AMERICAN REVOLUTIONS IN THE DIGITAL AGE

AMERICAN REVOLUTIONS IN THE DIGITAL AGE

Edited by Nora Slonimsky,
Mark Boonshoft, and Ben Wright

CORNELL UNIVERSITY PRESS   ITHACA AND LONDON

Thanks to generous funding from Iona University, the ebook editions of this book are available as open access volumes through the Cornell Open initiative.

Copyright © 2024 by Cornell University

The text of this book is licensed under a Creative Commons Attribution NonCommercial NoDerivatives 4.0 International License (CC BY-NC-ND 4.0). To use this book, or parts of this book, in any way not covered by the license, please contact Cornell University Press, Sage House, 512 East State Street, Ithaca, New York 14850. Visit our website at cornellpress.cornell.edu.

First published 2024 by Cornell University Press

Library of Congress Cataloging-in-Publication Data

Names: Slonimsky, Nora, 1986– editor. | Boonshoft, Mark, editor. | Wright, Ben, 1983– editor.
Title: American revolutions in the digital age / edited by Nora Slonimsky, Mark Boonshoft, and Ben Wright.
Description: Ithaca : Cornell University Press, 2024. | Includes bibliographical references and index.
Identifiers: LCCN 2023042168 (print) | LCCN 2023042169 (ebook) | ISBN 9781501771835 (hardcover) | ISBN 9781501771842 (paperback) | ISBN 9781501771859 (epub) | ISBN 9781501771866 (pdf)
Subjects: LCSH: Digital humanities—Research—Methodology. | History—Research—Methodology. | Historiography—Methodology.| Learning and scholarship—Technological innovations. | United States—History—Revolution, 1775–1783—Historiography. | United States—History—Revolution, 1775–1783—Research. | United States—History—Revolution, 1775–1783—Electronic information resources.
Classification: LCC E209. A523 2024 (print) | LCC E209 (ebook) | DDC 907.207301—dc23/eng/20231221
LC record available at https://lccn.loc.gov/2023042168
LC ebook record available at https://lccn.loc.gov/2023042169

Contents

Acknowledgments   vii

Introduction: North America, the United States, and Multiple Revolutions   *Mark Boonshoft, Nora Slonimsky, and Ben Wright*   1

Part I   **PUBLICS AND PEDAGOGY**

1. Digital Public History at Three Presidential Home Sites
Lindsay M. Chervinsky and Whitney Nell Stewart   17

2. New Media and Old Problems: Restoring Humanity in the Maryland Loyalism Project   *Kyle Roberts and Benjamin Bankhurst*   35

3. Discovering Revolution in Digital Sources: Other[ed] Colonial Voices   *Dorothy Berry*   50

4. Building a Relational Database to Explore Enslaved Midwives' Work in Early America   *Sara Collini*   65

Part II   **SPATIAL REVOLUTIONS**

5. Geographies of Emancipation: Geospatial Technology in Mapping Black Thought in the Age of Revolutions
Jessica M. Parr   83

6. Visualizing City-Spaces during the Age of Revolutions
Molly Nebiolo   103

7. Rethinking Enslaved Containment and Mobility in North Carolina's 1821 Insurrectionary Scare   *Christy Hyman*   119

8. Mapping Myaamia Landownership, 1795–1846 and Today
Cameron Shriver   136

Part III   **DATA AND DATABASES**

9. (Counter-)Revolutionary Discourse in the Age of Revolutions   *Brad Rittenhouse, Christian Boylston, and Afshawn Lotfi*   159

10. By Conversation with a Lady: Women's Correspondence Networks in the Founders Online Database   *Maeve Kane*   184

11. Identifying "A Slave": The Iona University Text Analysis Project Explores a Mystifying Letter to Thomas Jefferson
Gary Berton, Michael Crowder, Lubomir Ivanov, Smiljana Petrovic   201

12. Who Stands in the Digital Shadows?: "City of Refuge" at the Intersection of "Old" and "New" Media in the Age of the Digital Humanities   *Marcus P. Nevius*   220

Part IV ECHOES IN THE PRESENT

13. Media Literacy in Revolutionary America   *Jordan E. Taylor*   237

14. "A Busy, Bustling, Disputatious Tone": News Anxiety in the Age of Revolutions and Today   *Joseph M. Adelman*   252

15. Copyright and Historical Dangers of Licensing Regimes in the Digital Age   *Kyle K. Courtney*   267

Contributors   301
Index   307

Acknowledgments

In the acknowledgments of Leona Hudak's *Early American Women Printers and Publishers, 1639–1820*, essential reading for anyone interested in the history of media and communication in the long eighteenth century, she wished a "plague on all [the] pseudo intellectual houses" of those who had tried to stop her work. Very fortunately for us, *American Revolutions in the Digital Age* developed through the creativity and care of so many generous, supportive, and truly brilliant people.

All our thanks to each contributor for their thoughtful and engaged essays and for their patience and investment in the goals of the project. This book would not exist if not for the sharp eye and enthusiasm of Michael McGandy, now the editor in chief of the University of South Carolina Press, whose involvement from even before there was a day one made *American Revolutions in the Digital Age* possible, particularly as an open-access volume. Thank you to the entire team at Cornell University Press for seeing this book over the finish line, especially Mahinder S. Kingra, Bethany Wasik, Karen Carroll, and the marketing team.

Far less fortunately, for us and countless others around the world, the plague that Leona Hudak referenced was quite real. Due to the effects of the COVID-19 pandemic, *American Revolutions in the Digital Age* went through a few shifts in order to move forward amid the uncertainties of 2020 and 2021. We would like to thank the participants at the virtual symposium, "The Age of Revolutions in the Digital Age," which was rescheduled not once but twice amid COVID-19 safety precautions. Many of our participants felt the effects of the pandemic quite keenly, and we are deeply grateful to them for seeing this project through despite the circumstances.

That symposium, and indeed the volume itself, would not be possible without the unwavering support of the Iona University Provost's Office. Thank you to Tricia Mulligan, provost, senior vice president for academic affairs, and associate professor of political science, for being such an advocate for academic research, learning, and accessibility. Mulligan has been a steadfast and energetic supporter of all the work underway at the Institute for Thomas Paine Studies (ITPS) at Iona University, and this volume would not be possible without her. At Iona University, we were also encouraged by the departments of history and computer science, particularly around initiatives involving digital humanities. We were also fortunate to have the opportunity to draw on items in the Thomas Paine National

Historical Association (TPNHA) collection held at Iona. Our thanks to Rick Palladino and Natalka Sawchuk of Iona University Libraries for their commitment to the preservation of this collection. Our thanks as well to ITPS colleagues, past and present, including Michael Crowder, Alexi Garrett, Barry Goldberg, and Kellen Heniford, for their support of the volume in various ways.

American Revolutions in the Digital Age was meant to be accessible, both thematically and in the mode of publication. We would like to share our sincere appreciation for the support of the Robert David Lion Gardiner Foundation and its executive director, Kathryn Curran. An extremely generous grant from the Gardiner Foundation ensured that this volume could be published through an open access license and, as a result, remove barriers for students, researchers, educators, and interested members of the public who want to read it. The Gardiner Foundation's commitment to supporting the study of history in an inclusive way also ensured that we were able to provide honoraria for contributors so they could participate in a way that recognized their time and labor, and indeed made that participation financially feasible.

On a more personal note, the editors—Mark, Nora, and Ben—would also like to thank each other. In particular, Ben thanks Nora for fuzzy coats and Mark for Windows 95. *American Revolutions in the Digital Age* is truly a passion project, one we all deeply believe in for a cluster of related but unique reasons, and getting to work together on it over the last three years has been nothing short of awesome. That work has been made even better by the support of our partners—Jeremy and Whitney—and of course, family of the floofy variety, R2-D Dog, Teddy, and Frankie.

Introduction

NORTH AMERICA, THE UNITED STATES, AND MULTIPLE REVOLUTIONS

Mark Boonshoft, Nora Slonimsky, and Ben Wright

The creation of the United States unleashed revolutions that shook the worlds of politics, gender relations, religion, warfare, commerce, moral philosophy, diplomacy, and more. Many entrenched power structures wobbled; some toppled. Old inequities vanished, while others arose. Our present moment likewise appears unusually unsettled, as revolutions in technology, rising social divisions, and emerging public health and ecological crises wrack our politics, culture, and very sense of ourselves.

The chapters in this volume explore several revolutions in the past and present. They probe our understanding of the creation of the United States, the transformations the new nation spawned, the technological opportunities and challenges of our present moment, and our very ways of knowing. Then and now, there is more than one understanding of American independence and the origins of the United States and what it meant to those who lived through it and for generations since.

This is not a novel conclusion. The nature and scope of a singular American Revolution has remained unresolved and unsettled since independence. A host of complex events, movements, and legacies among several other events, movements, and legacies swirled out of what we now call the Age of Revolutions. It was an age of empire and of reason, defined by violence and dispossession alongside widespread politicization and calls for equality that are impossible to reconcile. The late eighteenth- and early nineteenth-century debates around revolution—in the United States, France, Haiti, Central America, and several other countries in Europe—are in certain ways timeless, yet they also reflect the circumstances in which they are studied.

As we approach these histories in our current moment, with the 250th anniversary of United States independence close at hand, the image of revolutions shifts yet again. In response to the changing communication technologies of the mid-twentieth century, the media theorist Marshall McLuhan coined the famous phrase "the medium is the message." McLuhan recognized that the presentation of ideas, information, arguments, events, and points of view—as much as the substance of the content—influenced audiences. A generation of scholars have applied these insights to a host of historical events, including the Age of Revolutions.[1]

The rise of new forms of communication in the twenty-first century—digital tools, spaces, and platforms that shape and reflect our day-to-day lives—are both distinct from and deeply connected to those of the past. For better or worse, revolutions are often talked about in terms of technology as much as ideas or movements. In that framework, modes of communication are a medium and a message, circuits and networks of revolt, wellsprings and restraints of change. *American Revolutions in the Digital Age* considers how the study of revolutions in the United States is likewise shaped by and reflective of digital history tools and methods. This volume does not consider the digital humanities as a replacement or corrective to classic questions and traditional approaches. It does quite the opposite. This collection examines what digital methods can tell us about American revolutions and in turn what American revolutions can tell us about our current digital age. It offers lessons about the past, the present, and the very ways of understanding both. Methodology, as much as argument, is the focus of this book.

In the late eighteenth century, oral, visual, and printed communication—from newspapers and pamphlets, to maps and prints—expressed pro- and anti-revolutionary arguments and were embodiments of those arguments themselves. The same can be said for classic and digital tools of historical study. For at least a generation, historians have recognized the transformative power of digital technologies and used them to revolutionize the study of early American history.[2] Inspired by that groundbreaking interdisciplinary work, our authors seek to bridge long-standing historical and historiographical debates with digital methods and the contemporary media landscape. Both timeless and of-this-moment interpretations of United States independence and the revolutions it enabled remain as complicated as any algorithm. This volume takes seriously both argument and method, so it is worth commenting on how argument and method have combined to shape how historians have understood the achievement of US independence.

For nearly a century, three paradigms—imperial, Progressive, and Whig, and their "neo" iterations—vied to explain the American Revolution.[3] Imperial historians situated the American Revolution as an event in the history of the British Empire and focused their attention on institutions of governance and trade. Progressive historians looked more closely at the internal structure of the colonies that revolted, emphasizing the importance of conflict—particularly class conflict—in bringing about the Revolution and shaping its course. Historians writing in a Progressive vein also argue that they are responsible for more recent attempts to incorporate non-white people and women into the literature on the Revolution.[4] Finally, there is the Whig interpretation, which prioritizes ideas, arguing that the Revolution was animated by a defense of constitutional principles. In British policies during the 1760s and 1770s, colonists saw a powerful conspiracy working to subvert their liberty and reacted accordingly.

All three paradigms began by trying to explain the origins of the Revolution. All three also, in John M. Murrin's memorable phrase, "self-immolated" by the 1980s.[5] More recently, historians have been divided over whether to explore lived experiences or ideological formulations of revolutionary-era peoples.[6] Still others ask whether these can ever be fully separated. At the same time, scholars have continued to expand the boundaries of what counts as revolutionary America, and who matters to the stories we tell about the period. The centrality of slavery, Native dispossession, the hardening of racial lines, and the resurgence of patriarchal householding emerge as key themes in this new work, especially on the war years.[7] This is coupled with a more cynical view of revolutionary politics, one focused on the deliberate manipulation of information and the mobilization of fear.[8]

Twenty-first-century scholarship on the Revolution also explicitly stresses continuity across the revolutionary divide, implicitly responding to Gordon Wood's assertion that the Revolution was a radical, paradigm-altering event.[9] But this recent work often actually reveals the tragically transformative importance of the long American revolutionary epoch, which created a new, less-restrained empire in North America.[10] This empire was not a repudiation of the Revolution's republican ideals but the product of the revolutionary search for self-determination by white Americans. These Americans believed that for the new nation to stake its place "among the powers of the earth," it either had to pacify or expel the other peoples who lived under its new dominion. This logic hastened dispossession, removal, the expansion of slavery, and the emergence of increasingly complex legal regimes to sustain inequality.[11] At the same time, this literature captures how Indigenous people and free and enslaved people of African descent fought against the new "Empire of Liberty" in pursuit of a broader vision of freedom, autonomy, and self-determination.[12] Many chapters in this book contribute to the ongoing effort to expand the temporal and geographic boundaries of the American Revolution, and to account for what it meant to the diverse peoples living primarily in regions that became part of the United States. The plural "revolutions" in the title of this collection, then, gestures toward how *the* American Revolution had multiple rounds of effects on many peoples and spawned other movements for self-determination within the expanding boundaries of the United States.

Our authors add to this literature by exploring how the explosion of digitized materials and digital methods have contributed to an overall reevaluation of the American Revolution. Early American history is one of the most digitized historical subfields, owing to the public and civic importance of the Revolution. One can access, for free, nearly 200,000 annotated documents pertaining to the Founding Fathers on founders.archives.gov. The federal government and private philanthropists have invested vast sums to produce the papers projects that are the root of this database. Nearly all the imprints and huge portions of the newspapers published in revolutionary America are available for a large fee on proprietary databases owned by the for-profit private company, Readex. Those databases came into existence thanks to partnerships with historical institutions that were founded in early America, maintain an interest in early America, and have preserved the original materials. Access to these paywalled databases have allowed our authors to trace the movement of information and misinformation over time and space. Yet there are constraints to even the most robust archive or the most impressive tool. *American Revolutions in the Digital Age*, then, also demonstrates how constraints of data and technologies, including archival silences and inflexible digital tools, shape the possibilities and limitations of scholarship.

The contributors to this volume have both used existing digital databases and created some of their own. They then deploy data to reconstruct networks of revolutionary-era Americans, posing questions about for whom and on what terms this period constituted a revolution. The subjects under consideration include recognizable political figures as well as understudied Americans, including enslaved people, women, Indigenous communities, and more. Other authors have worked to remediate the data within widely accessible sources to uncover the experiences of marginalized people during the period, without erasing their humanity. These chapters remind us to stay attuned to the way the structures of information shape historical scholarship on the Revolution, which is itself a productive intervention in the study of the revolutionary era.

The types of digital sources our authors use have shaped the questions they asked and the types of projects they undertook. In that regard, recent scholarship reveals the lingering influence of older historiographical paradigms, which were themselves distinguished at times by methodology and source base as much as analytical perspective. Progressive and neo-Progressive historians used the tools of social history to uncover the lives of "ordinary" people who left behind few written records. Several contributors to this book have used digital versions of similar sources to construct databases that allow us to recapture and humanize the experiences of comparatively voiceless historical actors. Imperial historians reimagined the map of early America, incorporating other parts of the British Empire and the Atlantic World, to reframe the Revolution's origins and consequences. Several chapters here join recent work by S. Max Edelson and Vincent Brown using sophisticated digital mapping and visualization tools to reframe the imperial history of revolutionary North America.[13] The study of print is not limited to cartographic publications but also includes multiple forms of texts. The great innovation that propelled the neo-Whig historians was the rediscovery of revolutionary pamphlets and print culture as repositories of revolutionary thought. Several of our authors have revisited similar corpora, now accessible in searchable and manipulable digital forms, to question long-standing paradigms of revolutionary intellectual history.

Just as understandings of the American Revolution have evolved, so too have the methods of the digital humanities. The most common origin story for the digital humanities involves the rise and evolution of humanities computing, mostly in the 1970s. In a project that stretched over three decades but culminated in 1979, Roberto Busa, with support from IBM, created the *Index Thomisticus*, a concordance of works by and relating to Thomas Aquinas.[14] Other text-centered projects dominated humanities computing in the 1980s.[15]

In the 1990s, *humanities computing* transformed into the *digital humanities*. The transition was not without tension, however, as the digital humanities brought a new focus that stretched beyond simply using technology to do the work of humanities. Indeed, digital humanists came to believe that the humanities, particularly the insights of intersectional feminism, were essential to make meaning of the digital itself.[16] Around the same time, *digital history* emerged through a separate genealogy, one more rooted in museum studies, archival practice, and public history.[17] From the beginning, digital historians demonstrated a greater interest in public-facing scholarship, and accordingly suffused their work with promises to democratize the study of the past.[18] Librarians, archivists, and public historians remain the vanguard of digital history, a reality that is reflected in this book.

Most digital humanities work completed in the first two decades of the twenty-first century falls into one of four categories. First, scholars in literature have continued developing the techniques begun in the era of humanities computing to perform increasingly complex forms of text analysis or—to use the phrase coined by Franco Moretti—"distant reading."[19] Historians, meanwhile, have embraced the possibilities of Geographical Information Systems (GIS) mapping as a means of understanding the past.[20] Third, digital humanists have applied themselves to the work of creating new tools to both create new knowledge and represent that knowledge in exciting ways. Some of these tools include data visualization, network analysis, stylometry and machine learning, virtual reality, 3D modeling, deformance, and digitization.[21] Finally, digital humanists have produced critical discourse that recognizes how the digital world has encoded and exacerbated inequalities.[22]

By 2016 Cameron Blevins had noticed that, despite the achievements of digital history, there had been relatively little work that endeavored to "advance academic claims about the past."[23] In other words, the medium had received more attention than the message, with the methods of digital history taking precedence over the conclusions those methods had drawn. This volume is a testament to how quickly the field is changing. Indeed, historians have recently produced numerous important projects that yield insightful historical as well as methodological arguments, and many of the chapters here do the same.

American Revolutions in the Digital Age begins with considerations of how we share historical knowledge with our students and with the wider public. Indeed, it is important to remember the special role that public history has played in the rise of the digital humanities, especially in the field of history. Before historians saw the digital world as an opportunity to create new arguments about the past, they recognized the democratizing potential of the internet and its ability to share history with new audiences. Lindsay Chervinsky and Whitney Nell Stewart delve into three presidential public history sites to evaluate their uses of digital tools. The President's House in Philadelphia, James Madison's Montpelier, and James Monroe's Highland all use digital tools to allow visitors to explore overlooked historical figures and offer visitors more agency in determining how they wish to relate to these historic sites. They further consider how these public history sites reflect changes in the scholarship on the history of the presidency.

Kyle Roberts and Benjamin Bankhurst reflect on their Maryland Loyalism Project, a collaborative endeavor to engage students in producing a digital archive that highlights the humanity of those on the losing side of the war for American independence. The chapter explores the history of Maryland in the American Revolution, the challenges of using quantified data to express complex lived experiences, and the opportunities of bringing this work into the classroom. Dorothy Berry similarly draws on her experience in creating Other[ed] Colonial Voices: Slavery and Indenture in New York. Berry explores the digital archival practice behind this project and ruminates on its applications for scholars and students. Sara Collini's work also involves database creation as a means of uncovering previously underrepresented histories. Collini deploys a relational database of enslaved midwives to help reveal the ways they supported their communities and how their work bolstered white freedom.

The next four chapters explore space, place, and our perceptions of both. Jessica Parr, Molly Nebiolo, Christy Hyman, and Cameron Shriver all employ the tools of digital mapping to better understand the history of the late eighteenth and early nineteenth centuries. Parr considers the opportunities and challenges of mapping Black history. As Parr notes, mapping has always been about power, and historians must untangle the inequalities that shape our archives, our tools, and our methods. Parr demonstrates how doing so can both reveal and move beyond imperial understandings of space following the American Revolution and beyond. Molly Nebiolo focuses on eighteenth-century urban spaces, which have long been key sites in work on the Revolution's origins. By comparing the lived experiences of eighteenth-century figures with georectified representations, Nebiolo brings historic American cities to life and reveals how space shaped identity, culture, and networks of power. This work also offers an important perspective on the very idea of the urban.

Christy Hyman continues the engagement with mapping but moves the focus from northeastern cities to southeastern swamps, where enslaved men and women relied on challenging geographies to resist their enslavement. Enslaved Americans built much of the United States, and this experience creating infrastructure informed the actions of the rebels in the 1821 North Carolina insurrection scare. Hyman's work unfolds the realities of slavery, the contingencies of creating American infrastructure, and the very idea of mapping as a means of knowledge. Cameron Shriver's chapter relays how he and other scholars at the Myaamia Center—a research collaboration between the Miami Tribe of Oklahoma and Miami University—created the "*Aacimwahkionkonci:* Stories from the Land" Project. This work is a process of recovery and creation, as members of the team work to combine and curate Myaamia land transfers as well as to connect present members of the community to important places in the past. The revolutions wrought by American independence were largely catastrophic for Indigenous people, and our digital world is haunted by the resulting traumas and inequities.

The next four chapters turn from explorations of space to analyses of texts to produce insights into the past. Brad Rittenhouse, Christian Boylston, and Afshawn Lotfi ask what we mean by the term "revolutionary." The team applied natural language processing tools to over one million works in English, French, German, Spanish, and Italian published between 1750 and 1875 with the goal of analyzing understandings of revolution. This chapter offers several insights about struggles for power in the age of American revolutions, the practical process of applying natural language processing as a research tool, and the importance of balancing the methods of distant and close reading. Maeve Kane uses the National Archives' "Founders Online" database to track women's correspondence networks. By analyzing over 165,000 records from 1730 to 1830, Kane shows that the American Revolution transformed the way women communicated with one another and with male political leaders. This network analysis intervenes in a classic literature about the Revolution's effects on women to reveal the importance of personal connections and the limitations of women's political participation.

A collection of interdisciplinary researchers—including an expert on Thomas Paine (Gary Berton), a public historian (Michael Crowder), and two computer scientists (Lubomir Ivanov and Smiljana Petrovic)—wrote chapter 11. These scholars, all affiliated with the Institute for Thomas Paine Studies at Iona University, employ text analysis to determine the authorship of an abolitionist letter sent to Thomas Jefferson from an anonymous person who identified as "a slave." Using both digital author attribution methodologies and careful historical analysis, they suggest that the letter was potentially dictated by Thomas Paine. Next, Marcus Nevius draws on his experience writing the monograph *City of Refuge: Slavery and Petit Marronage in the Great Dismal Swamp, 1763—1856* to compare the impressions of the past offered by digital sources and analog archives. As Nevius notes, our scholarly arguments flow from our sources. The limitations of digitalization offer another filter that colors our understanding of the past, especially of the revolutionary experience of Black Americans.

The final three chapters have direct messages for our modern readers. Jordan Taylor, Joseph Adelman, and Kyle Courtney all foreground ways that the history of the late eighteenth and early nineteenth centuries can help us make sense of the challenges and opportunities we face in our current digital world. Taylor highlights the idea of media literacy and considers how Americans, in a time of rapid social change, sought to identify and combat "fake news." Adelman focuses on the stress and fears of dislocation and confusion brought on by explosions of misinformation. The chapter chronicles how Americans in the early republic managed a deluge of information, an essential skill for our current media-saturated lives. As Adelman notes, managing information is critical for modern citizens and especially for research historians who must understand the information climate of the era they study if they are to do more than reflect the biases and inequalities of the archive. Finally, Kyle Courtney untangles the relationship between information and property by evaluating the opportunities and limitations of the new licensing climate of the digital world. This history and evaluation of copyright sheds light on contemporary debates surrounding censorship, capitalism, and authority.

Three common themes run through the chapters in *American Revolutions in the Digital Age*. The first and second are evident from the title itself: this book connects digital tools with the study of American revolutions. Woven throughout is a third thread that considers the omnipresence of what is not there. That is to say, the chapters attend to traditional archival silences and digital technical limitations. The subject of archival silences—gaps, fragments, deliberate erasure—is by no means a new one, nor is it exclusive to digital humanities. From the now-famous phrase of Laurel Thatcher Ulrich that "well behaved women seldom make history" to the groundbreaking work of Michel-Rolph Trouillot and Saidiya Hartman, scholars and students across disciplinary boundaries have long grappled with the challenge of how to understand histories with incomplete, disrupted, unconventional, or unseen evidentiary records.[24]

For the revolutionary histories described in this book, digital tools and methods provide a powerful way of considering those silences, yet they also risk creating others. Over the last decade, early American studies has focused more closely on gaps in the archives—from the work of Carolyn Steedman in *Dust: The Archive and Cultural History* and Marissa Fuentes in *Dispossessed Lives: Enslaved Women, Violence, and the Archive* to Jennifer Morgan's *Reckoning with Slavery: Gender, Kinship and Capitalism in the Early Black Atlantic*. Whether it is through GIS mapping and data visualization, computational text and linguistic analysis, text mining and machine learning, digitization and database building, network analysis and digital public history, recovery and multidisciplinary understanding is at the heart of digital knowledge production. Blending well-known historical figures like Thomas Paine with less-studied people like Pegg, Rachael, Nan, Nell, Kate, Jane, Peg, Sarah, and Lucy (all enslaved midwives), along with Cynthia Haycorn and Moses Judah, we gain an even wider range of who is considered a revolutionary actor.

Yet digital history is an imperfect solution. There are challenges with these methodologies, particularly involving issues of accessibility and resources. In learning from and referencing scholars of archival construction and source materials, we strive to be mindful that citation practices do not themselves become extractive by engaging with rather than appropriating methods to understudied voices and experiences. Much like the structures of authority that determined the development of archives in the Age of Revolutions, there are digital silences as well, and several components of the collection consider this, from the complex landscapes of the Great Dismal Swamp to the categorization of people in records designed to view them as anything but.

The contributors address these challenges in a multitude of ways. Most centrally, the volume examines the relationship between digital tools, archival silences, and what exactly is meant by a singular American Revolution by arguing that the stakes are not only in historical debates and circumstances but in contemporary ones as well. For example, Black digital humanities, data feminism, postcolonial digital humanities, and digital pedagogy all make explicit connections between approaches to the historical past (in our case, late eighteenth-century North America) and the current stakes of scholarship and research—not just to higher education but to a vast range of broader challenges in education, civic and political engagement, inequality, and other distributions of power. Much like the classic historiography of American independence and the fraught dynamics of the age of enlightenment and empire, *American Revolutions in the Digital Age* seeks to answer certain archival silences with digital amplification.

The chapters in this collection also lay bare how resources and organizations shape historical knowledge in the digital age. Institutions, particularly colleges and universities, set their own research priorities and respond to those of large grant-making agencies. For-profit database companies make decisions based on market considerations. Historic sites balance the interests of many constituencies. Those varied institutions helped make possible the chapters in this book, and inevitably shaped the topics our authors considered.

This volume is, itself, an example of how institutions enable and circumscribe the possibilities of digital historical scholarship. The conference from which it derives, "Revolutionary Texts in a Digital Age," was held at Iona University, home of the Institute for Thomas Paine Studies or ITPS, in the fall of 2018. Support from Iona, aided by a generous grant from the Robert David Lion Gardiner Foundation, provided the financial backing that allowed us not only to hold a subsequent symposium—though our in-person meeting was cancelled due to the COVID-19 pandemic—but also to provide contributors with honorariums for their work. We understand, based on conversations with many of them, that this funding made it possible to assemble a set of participants and authors from different career stages, statuses, and even professions. Moreover, that financial support allowed us to publish this collection with an open access Creative Commons license, which broadened the volume's ambitions and encouraged our authors to think about how to reach audiences beyond fellow scholars in their subfields. This license in turn allows interested readers—across economic, professional, and geographic categories—to access this book free of charge.

At the same time, collections like *American Revolutions in the Digital Age* are not immune from the challenges being faced across higher education. The fortunate background of its support, coupled with the enthusiasm and stewardship of Cornell University Press, also meant that this volume came out of a conference held in the Northeast, at an institute named for a Founding Father, and which has its greatest influence in American Revolution studies, as opposed to the broader Age of Revolutions. Recognizing these circumstances, with their benefits and limitations, we hope that ultimately *American Revolutions in the Digital Age* is not unlike the subjects that it treats, historical and contemporary.

This brings us, then, to a fourth thread. One of many, many efforts to reconsider revolutionary commemoration, the semiquincentennial was a contributing factor to creating this book. As the collection developed, we were struck by how relevant the stakes of responsible and accurate commemoration are to the silences of public and collective memory, particularly when it involves a period of such complexity and varied meanings. By applying digital tools and methods to both past and present interpretations of revolution, the authors in this book complicate whose perspectives are elevated and whose are overlooked. Their contributions are deeply relevant to how local, regional, and national communities approach the founding of the United States. *American Revolutions in the Digital Age* offers its readers—students, educators, researchers, public history professionals, archivists, librarians, and any interested person—historical accounts, digital tools, and varied methods through which they can find themselves amid the multiple legacies of American revolutions.

NOTES

1. Examples abound of scholars exploring the relationship between communication technologies and revolutionary change. See, for example, Elizabeth Eisenstein, *The Printing Press as an Agent of Change* (New York: Cambridge University Press, 1982); and Julius S. Scott, *The Common Wind: Afro-American Currents in the Age of the Haitian Revolution* (London: Verso, 2018).

2. In 2018 the premier journal of early American history, the *William and Mary Quarterly*, supported a workshop on the use of digital humanities methods in the field. See Sharon Block, "#DigEarlyAm: Reflections on Digital Humanities and Early American Studies," *William and Mary Quarterly* 76, no. 4 (October 2019): 611–48. See also Dan Edelstein, Paula Findlen, Giovanna Ceserani, Caroline Winterer, and Nicole Coleman, "Historical Research in a Digital Age: Reflections from the Mapping the Republic of Letters Project," *American Historical Review* 122, no. 2 (April 2017): 400–424; and Lauren Klein, *Data by Design: An Interactive History of Data Visualization, 1786–1900*, available at dataxdesign.io. Historians of the slave trade have long relied on the data available in the Slave Voyages Database, now hosted by Rice University and available at slavevoyages.org.

3. This historiographical summary primarily echoes Michael A. McDonnell and David Waldstreicher, "Revolution in the Quarterly? A Historiographical Analysis," *William and Mary Quarterly* 74 (October 2017): 633–66. See also Alan Taylor, "Introduction: Expand or Die: The Revolution's New Empire," *William and Mary Quarterly* 74 (October 2017): 619–32; and Serena R. Zabin, "Conclusion: Writing To and From the Revolution," *William and Mary Quarterly* 74 (October 2017): 753–64. For a perspective on the literature in distinct phases during which one approach or another predominated, see Michael D. Hattem, "The Historiography of the American Revolution," *Journal of the American Revolution*, last modified August 27, 2013, https://allthingsliberty.com/2013/08/historiography-of-american-revolution/.

4. Alfred Fabian Young and Gregory H. Nobles, *Whose American Revolution Was It? Historians Interpret the Founding* (New York: New York University Press, 2011).

5. John M. Murrin, "Self-Immolation: Schools of Historiography and the Coming of the American Revolution," in *Rethinking America: From Empire to Republic* (New York: Oxford University Press, 2018), 383–407. See also Patrick Griffin, introduction to *Between Sovereignty and Anarchy: The Politics of Violence in the American Revolutionary Era*, ed. Patrick Griffin, Robert G. Ingram, and Peter S. Onuf (Charlottesville: University of Virginia Press, 2015), 1–20, 1, 2–4.

6. Patrick Spero, "Introduction: Origins," in *The American Revolution Reborn*, ed. Michael Zuckerman and Patrick K. Spero (Philadelphia: University of Pennsylvania Press, 2016), 1–6, 3–4.

7. Claudio Saunt, *West of the Revolution: An Uncommon History of 1776* (New York: W. W. Norton, 2014); Kathleen DuVal, *Independence Lost: Lives on the Edge of the American Revolution* (New York: Random House, 2015); Alan Taylor, *American Revolutions: A Continental History, 1750–1804* (New York: W. W. Norton, 2016); Lauren Duval, "Landscapes of Allegiance: Space, Gender, and Military Occupation in the American Revolution" (PhD diss., American University, 2018).

8. Robert G. Parkinson, *The Common Cause: Creating Race and Nation in the American Revolution* (Chapel Hill: published by the Omohundro Institute of Early American History and Culture and the University of North Carolina Press, 2016); Holger Hoock, *Scars of Independence: America's Violent Birth* (New York: Crown, 2017).

9. Gordon S. Wood, *The Radicalism of the American Revolution* (New York: Alfred A. Knopf, 1992). See also "Forum: How Revolutionary Was the Revolution? A Discussion of Gordon S. Wood's *The Radicalism of the American Revolution*," *William and Mary Quarterly* 51, no. 4 (October 1994): 677–716.

10. Taylor, "Introduction," 630–31; Andrews Shankman, "Toward a Social History of Federalism: The State and Capitalism To and From the American Revolution," *Journal of the Early Republic* 37, no. 4 (Winter 2017): 615–53; and the chapters in *The World of the Revolutionary American Republic: Land, Labor, and the Conflict for a Continent*, ed. Andrew Shankman (New York: Routledge, 2014).

11. Eliga H. Gould, *Among the Powers of the Earth: The American Revolution and the Making of a New World Empire* (Cambridge, MA: Harvard University Press, 2012); Carroll Smith-Rosenberg, *This Violent Empire: The Birth of an American National Identity* (Chapel Hill: published by the Omohundro Institute of Early American History and Culture and the University of North Carolina Press, 2010); Gregory Ablavsky, "The Savage Constitution," *Duke Law Journal* 63, no. 5 (February 2014): 999–1089; Matthew Spooner, "The Problem of Order and the Transfer of Slave Property in the Revolutionary South," in *American Revolution Reborn*, ed. Patrick Spero and Michael Zuckerman, 231–47; Samantha Seeley, *Race, Removal, and the Right to Remain: Migration and the Making of the United States* (Chapel Hill: published by the Omohundro Institute of Early American History and Culture and the University of North Carolina Press, 2021).

12. Paul J. Polgar, *Standard-Bearers of Equality: America's First Abolition Movement* (Chapel Hill: published by the Omohundro Institute of Early American History and Culture and the University of North Carolina Press, 2019); Sean Gallagher, "Black Refugees and the Legal Fiction of Military Manumission in the American Revolution," *Slavery and Abolition* 22, no. 1 (August 6, 2021), 140–59; Eliga Gould, "Independence and Interdependence: The American Revolution and the Problem of Postcolonial Nationhood, circa 1802," *William and Mary Quarterly* 74, no. 4 (October 2017): 729–52; Michael Witgen, *Seeing Red: Indigenous Land, American Expansion, and the Political Economy of Plunder in North America* (Chapel Hill: published by the Omohundro Institute of Early American History and Culture and the University of North Carolina Press, 2021); Jean O'Brien, *Firsting and Lasting: Writing Indians Out of Existence in New England* (Minneapolis: University of Minnesota Press, 2010); Elizabeth Ellis, "The Natchez War Revisited: Indigenous

Diplomacy, Multi-Directional Slave Trades, and Violence in the Lower Mississippi Valley," *William and Mary Quarterly* 77, no. 3 (Summer 2020): 441–72.

13. S. Max Edelson, *The New Map of Empire: How Britain Imagined America before Independence* (Cambridge, MA: Harvard University Press, 2017); Vincent Brown, *Tacky's Revolt: The Story of an Atlantic Slave War* (Cambridge, MA: Harvard University Press, 2020).

14. Roberto Busa, "The Annals of Humanities Computing: The Index Thomisticus," *Computers and the Humanities* 14, no. 2 (1980): 83–90.

15. Susan Hockey, "The History of Humanities Computing," in *Companion to Digital Humanities*, ed. Susan Scheibman, Ray Siemens, and John Unsworth (Oxford: Blackwell, 2004), 3–19.

16. Patrik Svensson, "Humanities Computing as Digital Humanities," *Digital Humanities Quarterly* 3, no. 3 (2009), http://www.digitalhumanities.org/dhq/vol/3/3/000065/000065.html; Catherine D'Ignazio and Lauren F. Klein, *Data Feminism* (Cambridge, MA: MIT Press, 2020); Elizabeth Losh and Jacqueline Wernimont, eds., *Bodies of Information: Intersectional Feminism and the Digital Humanities* (Minneapolis: University of Minnesota Press, 2018); Safiya Umoja Noble, "A Future for Intersectional Black Feminist Technology Studies," *Scholar and Feminist Online*, 13.3–14.1 (2016): 1–8; and Kim Gallon, "Making a Case for the Black Digital Humanities," in *Debates in the Digital Humanities 2016*, ed. Matthew K. Gold and Lauren F. Klein (Minneapolis: University of Minnesota Press, 2016), 42–49.

17. See Stephen Robertson, "The Differences between Digital Humanities and Digital History," in *Debates in the Digital Humanities 2016*, ed. Matthew K. Gold and Lauren F. Klein (Minneapolis: University of Minnesota Press, 2016), 289–307; and Tom Scheinfeldt, "The Dividends of Difference: Recognizing Digital Humanities' Diverse Family Trees," Found History, April 7, 2014, https://foundhistory.org/2014/04/the-dividends-of-difference-recognizing-digital-humanities-diverse-family-trees/.

18. Joseph L. Locke and Ben Wright, "History Can Be Open Source: Democratic Dreams and the Rise of Digital History," *American Historical Review* 126, no. 4 (December 2021): 1485–1511.

19. Franco Moreti, *Distant Reading* (London: Verso, 2013); Ted Underwood, *Distant Horizons: Digital Evidence and Literary Change* (Chicago: University of Chicago Press, 2019); Katherine Bode, *A World of Fiction: Digital Collections and the Future of Literary History* (Ann Arbor: University of Michigan Press, 2018); Matthew L. Jockers, *Macroanalysis: Digital Methods and Literary History* (Champaign: University of Illinois Press, 2013).

20. For overviews of how historians have used digital mapping, see Ian Gregory, Don DeBats, and Don Lafreniere, *The Routledge Companion to Spatial History* (London: Taylor and Francis, 2018); Ian N. Gregory and Paul S. Ell, *Historical GIS Technologies, Methodologies, and Scholarship* (New York: Cambridge University Press, 2007); and Anne Kelly Knowles, *Placing History: How Maps, Spatial Data, and GIS Are Changing Historical Scholarship* (Redlands, CA: ESRI, 2008).

21. For a sample of tools that historians have used as well as a guide on how to use them, see The Programming Historian (https://programminghistorian.org) and The Digital Humanities Literacy Guidebook (https://cmu-lib.github.io/dhlg/topics/).

22. Alexis Lothian and Amanda Phillips, "Can Digital Humanities Mean Transformative Critique?," *Journal of e-Media Studies* 3, no. 1 (2013): 1–25; Noble, "Future for Intersectional Black Feminist"; Kim Gallon, "Making a Case for the Black Digital Humanities," in *Debates in the Digital Humanities 2016*, ed. Matthew K. Gold and Lauren F. Klein (Minneapolis: University of Minnesota Press, 2016), 42–49; Jessica Marie Johnson and Mark Anthony Neal, "Introduction: Wild Seed in the Machine," *Black Scholar: Journal of Black Studies and Research* 47, no. 3 (2017): 1–2.

23. Cameron Blevins, "Digital History's Perpetual Future Tense," in *Debates in the Digital Humanities 2016*, ed. Matthew K. Gold and Lauren F. Klein (Minneapolis: University of Minnesota Press, 2016), 308–24.

24. Michel-Rolph Trouillot, *Silencing the Past: Power and the Production of History* (Boston: Beacon Press, 1995); Saidiya Hartman, *Lose Your Mother: A Journey along the Atlantic Slave Route* (New York: Farrar, Straus and Giroux, 2007).

Part I
PUBLICS AND PEDAGOGY

1
DIGITAL PUBLIC HISTORY AT THREE PRESIDENTIAL HOME SITES

Lindsay M. Chervinsky and Whitney Nell Stewart

Digital history offers the potential to reconceptualize how we make and present historical knowledge. Digital tools and networks present the opportunity to expand who we include in the historical narrative, the historical process, and consumption of public history. Digital technologies, in particular, provide the potential to build public history projects—including the three examined in depth in this chapter—that go beyond the Founders-focused Revolution that most Americans already know.

Recent literature on the Revolution—including works by Annette Gordon-Reed, Erica Armstrong Dunbar, and Robert Parkinson—examines the lives of the Founders in their entirety, often emphasizing the pervasive and interwoven nature of slavery with every aspect of life.[1] Though the challenges are many, digital public history offers another, complementary way to broaden the Founders narrative, include the vast array of voices present in the revolutionary age, and make the American Revolution more relevant to the diverse people who now populate the nation that it created. Digital public history demonstrates that both women and enslaved individuals were ever present, even during political events traditionally considered to be the realm of white men. The projects reveal the constant overlap between public and private spaces, and the people that occupy them.[2] Political negotiations, diplomacy, and appointments rarely just occurred in the halls of Congress. Instead, the Founders hosted visiting dignitaries, congressmen, and local elites at their homes to socialize—and conduct business. These blended spaces were essential to the Revolution and the political project of creating a new nation.

Finally, whether the American Revolution was a radical event or a retention of the status quo has remained a lively debate within the scholarship for the last thirty years.[3] These projects offer an interesting lens through which to consider this question. At the President's House in Philadelphia, James Madison's Montpelier, and James Monroe's Highland, the social hierarchy, gender norms, and enslaved status of many residents remained the same both before and after the Revolution. Indeed, often the only changes were the citizenship of the white owners and to whom they sold the products produced by the enslaved community. Therefore, these projects force us to consider that if the American Revolution brought radical changes, who was creating that change and whose labor was required to support the new radical systems?

In this chapter, we examine three digital projects that focus on presidential domestic sites of the revolutionary era, and we do so in chronological order of the presidents who lived in them: the President's House, where George Washington and John Adams lived in Philadelphia from 1790 to 1800; Montpelier, James Madison's Virginia plantation; and Highland, James Monroe's Virginia plantation. Each of the projects builds on the scholarly debates around the Revolution and the Founders and contributes to discussions about the best practices in digital public history. Additionally, the three sites are in different states of preservation and are using different digital tools to interpret places, objects, and people, thus providing both common ground and diversity. Through these case studies, we consider the promises and challenges of representing and interpreting the built environment through digital media, and we reflect on how these projects can tell broader stories, engage wider communities, and expand the public's participation in the creation and consumption of history. All three projects encourage the public to understand history in new and innovative ways, whether by highlighting previously overlooked historical actors or by providing a user-led experience for each visitor to chart their own path.

Best Practices of Digital Public History

Before exploring the digital public history projects in depth, it is necessary to first understand what digital public history is. Breaking it into its component parts, it would seem to be part digital history and part public history. But both have a similar problem: there is no agreed-upon definition of either, despite scholars' best efforts to define them.[4] Are they fields, subjects, methods, or all of the above? Is the digital in digital history the subject matter or the tool? Is the public in public history simply an audience or an active participant? The debates over definition have pushed many to the same exasperated conclusion: like Justice

Potter Stewart said about pornography, "I know it when I see it." Unfortunately, this would seem to leave us in the same definitional quandary for digital public history. Indeed, the basic definition of digital public history—the use of digital tools as a means of "doing" public history—is vague to the point of glossing over major debates.[5]

Rather than providing a single definition of what digital public history *is*, we propose instead to define what digital public history *should be*. Scholars and practitioners assert that the core tenets of digital public history are about one thing: involving many voices, both past and present. The digital is the tool or technology for engaging with various audiences who hold a stake in the project. The field's best practices emphasize how crucial inclusion, shared authority, collaboration, and user-centered history are to "good" digital public history.

These terms overlap, as they stem from the desire to incorporate all relevant communities in the process. Inclusion can indicate both an inclusive historical narrative that centers the experiences of frequently overlooked historical actors and an inclusive process involving different stakeholders. Additionally, collaboration often entails the same kind of inclusive process but at a higher level of engagement, while it can also refer to collaboration across disciplinary boundaries (similar to other kinds of digital humanities work). Though sometimes imprecise, these best practices provide a clear indication of what practitioners and scholars believe are the core objectives of digital public history.

Digital public history offers the chance to reveal and highlight the experiences of people beyond the great-man narrative of American history (an especially important goal at places like presidential sites) as well as embrace the voices of various stakeholder communities—not just in the consumption but also in the process of creating this history. Public history and digital public history must center the stories of people who are too often seen as peripheral to history. Historic sites often replicate the injustices that took place there in the past, interpreting enslaved people only through the lens of labor or women only as handmaidens of their husband. This same exclusion is replicated in digital history; as Sharon Leon contends, the historiography of digital history too often overlooks the contributions of women to the field.[6] Digital tools, though, provide the possibility of building more inclusive histories. In her work with the Parkland History Project, Lara Kelland has blended traditional oral histories with 3D re-creations to give prominence to the people of color who inhabited this neighborhood of Louisville, Kentucky.[7] Even as the projects highlighted in this chapter are in some way about "great men" of American history, they also emphasize the diverse peoples who were an integral part of these places. Indeed, all three projects or institutions explored here are women-led, and all three seek to center and prioritize the non-white, non-male actors in their stories.

Along with inclusion of diverse historical voices, public historians have long touted the necessity of shared authority and collaboration, both of which encourage community engagement. Shared authority means that historians do not have exclusive right to the past but must enter into dialogue with the experiences/interpretations of the public.[8] Collaboration is about actually working with that public, about including them in the conversation and process.[9] Historians should not relinquish their expertise; rather, as Wendy Hsu has argued, collaboration and engagement allow historians, stakeholders, and others to bring their own expertise to the table.[10]

User-centered history arises from projects that focus on the user in multiple ways, both within the consumption of history and its creation.[11] As Roy Rosenzweig and David Thelen showed in the late 1990s, Americans want their museums and historic sites to include them in the intellectual process.[12] Rather than telling visitors the interpretation, digital public history offers the possibility of leading them through the interpretive process. Offering narratives and primary sources, including digital re-creations of places and buildings long since gone, helps the user feel more a part of the process. But the term "user-centered" can also denote foregrounding the user in the entire digital public history process, from conception to design to implementation.[13]

All these best practices are touted by public historians working at traditional, physical sites. What is unique about digital public history is the possibilities that the digital brings to the table. Digital technologies provide practitioners with far more opportunities to center important stakeholders. By stakeholders, we mean those with a financial or professional stake in the project as well as groups with a personal, communal, and historical stake in the history being presented.[14] It has not always been easy to interface with these groups, but the internet provides a clear way to do so. Websites, blog posts, and social media allow public historians the opportunity to communicate with and engage various stakeholders.[15]

Engagement, however, need not only come after a project is complete. The promises of digital public history will only be realized by prompting stakeholders to participate in all stages of the process.[16] Project managers should ensure that the community helps inform the decision making from inception to completion. Digital history provides tools—such as crowdsourcing the acquisition of materials—to promote early and frequent inclusion of the public in the process.[17] This kind of participation also helps define exactly who is the "public" for this project. If we create projects for an amorphous "public," who are we really creating them for but ourselves, with our interests and goals? But more than just ensuring there is an audience *for* this project, digital public history should seek to ensure that the project is created *with* that public.[18]

Best practices are easy to write about, but they are much more difficult to realize. There are many obstacles to succeeding in building shared authority and practicing collaboration for digital public history practitioners, starting with the continued digital divide. While the digital divide first developed around disparities in access to computers and the internet, more recently scholars and activists have been focused on the disparity of abilities in actually using those technologies. If individuals feel overwhelmed by the technology, unable to navigate or use it, they simply will not participate. "As practitioners gain more in-the-field experience with electronic media," Andrew Hurley notes, "they are discovering it is not the case that 'if we build it, they will come.'"[19] Other challenges are reminiscent of those experienced by many public historians: a lack of resources and staff with the right skill set, or a board or administration that does not fully understand or support the core tenets of shared authority and collaboration.[20] Additionally, some digital projects meant to break down barriers between the public and public historians might actually devalue the historical process—like Museum Selfie Day, which some believe fetishized objects and eliminated historical context.[21]

Actually practicing "best practices" is not easy, but making inclusion, shared authority, collaboration, and a user-centered experience key to digital public history projects will make them more relevant, engaging, and meaningful. The three presidential domestic site projects demonstrate this potential. Each of the following case studies explores the project's goals, its development, and its challenges, and shows how one or more of the best practices advanced by practitioners and scholars plays out in the real world.

President's House

In the fall of 1790, President George Washington moved into a large brick house on the corner of Sixth and Market (or High) Streets in Philadelphia. He hosted visiting dignitaries, dined with friends and colleagues, and oversaw the work of the executive branch from this building. After taking the oath of office, John Adams, now president of the United States, moved into the same house, and lived and worked there until November 1800, when the seat of government moved to Washington, DC. But Washington and Adams did not live in these homes alone. Their wives, children, and their children's families frequently joined them. The large house required an enormous staff to clean the building, cook meals for the family and guests, tend to the horses and carriages, deliver messages, and more. During Washington's presidency, up to thirty-five enslaved and free people occupied the space.²² Adams employed a slightly smaller staff, including a mix of white and Black, free and hired enslaved individuals.²³

Unfortunately, this house no longer exists.²⁴ Archaeological excavation on the site began in 2007 and a National Park Service site was completed in 2010 with a dedication to the enslaved individuals who had labored at the site.²⁵ The existing NPS site depicts part of the floorplan of the first floor, but the experience is limited. Visitors have a hard time visualizing what the completed space would have looked like, and it is even harder for them to imagine the sights, smells, sounds, and material culture that would have filled the rooms. As a user-centered experience, the site often leaves the public wanting more information and context, and there are few options for individual exploration.

"The Virtual White House" plans to provide a virtual, 3D re-creation of the President's House in Philadelphia, then expand the project to include other early presidential sites, including the New York presidential homes and the early White House in Washington, DC. Lindsay Chervinsky, one of the coauthors of this chapter, started this project in 2017 with the support of the Center for Presidential History at Southern Methodist University. She has since partnered with Whitney Nell Stewart (her coauthor for this chapter) at University of Texas at Dallas to continue the project. Currently, work on the president's private study and the exterior of the home as well as the scaffolding of the interior are finished. Once the rest of the house is "constructed," the virtual President's House will link archives with public history sites, classrooms, and scholars. To present a more inclusive history, the digital re-creation of the President's House reveals the lived experiences of the white, enslaved, and servant communities living in the home, which tend to be less documented in the written record.

This project will serve three audiences for the user-centered experience. First, the President's House will help researchers utilize an interdisciplinary approach in their scholarship. Chervinsky started this project by re-creating the president's private study so that she could imagine the room where the first cabinet met.²⁶ After visualizing the space, she realized that the secretaries would have felt confined and stifled by the room's narrow dimensions and excessive furniture. The story of the first cabinet cannot be told without first understanding their meeting space. The President's House will provide other researchers with similar opportunities to observe the physical spaces where their actors interacted and events transpired, and draw appropriate conclusions for their work in diplomatic, social, and economic history.

The second audience consists of teachers and students. Many history and social studies teachers are required to complete units on Washington, the presidency, and slavery. Accordingly, many teachers merge these conversations by assigning Erica Armstrong Dunbar's excellent biography of Ona Judge, *Never Caught*,

Teachers have shared with Chervinsky that the images and physical space of the President's House would help students understand the lived experience of Ona. For example, Judge likely slept in one of the rooms set aside for enslaved female workers. To get to this room, she had to walk past the Washingtons' bedrooms and George Washington's private study.[27] These details reinforce the fact that enslaved individuals, especially in urban dwellings, lived under nearly constant surveillance and privacy was at a premium. Teachers' contributions are pivotal to the project's development, leading the codirectors to reaffirm their commitment to highlighting the lives and labor of individuals like Ona Judge.

The final audience would be the history-inclined public, especially those who visit the National Park Service site. This digital project can support and enhance the work done at a traditional public history site by complementing the existing physical structures, helping the public learn more about the early presidencies, revealing details about all the people who lived and worked in the space, and providing an urban alternative to the plantation-set narrative that dominates the public's understanding of slavery.

By encouraging user experimentation, this project encourages the public to serve as active participants. Each user will be able to explore the home on their own, create their own experience, and see the rooms in their own way. The President's House also honors the authority of other museums, archives, and historic sites over this history, including Mount Vernon and the NPS site in Philadelphia. We have consulted with experts at these sites in recognition of their shared authority. The house will include "Points of Interest" on furniture, art, décor items, and architectural elements that survive. These POIs will link to the items in collections and sources in archives that provide documentation for the re-creation. For example, before moving in, Washington added a bow window in the first- and second-floor parlors. The house will have a POI on the windows and link to the letters that discuss this architectural addition.[28] Similarly, the globe that stood in a corner of Washington's private study will be linked to the globe in the Mount Vernon collections.[29] These links will invite students and other members of the public to engage with these institutions and recognize archive and museum collections' shared authority over this history.

The President's House will run online to make it as accessible as possible. The users will not be required to download additional software, as we wish to eliminate all obstacles to public use. Long term, we would love to partner with the NPS to create a virtual reality mobile application that will allow users to walk through the 3D site on their phones, while walking through the physical site in person.

Collaboration has proved to be the biggest challenge with the President's House project. While we have consulted with many enthusiastic community-based partners and other stakeholders of this history, we still have work to do.

We plan to consult with local communities, especially Black descendants, to hear what they need and want to see from this project. Secondarily, as historians, we do not have the technological skills to build out the digital platform. Finding the appropriate graphic artists and computer scientists, and the required funding to compensate them for their work, has been challenging.

James Madison's Montpelier

A few months after George Washington left Philadelphia in March 1797 for his home in Virginia, another Virginian would do the same. Accompanied by his wife, Dolley, James Madison traveled to Montpelier, his father's plantation in Orange County, Virginia. Though he would be called to political service far from Montpelier again and again, he constantly talked, wrote, and thought about building, farming, and enslaving at the plantation.[30] By the time Madison last retired to Montpelier after the end of his two terms as president, approximately 112 enslaved people would also call that place home.[31] They were the individuals who built and beautified the Madisons' house and grounds. They polished the mahogany furniture and the silver, smoked meats and baked biscuits, renovated the mansion, harvested the crops, and labored in countless other ways to make possible the Madisons' life at Montpelier.[32]

These are the individuals the contemporary Montpelier historic site center in their exhibition: "The Mere Distinction of Colour." Opened in June 2017, the exhibit tells the story of all the people who lived at Montpelier through the voices of descendants, archaeological finds, archival materials, and material culture items. In the process of creating this exhibit, the Montpelier team, led by Elizabeth Chew, realized that their current collections databases impeded their collaborative efforts, both within the estate and with the broader community. Since then, the archaeological, curatorial, history, and education team at James Madison's Montpelier has been working to create a new digital platform, titled Legacies of Montpelier Digital Collections Project, that will be available to the site's employees, researchers, and the wider public. The goal of the new platform is to tell the full, complicated story of Montpelier from Madison and the Constitution to the du Pont family's ownership in the twentieth century, enhance accessibility and public engagement, and center the shared authority of the Montpelier descendant community.

The new platform will eventually house the records of 3 million archaeological objects, 4,000 decorative arts items, 1,100 architectural items, and 35,000 historic documents (or links to documents in other collections) in one place. The need to search and compare all these items in one database became clear while the team was completing "The Mere Distinction of Colour." Mary Minkoff, Montpelier's curator of archaeological collections, recalled that she had to search through seven databases to find the appropriate selections for one panel on buttons created and worn by enslaved individuals.[33]

To complete the new database, the Montpelier team received a planning grant from the National Endowment for the Humanities (NEH) to convene a working group consisting of digital humanists, historians, descendant community representatives, and site volunteers, for a two-day intensive planning session. The participants formed a governing committee to guide the project going forward. By organizing this initial workshop, the staff at Montpelier embraced collaboration, inclusive history, and shared authority from the very beginning. It is not hyperbole to say that their design process models the best practices for digital public history.

After the initial planning workshop, the governing committee applied for an NEH completion grant to fund the design of the future platform and the populating of the database with an initial sampling of items. The Montpelier team is partnering with the Matrix program at Michigan State University to use the KORA platform for this project.[34] This collaboration has proved beneficial for both sides. A few years ago, the Michigan State team received a large NEH grant to build out the Enslaved database, which collects data and research on enslaved individuals from historians and archives across the country.[35] In the coming years, the Matrix team at Michigan State is planning to expand the Enslaved database to include material culture, archaeological, and architectural items. Montpelier can provide these items, and in return the Michigan State team will provide the technological expertise required to build the Montpelier platform. This partnership solves one of the challenges that stymies many public historians and historic sites—most historians do not have the technical skills to build a database, while most computer scientists do not have the historical knowledge to offer crucial historical context.

The final platform will allow users to access material anywhere they have connection to the internet, as the current databases can only be used on site at Montpelier. Certain details, like collection management data, will be accessible to Montpelier curators only. The rest of the platform will feature analysis and interpretation from curators, historians, and descendants. To design the platform, the staff drew inspiration from the Mukato program utilized by Native communities, which permits communities to share their own oral histories and interpretations with equal weight. The descendant community, represented by the Montpelier Descendants Committee, will select their own representatives to input data, which offers one more level of shared authority.[36]

The completed platform will also facilitate the user experience by linking archival, material culture, and archaeological records to a 3D re-creation of the mansion house, landscape, enslaved dwellings, and other outbuildings. Users will also be able to search by individual name to browse items and documents linked to that person, which was particularly important to the descendant community, who emphasized the necessity of centering enslaved individuals' names as often as possible. These different search methods also encourage a user-based experience, recognizing that not all users will have the same research needs.

Finally, the governing committee collaborated with local teachers to determine what resources would be most helpful. The NEH grant will provide funds for Montpelier to hire an education specialist to create blueprints for teachers to incorporate into their classrooms. These blueprints will focus on an inquiry-based model that encourages students to self-discover primary sources, which is both the recent focus of secondary education pedagogy and Montpelier's education programs. The blueprints are easier for teachers to utilize than structured lesson plans, which teachers explicitly stated they did not need. These teachers were not only providing feedback after the fact; they were integral to the blueprints' development. Once again, we see how Montpelier incorporates the best practices of public history; teachers are engaged actors in the project, not simply a receptive audience.

The Montpelier team is hopeful the NEH will provide the needed funding to complete this valuable project. Even with the necessary funding, the team has encountered a few challenges, mostly posed by the enormity of the collection. To manage the sheer number of items, the team needed to articulate a triage strategy. After the opening of "The Mere Distinction of Colour" exhibit, many visitors had the most visceral, meaningful reaction to the final installation, which analyzes the Constitution and the complicated legacy of slavery up through the present moment.[37] This discussion offers a natural focus for the initial round of items loaded into the system. The final product will be a remarkable example for other digital public history projects on how to practice collaboration, inclusive history, and shared authority in their own work.

James Monroe's Highland

Some thirty miles from Montpelier stands another presidential home site: James Monroe's Highland. By 1799 construction on this homeplace was underway, the centerpiece of what would become a 3,500-acre estate in Albemarle County, Virginia, that Monroe hoped would be both a profitable plantation and a comfortable home.[38] Highland served as the Monroe homestead for nearly three decades, but it also served as sites of living and laboring for dozens of enslaved people. In 1828, with his debts swelling, Monroe sold Highland and many of the enslaved laborers living there to a plantation owner in Florida.[39] Over the course of the next century and a half, Highland went through several owners until the College of William and Mary acquired it in 1974.[40]

Much like the man himself, James Monroe's Highland has sat in the shadows of other presidential home sites in Virginia and beyond. Just next door, Thomas Jefferson's Monticello attracts far more scholarly attention, public interest, and monetary contributions. The lack of a coherent plan for Highland contributes to this oversight. While the site had been open to the public before its acquisition by William and Mary, only in the last decade has preservation, interpretation, research, and visitors' experience received the attention they deserve. Under the guidance of the site's second director, Sara Bon-Harper, the expanded staff has debunked much of the assumed history of the buildings and people who inhabited them.[41] Bon-Harper and the Highland staff are committed to changing the narrative of the "forgotten presidential house" including the innovative work of research, and centering the voices of those like the enslaved people sold to Florida in 1828, and they are using digital technology to do it.[42]

In 2018 Highland began using augmented reality in their site tours, which brings recent research discoveries and silenced voices to the forefront.[43] The wearable AR technology—which include glasses, headphones, and a handheld device—adds elements onto a visitor's field of vision, thereby augmenting the reality that they are experiencing. The AR tour at Highland, which staff developed with Richmond-based ARtGlass US, is meant to bring into "reality" two elements that are not currently a part of the landscape: the original 1799 house and the people who inhabited Highland in the 1810s. Rather than physically reconstructing the original house, the AR technology allows a 3D rendering of it to be superimposed onto the current landscape. The technology makes it easy to implement changes to the tour, allowing staff to keep interpretation up to date with the most recent research and archaeological finds. Indeed, Bon-Harper believes that "the potential to alter elements when new information becomes available" is one of the main strengths of digital technology projects like this one.[44]

Additionally, at various points throughout the site, the visitor can hear from different residents of Highland. This narration provides the inclusion of many diverse voices that made up Highland during Monroe's time. While the core of the traditional site tour continues to focus on the Monroes, the AR tour prioritizes the individuals who labored and lived there in bondage. The Highland team intentionally chose to make these figures animated rather than use live actor video, a decision that emphasizes the difficulty in representing people who left so little documentation. As Bon-Harper notes, these animated figures are "meant to evoke the imagination, to allude to the clear knowledge that we can't entirely represent those individuals, enslaved and free, yet we can invite visitors to learn about them, their experiences, and their roles in our shared history."[45]

For example, in front of the guest house visitors can meet the animated figures of Hannah (an enslaved cook) and Nelson (an enslaved blacksmith), who ruminate on the pain of being separated from loved ones. The possibilities of presenting history as overlapping multivocal narratives is greater with technologies like AR that can literally superimpose various layers over others. Patrick Gallagher has argued that visitors "comprehend a concept in more depth when the spaces they are in emulate the reality of the situation."[46] Especially when discussing difficult history, like slavery, this kind of emulation of reality is useful for visitors.

Highland is also working to engage and collaborate with relevant communities today, including those stakeholders who are descendants of individuals like Hannah and Nelson. Around the same time that archaeologists discovered the original 1799 house, Highland staff members also began communicating with a descendant community just four miles away, known as Monroe town.[47] Projects are underway to collect oral histories and include these descendants and their stories in future projects and interpretations.[48]

Inviting visitors into the experience and process of history reflects the user-centered nature of the Highland AR tour.[49] Primary sources and evidence from archaeological digs mix with animation and ambient sounds, combining the historical with the imagined. The visitors are included in the historical process as they learn about how these sources changed the historical narrative, thereby prompting deeper thinking about the processual nature of history. Bringing visitors "behind the curtain" of history, making them feel a part of the intellectual process, encourages a higher level of engagement than simply talking at someone.[50]

Visitors are required to engage with the site and the material to a higher degree than other tours, but they also hold the power to decide how they will move through the landscape, unlike the more formal guided tours typical at most historic sites. A camera in the AR glasses identifies where the visitor is and starts a sequence of images and script, and there is a suggested route for visitors to take, but they hold the reins to decide where they want go and how much time they want to spend in different parts of the site.

Centering the user also offers challenges, as does the technology. Like other self-guided tours, visitors may feel overwhelmed by all their options. When you add new technology, visitors may not feel they have the skills to engage with the tour, reflecting the latest conversation over the digital divide. Indeed, the wearable technology is cumbersome and a tad uncomfortable, which might be an additional deterrent to certain visitors. Like all outdoor tours, they rely on good weather, yet a particularly bright and sunny day is a negative for AR tours, as it impedes one's sight of the augmented landscape. Still, like all the projects surveyed here and discussed in the relevant literature, the challenges of digital public history do not outweigh the possibilities. A digital public history project like the Highland AR tour provides us with an incredible opportunity to use technology to superimpose the past over the present, thereby providing visitors with a unique visual understanding of how the built environment affected the lived experiences of free and enslaved persons.

These three projects offer an introduction to best practices of digital public history: inclusion, collaboration, shared authority, and user-centered history. But they are also, just like any good public history project, solid pieces of historical scholarship that contribute to the historiographical debates about the American Revolution and the Founders. The President's House, James Madison's Montpelier, and James Monroe's Highland all employ digital tools to center the lives of enslaved and previously overlooked individuals. All three projects also demonstrate the diverse ways digital public history can collaborate and respect the shared authority of multiple communities. In public history literature, collaboration and shared authority are typically described as historians or academics working with descendant and/or local communities.[51] These projects demonstrate that collaboration and shared authority are much more complex and multifaceted than previously described. Shared authority *can* be engaging descendant communities in the planning and interpretation of a project, as modeled by Montpelier, but user-centered digital public history projects also offer authority to each viewer to choose their own experience. Similarly, collaboration can exist between professional historians, local communities, and enthusiasts, like at Montpelier. But it can also be an enriching, interdisciplinary partnership across several professional fields, like the President's House and Highland AR projects. If these projects reveal anything about digital public history, it is that the field is continuously evolving and expanding on how to deploy these best practices.

When we conceived of this chapter, we expected to write about the funding challenges of digital public history projects, or perhaps the technological skills required to build online platforms. While those challenges continue to plague the public history field, the COVID-19 pandemic has shed new light on the benefits and obstacles of digital public history. In the early years of digital humanities, many practitioners wrote about the inequalities inherent in the field, particularly as they related to the digital divide mentioned earlier. Access to the internet, computers, and databases exacerbated existing disparities along race and class lines. In recent years, many scholars have assumed that increased internet access and decreasing prices for technology have had a democratizing effect.[52] Yet the COVID-19 pandemic has revealed that these inequalities are far from eradicated.

Despite the incredible challenges, the pandemic has also presented new opportunities to the field of digital public history. Now more than ever, public history sites and scholars are embracing the opportunity to connect with audiences across the globe using a digital medium. However, as Thomas Cauvin wrote in *Public History: A Textbook of Practice*, not all digital history projects are public history.[53] As more traditional academic historians turn toward digital platforms, the greatest challenge going forward will be to reimagine their scholarship as public history projects, especially if they are new to the field and best practices of digital public history.

NOTES

1. Annette Gordon-Reed, *Thomas Jefferson and Sally Hemings: An American Controversy* (Charlottesville: University of Virginia Press, 1998); Annette Gordon-Reed and Peter Onuf, *"Most Blessed of Patriarchs": Thomas Jefferson and the Empire of Imagination* (New York: Liveright, 2016); Erica Armstrong Dunbar, *Never Caught: The Washingtons' Relentless Pursuit of Their Runaway Slave, Ona Judge* (New York: 37Ink, 2017); Robert Parkinson, *The Common Cause: Creating Race and Nation in the American Revolution* (Chapel Hill: University of North Carolina Press, 2016); Robert Parkinson, *Thirteen Clocks: How Race United the Colonies and Made the Declaration of Independence* (Chapel Hill: University of North Carolina Press, 2021).

2. These projects, therefore, continue the pioneering work of women's historians such as Linda Kerber who have for decades emphasized the blurred boundaries between public and private spheres.

3. The central work in this debate is Gordon Wood, *The Radicalism of the American Revolution* (New York: Alfred A. Knopf, 1991). Responses include Gary Nash, *The Unknown American Revolution: The Unruly Birth of Democracy and the Struggle to Create America* (New York: Penguin Books, 2006); Parkinson, *Common Cause*; Parkinson, *Thirteen Clocks*; and Christopher R. Pearl, *Conceived in Crisis: The Revolutionary Creation of an American State* (Charlottesville: University of Virginia Press, 2020).

4. For digital history, see Daniel J. Cohen, Michael Frisch, Patrick Gallagher, Steven Mintz, Kirsten Sword, Amy Murrell Taylor, William G. Thomas III, and William J. Turkel, "Interchange: The Promise of Digital History," *Journal of American History* 95, no. 2 (2008): 452–91; for public history, see National Council on Public History, https://ncph.org/what-is-public-history/about-the-field/.

5. Jesse Stommel maintains that "the public digital humanities is a Venn diagram at the point where public work, digital work, and humanities work intersect." Jesse Stommel, "Public Digital Humanities," in *Disrupting the Digital Humanities*, ed. Dorothy Kim and Jesse Stommel (New York: Punctum Books, 2018), 79–90, 81.

6. Sharon M. Leon, "Complicating a 'Great Man' Narrative of Digital History in the United States," in *Bodies of Information: Intersectional Feminism and the Digital Humanities*, ed. Elizabeth Losh and Jacqueline Wernimont (Minneapolis: University of Minnesota Press, 2018), 344–66. Leon notes that most digital history scholars focus on the work of other academics, not the digital history work being done at museums, libraries, and other cultural institutions outside of academia. This further masks the work of women, who are the core of the staff at these kinds of institutions.

7. Lara Kelland, "Digital Community Engagement across the Divides," *History@Work* (blog), April 20, 2016, NCPH, https://ncph.org/history-at-work/digital-community-engagement-across-the-divides/.

8. Jason Heppler, Rebecca Wingo, and Paul Schadewald make a distinction between shared authority and "a shared authority," emphasizing that the former privileges the historian's power while the latter emphasizes engagement. Jason Heppler, Rebecca Wingo, and Paul Schadewald, introduction to *Digital Community Engagement: Partnering Communities with the Academy*, ed. Jason Heppler, Rebecca Wingo, and Paul Schadewald (Cincinnati, OH: University of Cincinnati Press, 2020).

9. Cherstin M. Lyon, Elizabeth M. Nix, and Rebecca K. Shrum propose that shared authority and collaboration (along with reflective practice) are the distinguishing characteristics between public history and other types of historical practice. *Introduction to Public History: Interpreting the Past, Engaging Audiences*, ed. Cherstin M. Lyon, Elizabeth M. Nix, and Rebecca K. Shrum (Lanham, Md.: Rowman and Littlefield, 2017), chapter 1.

10. Wendy F. Hsu, "Lessons on Public Humanities from the Civic Sphere," in *Debates in the Digital Humanities 2016*, ed. Matthew K. Gold and Lauren F. Klein (Minneapolis: University Minnesota Press, 2016), 280–86. Richard Weible, among others, has argued that, while shared authority is important, some public historians cede *too* much authority to the public. He contends that not everyone has the same ability to interpret the past and that should be recognized. This, however, seems an oversimplification of the shared authority that public historians actually propose, which recognizes and values the expertise of historians (as well as other participants). Richard Weible, "Defining Public History: Is It Possible? Is It Necessary?," *Perspectives on History*, March 1, 2008, http://www.historians.org/publications-and-directories/perspectives-on-history/march-2008/defining-public-history-is-it-possible-is-it-necessary.

11. Sharon Leon, "User-Centered Digital History: Doing Public History on the Web," *Bracket*, March 3, 2015, https://www.6floors.org/bracket/2015/03/03/user-centered-digital-history-doing-public-history-on-the-web/.

12. Roy Rosenzweig and David Thelen, *The Presence of the Past: Popular Uses of History in American Life* (New York: Columbia University Press, 1998).

13. Nina Simon, *The Participatory Museum* (Santa Cruz, CA: Museum 2.0, 2010).

14. Lyon, Nix, and Shrum define stakeholders as "those who have a specific interest or a stake in the topics we study, the communities about which we write, or the institutions or places where we work." *Introduction to Public History*, 2.

15. Tim Grove, "New Media and the Challenges for Public History," *Perspectives on History*, May 1, 2009, https://www.historians.org/research-and-publications/perspectives-on-history/may-2009/new-media-and-the-challenges-for-public-history.

16. Simon has shown how the digital can reconnect the public to museums and history through encouraging "active engagement." Simon, *Participatory Museum*.

17. Wendy Hsu had luck with live-tweeting events, while Faye Sayer found crowdsourcing to be a good way of engaging communities. Hsu, "Lessons on Public Humanities"; Faye Sayer, *Public History: A Practical Guide* (London: Bloomsbury, 2015).

18. Leon, "User-Centered Digital History."

19. Andrew Hurley, "Chasing the Frontiers of Digital Technology: Public History Meets the Digital Divide," *Public Historian* 38, no. 1 (February 2016): 69–88, 87.

20. Sheila A. Brennan, "Public, First," in *Debates in the Digital Humanities 2016*, ed. Matthew K. Gold and Lauren F. Klein (Minneapolis: University Minnesota Press, 2016), 384–90.

21. Meg Foster, "Online and Plugged In? Public History and Historians in the Digital Age," *Public History Review* 21 (December 2014): 1–19.

22. Lindsay M. Chervinsky, "The Enslaved Household of President George Washington," Slavery in the President's Neighborhood Initiative, White House Historical Association, https://www.whitehousehistory.org/the-enslaved-household-of-president-george-washington.

CHAPTER 1

23. Lindsay M. Chervinsky, "The Households of President John Adams," Slavery in the President's Neighborhood Initiative, White House Historical Association, https://www.whitehousehistory.org/the-households-of-john-adams.

24. Edward Lawler, "The President's House in Philadelphia," US History.org, https://www.ushistory.org/presidentshouse/history/pmhb/ph1.php.

25. City of Philadelphia, https://www.phila.gov/presidentshouse/pdfs/Presidents_House_Site_December152010.pdf.

26. Lindsay M. Chervinsky, *The Cabinet: George Washington and the Creation of an American Institution* (Cambridge, MA: Harvard University Press, 2020).

27. Dunbar, *Never Caught*; Lawler, "President's House in Philadelphia."

28. George Washington to Tobias Lear, September 5, 1790, *The Papers of George Washington*, Presidential Series, ed. Mark A. Matromarino (Charlottesville: University Press of Virginia, 1996), 6:397–401.

29. "Terrestrial Floor Globe," Mount Vernon Ladies Association, W-166, https://www.mountvernon.org/preservation/collections-holdings/browse-the-museum-collections/object/w-166/.

30. Letters to and from James Madison during this era (including dozens in *The Papers of James Madison*) show his deep interest and involvement in the building and landscaping activity of the entire plantation.

31. Although only 53 individuals were tithable/taxable (16 years or older), 112 were known to be living at Montpelier around 1820. See appendix B, "Madison Family Slave Population, 1720–1850," in Douglas B. Chambers, *Murder at Montpelier: Igbo Africans in Virginia* (Jackson: University Press of Mississippi, 2005), 199–208, 204.

32. Galliard Hunt's 1902 biography of James Madison almost completely elides the labor of enslaved people, but the extensive list of daily chores, beautification projects, and renovations around Montpelier during James and Dolley's residence is an indicator of just some of the work enslaved people engaged in for the benefit of the Madisons. Galliard Hunt, *The Life of James Madison* (New York: Doubleday Page, 1902), esp. chapter 36.

33. Mary Furlong Minkoff, Elizabeth Chew, and Matt Reeves, conversation with authors, August 7, 2020.

34. Legacies of Montpelier Digital Collections Project Narrative, National Endowment for the Humanities grant application, provided by Mary Minkoff, August 2020.

35. Enslaved database, https://matrix.msu.edu/enslaved.

36. Legacies of Montpelier Digital Collections Project Narrative, National Endowment for the Humanities grant application; Mary Furlong Minkoff, Elizabeth Chew, and Matt Reeves, conversation with authors, August 7, 2020. For more on the Montpelier Descendants Committee (MDC), see https://montpelierdescendants.org/. In 2021 the Montpelier Foundation made the enormously important step of committing to structural parity, thereby creating a shared governance structure with descendants. But in March 2022, the Montpelier Foundation's board of directors reversed their decision and soon after fired several staff members, including Elizabeth Chew and Matthew Reeves, who spoke out against this unacceptable reversal. Thanks to the persistent work of the MDC, who launched a campaign that drew thousands of signatures and widespread media attention, the board recommitted to structural parity in May 2022 and rehired the unjustifiably fired staff members soon thereafter. Montpelier seems to once again be on a promising path. See James French, "Montpelier: A Model for Reconciliation in Peak Polarization," *Washington Post*, May 16, 2022, https://www.washingtonpost.com/opinions/2022/05/16/montpelier-model-reconciliation-peak-polarization/.

37. Amber Galaviz, "Montpelier: The Mere Distinction of Colour," *Daily Progress* (Charlottesville, VA), May 25, 2017, https://dailyprogress.com/orangenews/news/montpelier-the-mere-distinction-of-colour/article_b6afc4b4-415f-11e7-b93d-37b395d75de5.html#:~:text=%E2%80%9CWe%20have%20seen%20the%20mere,man%2C%E2%80%9D%20the%20quote%20reads.

38. In July 1799, Monroe wrote to Madison asking him to visit and give advice on "farming and house building." James Monroe to James Madison, July 13, 1799, in David Preston, *A Comprehensive Catalogue of the Correspondence and Papers of James Monroe* (Westport, CT: Greenwood Press, 2001), 73. For more on Monroe's landholdings, see Christopher Fennell, An Account of James Monroe's Land Holdings, http://www.histarch.illinois.edu/highland/ashlawn5.html.

39. James Monroe to James Madison, March 28, 1828, Take Them in Families, https://taketheminfamilies.com/.

40. The Ash Lawn–Highland Records in the Special Collections of Swem Library at the College of William and Mary include architectural and archaeological surveys and reports dating back to 1975.

41. For much of its history, it was assumed that the Guest House, built in 1818, was the original house. An archaeologist by training, Sara Bon-Harper led the way in debunking this, discovering the foundation of the original 1799 house in front of the guest house. See "'A Plan of Comfort from the Unquiet Theatre': Phase I and II Archaeological Investigations at James Monroe's Highland (44AB0398), Albemarle, Virginia," Highland archive, Rivanna Archaeological Services (Charlottesville, VA: Rivanna Archaeological Services, 2016).

42. Sara Bon-Harper, conversation with authors, June 4, 2018.

43. Augmented reality and its digital sibling virtual reality have a long history, going back to the late 1960s, but it has been just since the 2010s that museums and historic sites have begun exploring the technology. The American Association for Museums, for instance, included an article on the "promise and peril" of AR/VR in their 2016 *TrendsWatch* magazine, an annual report that highlights the most important trends happening in the museum world. Center for the Future of Museums, "Me/We/Here/There: Museums and the Matrix of Place-Based Augmented Devices," *TrendsWatch 2016*, 23. For several examples of how museums are currently using AR/VR technology, see Charlotte Coates, "How Museums Are Using Augmented Reality," *Museum Next*, July 17, 2021, https://www.museumnext.com/article/how-museums-are-using-augmented-reality/. Many of the museums at the forefront of the AR/VR trend are art museums, but the possibilities for historians, teachers, and historical institutions are just as great. See John Bonnett, "Following in Rabelais' Footsteps: Immersive History and the 3D Virtual Buildings Project," *Journal of the Association for History and Computing* 6, no. 2 (September 2003): 107–50. For more on public history and AR/VR, including some of the challenges, see Christian Bunnenberg, "Virtual Time Travels? Public History and Virtual Reality," *Public History Weekly*, February 1, 2018, https://public-history-weekly.degruyter.com/6-2018-3/public-history-and-virtual-reality/. AR/VR is beneficial not only for educational purposes but also especially for preservation. See Megan Crutcher, "Artificial Reality Is the Likely Future of Preservation," *History@Work* (blog), June 11, 2020, NCPH, https://ncph.org/history-at-work/artificial-reality-likely-future-of-preservation/.

44. Sara Bon-Harper, interview by authors, March 30, 2020.

45. Sara Bon-Harper, "New History at James Monroe's Highland," *VAM Voice*, Winter 2018 (February 7, 2019), https://issuu.com/vamuseums/docs/winter_2018_vam_voice_.

46. Cohen et al., "Interchange," 470.

47. Nancy Stetz, education manager, conversation with authors, June 5, 2018. See more on Highland's Council of Descendant Advisors at Monroe's Highland, https://highland.org/descendant-advisors/.

48. "W&M secures $1 million for research, teaching about those enslaved on campus, Monroe's Highland," *Williamsburg Yorktown Daily*, July 31, 2019, https://wydaily.com/local-news/2019/07/31/wm-secures-1-million-for-research-teaching-about-those-enslaved-on-campus-monroes-highland/.

49. Recent surveys of museums and historic sites have shown how AR can improve a visitor's engagement with the past, though it requires far more than simply having

technology available in the welcome center. As Stella Ress and Francesco Cafaro have shown, it must be incorporated into the site and into the experience. It is still important to know why visitors come to historic sites, what they expect, and what they already enjoy, for only then can we use technology as a means of engaging visitors further and enhancing their experience. Stella A. Ress and Francesco Cafaro, "'I Want to Experience the Past': Lessons from a Visitor Survey on How Immersive Technologies Can Support Historic Interpretation," *Information* 12, no. 1 (2021): https://doi.org/10.3390/info12010015.

50. See Rosenzweig and Thelen, *Presence of the Past*.

51. Heppler, Wingo, and Schadewald, introduction.

52. Shawan Graham, Guy Massie, and Nadine Feuerherm, "The HeritageCrowd Project: A Case Study in Crowdsourcing Public History," in *Writing History in the Digital Age*, ed. Jack Dougherty and Kristen Nawrotzki (Ann Arbor: University of Michigan Press, 2013), 222–32.

53. Thomas Cauvin, *Public History: A Textbook of Practice* (New York: Routledge, 2016), 177.

2

NEW MEDIA AND OLD PROBLEMS
Restoring Humanity in the Maryland
Loyalism Project

Kyle Roberts and Benjamin Bankhurst

Following the American Revolution, Loyalist refugees lined up in New York, London, and Halifax, Nova Scotia, to share their experience of upheaval in the hope that the British state would recognize their political allegiance and compensate them for their losses through evacuation, land, a pension, or reimbursement. The state recorded the narratives of this refugee population, one of the eighteenth-century Atlantic's largest, in two series of documents: the Inspection Roll of Negroes (known also as the Book of Negroes) and the papers of the parliamentary Loyalist Claims Commission (LCC). Both serve to establish the political allegiance and right to confiscated property of women and men who supported the Crown during a civil war, in the face of betrayal, persecution, and loss of family and friends. They are filled with firsthand accounts of wartime experiences, biographical details, and evidence of networks and geographic movements of a displaced people.

Nearly 250 years later, the Maryland Loyalism Project engages undergraduate and graduate students to use digital platforms to make these poignant stories and revealing data available to scholarly and descendant communities.[1] Scholars have long used these sources, but too often independently of each other, focused either on the experience of white Loyalism or Black self-emancipation.[2] Rarely is the history of Loyalism multihued.[3] In telling this broader history, a guiding principle is to document while not reproducing the inhumanity often embedded in the construction and content of these historical records. Even as white Loyalists emphasized their suffering and the inhumanity of American rebels, they submitted financial claims that chronicled their own denial of humanity to enslaved women and men. Most enslaved people were recorded in aggregate counts ("95 Negroes") and denied the dignity of being named.[4] As Black Loyalists told their narratives of flight from Patriot enslavers and service in the British army, British and American commissioners, who stood between their return to enslavement in the United States and their freedom elsewhere in the British Empire, reproduced in manuscript ledger books the data categories and language of runaway slave advertisements, perpetuating the commodification of the newly emancipated.

This chapter reflects on the challenges and opportunities the Maryland Loyalism Project offers through the use of digital platforms and an engaged pedagogy grounded in restoring the humanity of all those who were impacted by the British loss in the American Revolution. The authors take seriously calls by Jessica Marie Johnson and other scholars not to perpetuate the datafication of enslaved people in creating this new digital project.[5] They seek to construct antiracist ontologies for their digital archive and database to represent data extracted from these rich archival sources while recognizing past interpretations have the potential to influence present construction of these ontologies. Yet the authors have found that engaging students to study these disparate sources can reveal latent assumptions that might otherwise limit possibilities for new interpretations. This chapter reflects on the affordances provided by different platforms for digital humanities work and reveals possible ways to surmount silences within the archival record when creating a digital project that embodies eighteenth-century lives and experiences for the education and edification of twenty-first-century scholarly and popular audiences.

Origins of the Project

The authors decided to create the Maryland Loyalism Project as a public-facing digital archive and biographical database after team-teaching a synchronous online course for Loyola University Chicago and Shepherd University students on digital approaches to the American Revolution in the spring semesters of 2017 and 2019. The course—open to undergraduates, honors students, and public history graduate students—instructed students in new approaches to digital history through the lens of a pivotal historical event. Assignments throughout the semester introduced students to different digital skills as a means of exploring primary sources and learning a variety of methodological approaches, such as data creation, textual analysis, and spatial visualization.[6]

A popular assignment involved creating a timeline to highlight the experience of a Loyalist refugee based on the Loyalist Claims Commission records. The Loyalist Claims Commission was created by an act of Parliament in July 1783.

With the loss of the colonies all but inevitable, the commission sought to provide compensation and financial support for Loyalist refugees. The LCC produced the largest collection of government sources on American Loyalism. Housed principally in the Audit Office Papers (AO) at the British National Archives, the LCC records are divided into two series, AO12 and AO13. The first series, AO 12, is the official record of the LCC in 146 volumes composed between 1776 and 1831. These volumes consist of compiled copies of original material submitted to the commissioners, including Loyalist memorials, evidence given by witnesses attesting to the validity of Loyalist claims and character references in support of individual claimants, and commission notes of the hearings held in either London or Halifax, Nova Scotia. The number of Loyalists who submitted claims to the commission totaled 3,225 seeking £8,216,126 in restitution. The commissioners deemed 2,291 claims worthy of compensation totaling £3,033,091.[7]

The data-rich LCC volumes are difficult to use in their analog form but ideal for the digital medium. Provided with digital images of the memorials of a dozen or so Maryland Loyalists, students had the opportunity to view the American Revolution from a different perspective, which helped them see the particular complexities of life during the Revolution and more universal issues around wartime displacement. Students transcribed a few pages of a digital surrogate of a memorial, constructed a dataset from that memorial (based on a template), and learned how to design a web-based story map of the historic subject's experience. The class culminated in students performing their Loyalist's claims for their classmates, themselves playing the role of the Claims Commission, and learning what the state ultimately awarded the person. Empathy was built through reconstructing the lives of refugees, while also providing an opportunity for understanding the rhetoric of petitioning and the material worlds these individuals and their families lost.

The authors' experience teaching this course made them realize the importance of this material not only for their students' education but also for the field in general. They determined to pilot a project to digitize, transcribe, index, and make available the original LCC records. They chose to focus on Maryland for four reasons. First, many of the Shepherd University students hail from western Maryland. Local stories of Maryland refugees offered specific relevance to students familiar with the state's geography and history. Second, the subset of records on the Maryland colony was of a manageable size.[8] Third, Maryland's different economic regions, ranging from the tobacco plantations of the Eastern Shore to the wheat fields of the mountainous west, coupled with the colony's ethnic diversity, ensured an array of Loyalist experiences. The Maryland claims include testimonies from German freeholders working relatively modest farms, women tavern keepers, soldiers, and owners of large estates. Fourth, Maryland

Loyalists remain understudied in comparison to other colonies, allowing students to make a valuable contribution to scholarly research.[9] The authors were not initially aware, but have come to appreciate, how using Maryland as a focus empowered their students to tell a more complex story of Loyalism, one both white and Black, free and enslaved.

Between 2019 and 2021, the Maryland Loyalism Project transformed from a classroom assignment into a standalone digital project through the work of student interns. A Lapidus Digital Collections grant from the Omohundro Institute for Early American History and Culture supported digitization of manuscript volumes at the National Archives, employed student interns to develop the project, and covered initial hosting expenses. In the summer of 2019, Zachary Stella, a student in the Digital Humanities MA program at Loyola University, worked with the authors to design the site. Over the 2019–2020 academic year, the Shepherd University digital interns Claire Tryon and Michael Mastrianni revealed the research possibilities for such a site while documenting particular challenges in representing those who were marginalized in the LCC records—namely, white women and the enslaved. In fact, Mastrianni's work was crucial in convincing the authors to include the Marylanders listed in another state-created document, the Inspection Roll, on an equal footing in the digital project.

Even before the commencement of the Loyalist Claims Commission, representatives of the British state created the Inspection Roll to record Black Loyalists who sought freedom beyond the United States in return for their service during the war. In 1783 Sir Guy Carleton, commander-in-chief of His Majesty's Armies in North America, was tasked with the responsibility of orchestrating the removal of British military forces and evacuation of Loyalists who did not want to remain in the United States.[10] As the war wound down, Loyalists fled to Savannah, Charleston, and New York. Among their number were tens of thousands of women and men who had freed themselves from bondage by escaping their Patriot enslavers.[11] The Inspection Roll emerged from Carleton's refusal to adhere to George Washington's expectation that escaped slaves would be returned to fulfill the terms of the preliminary Treaty of Paris.[12] Carleton insisted that anyone who served the British army for a year or more was now free and no longer considered the "property" of Americans. He ordered his officers to register all self-emancipated women and men who could prove their service to the British forces and wished to leave the new United States. The three-volume set of detailed ledgers represents nothing less than, in the words of the novelist Lawrence Hill, "the first massive public record of blacks in North America."[13] The ledgers record nearly 3,000 men, women, and children who shipped out on 219 different voyages from New York between April and November 1783.[14] Transcriptions of the LCC and Inspection Roll materials for Black and white Maryland

Loyalists was largely completed in the summer of 2020 by Jillian Curran and Elizabeth Lilly, C. V. Starr Center for the American Experience at Washington College interns at the American Philosophical Society.

When the first iteration of the Maryland Loyalism Project site went live in late summer 2020, it showcased the testimonials of Black and white Loyalists who had actively sought recognition, and often compensation, from the British Government. The site did not yet, however, give voice to the silenced women and men listed as property confiscated by the Maryland legislature in the LCC archive. It was clear that another source base was needed to rectify this injustice and to fill the absences in the parliamentary documents. To that end, the project team digitized the records of the Maryland Commission to Preserve Confiscated British Property and asked Shepherd University undergraduate students to transcribe the documents as part of their coursework in the spring 2021 semester.

Representing Eighteenth-Century People on Twenty-First-Century Digital Platforms

The decision to create a digital archive and database of the LCC and Inspection Roll materials represented one more in a long line of new media remediations, this time in a digital format. Previous remediations—whether as transcription, microfilm, print, or electronic database—were designed to provide different modes of access for audiences who could not use the original manuscripts with ease. Each remediation, however, contributed new layers of meaning with which scholars must contend.[15] The authors recognized their efforts similarly ran the risk of perpetuating the interpretations, interventions, and legacies of the past, especially the marginalization of certain voices. Extracting information from eighteenth-century manuscript records and representing it as tabular data in the database are not neutral acts, each requiring specific decisions to be made about what is recorded, what is not, and what is inferred. Scholars in Black digital history and critical race studies have spoken powerfully to issues of representation and the necessity of care that arise as analog sources are translated to digital platforms.[16] As the authors and students discovered, the act of designing a database requires a variety of decisions about the naming, definition, and representation of categories, properties, and relations. Building off nearly 250-year-old sources (and their subsequent translation and interpretation by editors and scholars) can bring with it many assumptions that intentionally or inadvertently are biased.

But remediation can also offer an opportunity to approach the past with a new sensitivity, a new humanity, to begin to remedy the failures of the past. Digitization, as Michael Kramer paradoxically notes, might actually ask us to slow down (to excavate previous remediations and to think through how to represent analog data on a new digital platform) rather than to simply speed up (through increasing the availability and accessibility of these documents—and, by implication, the harmful practices they maintain).[17] All creators need to find new modes of correcting the record. The manuscript ledger is frozen in time, but the digital archive and database need not be.

Affordances of Platform

A variety of platforms are available today for supporting digital humanities projects. The challenge is matching design intention and platform affordances with the data available, the mode of representation desired, and the expected audience. Selecting the right platform is particularly important for student-centered projects. Such platforms should be familiar to students from their coursework as well as accessible for their use after graduation. In the long run, proprietary software that is only available in higher education is not helpful for a public history student working at a small historical society or museum.

The idea of affordances has been deployed from a variety of disciplinary perspectives, but at its core, it usefully denotes the range of circumstances in an environment that allows for something to happen.[18] A digital platform's affordances might provide for different modes of representation, capacities for storage and preservation, and modes of analysis. The authors and site developer thought carefully about these concerns when they created a standalone digital project. In addition to being student friendly, the project had to make eighteenth-century documents accessible to twenty-first-century audiences. The documents required an archive, but users needed a database to locate individuals. The project team had to find a platform or platforms that would achieve those ends.

The project team went back and forth debating Scalar and Omeka as possible platforms and, in the end, decided to use both. Scalar (https://scalar.me/anvc/) is an online digital publishing platform modeled on the idea of a digital edition, an orientation reflected not only in its structure but also in its terminology. (For example, the site refers to projects as "books.") Scalar's structure works particularly well at representing volumes digitally. A stipulation of the Lapidus grant was that all digitized materials would be made available for public use; Scalar was an obvious choice for the digital archive.[19] Omeka (https://omeka.org/), on the other hand, is a web publishing platform modeled on a database that excels at representing tabular data. Omeka S expands beyond the original use of Dublin Core to include a range of other standard and customizable ontologies.[20] This allows for the creation of resource templates that accommodate a wider variety of information. Permitting individual items, the basic building blocks within the program, to be associated with others in item sets makes Omeka a useful relational database, correlating connections between various data.

Representations of Individuals as Data

Having acknowledged the advantages and limitations of Scalar and Omeka for indexing individuals named in the manuscript volumes, the project team had to decide *who* would be represented in the database, *how* that would be done, and *what* aspects of their lives could or should be converted into tabular data. The nature of the source required the team to make careful decisions about how to translate from an analog to a digital format. The history of remediations of these sources made the team aware that their work was not neutral; the line between objective and subjective data choices always had to be considered.

Construction of the Omeka database began with the design of a people resource template based on the LCC records outlining the fields to be recorded for individuals. Classroom exercises and internship assignments had already introduced students to the steps of identifying and converting specific types of biographical information into tabular data. For example, students had been asked to identify biographical, financial, and social network information for a Loyalist in the LCC records. A follow-on step asked them to think about the type of field format in which that information was represented, with an eye toward standardization and interoperability.[21] Inspiration came from looking at other digital projects specifically on Loyalism, such as Loyalist Migrations and New Brunswick Loyalist Journeys.[22]

Before the project team could determine what data to record about individuals, it had to decide on whom, out of the thousands of names in the LCC records for Maryland, to focus. Omeka allowed the project team to group individuals into item sets based on shared characteristics. In Scalar, the presentation of manuscript ledgers was organized around memorialists, the individuals who "presented claims before the commission." This follows the practices set by the clerks of the Loyalist Claims Commission. Memorialists thus provided the first item set in Omeka. The team could have stopped there, providing an index of those who pursued recognition and reward for their political allegiance. This would have been consistent with earlier remediations, such as those by Gregory Palmer and Peter Wilson Coldham.[23] But this practice posed two ethical problems. First, it reproduced an often white, patriarchal, heteronormative household structure. White women are infrequently named as memorialists in the LCC volumes. White married women almost never are named, even when the property they brought into the marriage was the subject of the dispute. Unmarried or widowed white women did make claims, but largely because of their expanded property rights. Second, it missed the opportunity to explore the networks—economic, political, religious, social—that connected Loyalists in the colonies and the diaspora.

Reading through the memorials quickly reveals that these accounts are heavily populated with a range of people—male and female, Loyalist and Patriot, free and enslaved—whose lives intersected with the memorialists. Testimonies—some written, more often oral—from other members of the Loyalist diaspora were utilized to verify their claims, either of political allegiance or property ownership. Those witnesses, as the project team labeled them, admittedly looked a lot like the memorialists—white, predominantly male, also living in exile. While memorialists and witnesses both speak directly through the records, a third, and much larger, group speaks through the mediation of others. The name the project team assigned to this third group, the mentioned, reflects as much their auxiliary relationship to the proceedings as their own lack of agency in the created record. They neither actively presented a memorial nor witnessed someone else's allegiance, but were passively invoked for a range of reasons: for being family members; Patriot persecutors; Loyalist neighbors; or past, present, or future property owners. Nearly eight hundred had been enslaved by memorialists and claimed as lost property, tallied alongside land, household furnishings, and livestock.

The decision in the summer of 2020 to include Marylanders from the Inspection Roll in the digital archive and database aimed to provide a truer sense of the Loyalist experience but also required rethinking the assumptions that undergird the database structure in two ways. First, the project team had to determine—within a structure of memorialists, witnesses, and mentioned—where to locate the self-emancipated who presented themselves to the commissioners for transportation out of the new United States. The project team determined to put the 56 Marylanders from the Inspection Roll in the primary category of memorialist, alongside the 73 Marylanders from the LCC records, because their role involved essentially presenting evidence of their allegiance to the state. In fact, those in the Inspection Roll often had to cross a higher threshold than white Loyalists, typically one year's active duty. Witnesses for memorialists from the Inspection Roll tended to be either General Samuel Birch or General Thomas Musgrave, who had earlier issued certificates of allegiance.[24] The mentioned played a range of roles in their lives—as former owners, family members, and so on.

Second, the people resource template had to be updated to reflect the new information provided by the Inspection Roll. Derived from the LCC records, the original resource template had thirty-nine fields assembled in Omeka S from Bibliographic, Dublin Core, FOAF, and military ontologies, among others. The historical data in the Inspection Roll suggested fifteen additional categories of information. Given its specificity, our developer decided to create a custom ontology in Omeka S for this information.[25] These fields allowed the project team to record information about how and when freedom was obtained, names of former owners and claimants, and even the ship, shipmaster, and their destination. The creation of these fields reflected the historic record but also forced the team to rethink some of the basic assumptions about what was necessary to record about memorialists in the LCC records. For example, why was it seen as necessary to document the race of the self-emancipated but not the LCC claimant? Does not forcing the team to record the legal status of LCC claimants reinforce contemporary assumptions about who should be free and who should not? Fields for military service and destination in the Loyalist diaspora are explicitly spelled out in the Inspection Roll, but equally apply to the LCC records, even if they are not always consistently recorded. Putting these two historical sources into conversation not only revealed latent assumptions but also improved the information in the database.

The decision to include both LCC and Inspection Roll petitioners began to restore the complexity of the Loyalist experience, fleshing out the humanity of those who remained loyal to the Crown amid a civil war. Yet there was one group of people who remained silenced in the records: the women and men enslaved by Loyalists. Thinking about how to represent this group—who appeared exclusively in the catchall category of the mentioned where they were consistently reduced to property—raised a new set of considerations, one that straddles historic reporting and contemporary responsibilities.

Naming the Unnamed

Enslaved women and men permeate the LCC records, just as they did in colonial Maryland. Yet only one of the seventy-three white memorialists bothered to enter the names of the people he enslaved. In his memorial, James Chalmers listed Ben, Plymouth, James, Alfred, Bob, Sam, Tom, Queen, Ipheginia, Christian, Sarah, Monimia, Hanah, Rene, and Juda, valued at £525, above his twenty-two horses and below his "House & Lotts in the County Town."[26] The rest merely provided numbers. Hugh Dean claimed he had been "possessed of four Negroe slaves valued at £40 each."[27] The former royal governor Robert Eden laid bare even more blatantly the status of those he enslaved as property by lumping them under his schedule of losses as "House Servants (black) Furniture plate Books Linnens wine 1500" and "Slaves, Horses, Cattle on the Homing pot 1000."[28]

The project team recognizes the contemporary imperative to restore the humanity of these enslaved women and men. In restoring a voice to the most inarticulate, the project looks to the work of archivists attempting to decolonize the archive by using metadata to address "the experiences, needs and aspirations of marginalized and under-represented groups as well as addressing the wider social imperative to ensure recordkeeping can help to document, empower and enfranchise."[29] Naming—of both people and places—wherein the denial of authority by the colonial apparatus, especially of the enslaved, is rooted as much in the programmatic activities of the claimants and scribes as in the bureaucratic worldviews of the time. The challenge to such an approach is the paucity of participation by or consultation of marginalized communities for their views on representation in the historical record.

Realizing the needs of twenty-first-century audiences, the project takes two basic steps to begin to address silences in the eighteenth-century record. The first is to grant each enslaved person her or his own record in the database. This seemingly simple step has a radical implication: it refuses to perpetuate the reduction by eighteenth-century memorialists—as well as the state—of the enslaved to lost property. The challenge, which struck the project team early in the works, is how to identify them. To label them "Unnamed Enslaved Person #397" does the work of embodying an individual in the database but is an inelegant formulation at best and runs the risk of perpetuating dehumanization at worst. The second step in the process is to try to identify those who are rendered nameless in the records. This involves looking to other primary sources, such as those created by Americans at the same time. To this end, the project team turned to another source base: the records the new state governments created as they sold confiscated Loyalist property to pay debts and raise revenue in support of the war. In 1781 the Maryland legislature established the Commission to Preserve Confiscated British Property to oversee the confiscation and sale of local Loyalist property. "Property" sold at auction by the state included enslaved women and men living on seized estates. The journals and sales books of the commission are housed today in the Maryland State Archives.[30]

As a case study to see if it was possible to recover the names of enslaved people listed but unnamed in the LCC claims, Jillian Curran and Elizabeth Lilly consulted the confiscation records for one of the largest Maryland LCC claimants, the Principio Company, in the summer of 2020. The Principio Company was a major commercial enterprise in colonial Maryland. Chartered in 1724 to establish an iron furnace on the outskirts of Baltimore, the venture proved a success and by the 1770s the operation had expanded to include several furnaces across Maryland and Virginia. Nonresident trustees owned a majority share, making it an easy target for the Maryland legislature seeking to expand revenue. The assets of the company—the forges, wooded lands, and enslaved laborers—were seized and sold at public auction.[31] The trustees submitted a claim for lost property to the LCC in May 1786 that included "95 negroes" valued at £4,750 in their list of lost property.[32] The LCC claim list offers no other identifying information.

Curran and Lilly, however, discovered the names of ninety-nine people living on Principio estates in the listing of several public auctions held by the commission between September 1781 and August 1782. In addition, the auction records include demographic information missing from the LCC claim, such as age data. Unlike the Inspection Roll, however, occupation and other defining details are not listed. Putting the two sets of records in dialogue with each other offers a way to transcend the forced anonymity of the enslaved in the LCC records.

The confiscation records offer a possible way forward for restoring names to the hundreds of unnamed enslaved people in the Maryland Loyalism Project database, although they must be used with caution. First, the numbers in the two sources do not always exactly match. For example, there are four people listed in the auction sales who are not included in the LCC claim. This is likely due to the fact that seven children had been born on the estates after confiscation, a situation perhaps unknown to the claimants in London and their trustees in America. Second, not all confiscation records list enslaved people by name. During the February 1782 auction of assets of the Nottingham Company, for example, thirty-four unnamed people were sold to ten purchasers. In this instance, gender and age were indicated ("1 Woman & 2 Children," "1 Old Negro"), but no other information other than sale price was listed.[33] Further work by undergraduate researchers into these records during the spring of 2021, however, revealed that names of enslaved laborers missing in one confiscation document may be listed in another. Noah Biedrzycki found the names and ages of twenty-eight people, a majority of whom were children, in the auction lists for the Whitemarsh Plantation, a property owned by the Nottingham Company before the Revolution. These lists were more detailed than both the Nottingham Company claim submitted to the LCC and the auction lists included in state auction records.[34] Although the confiscation records will not provide the names of all the anonymous people included in LCC claims, they do provide us with many of them. They also provide future students with potential leads to discover what happened to the enslaved on confiscated estates in the years that followed the end of the Revolution. The auction lists include the names of those who purchased enslaved women and men alongside the lands confiscated by the state. Probate records of these buyers, then, might be helpful in generating further leads. All of this serves as a reminder that to restore humanity the team needs to keep creatively mining the archive to fill in the gaps.

Not only does the Maryland Loyalism Project offer the opportunity for students to restore the identities of a world of revolutionary-era Marylanders—free and enslaved, rich and poor, Loyalist and Patriot—it also provides a means by which the state can reverse a silence imposed nearly two and a half centuries ago. The

UK National Archives today recognizes this silence. A stipulation they put on providing the scans for the project was that certain types of metadata would be returned to them that could be used to expand accessibility. The Discovery portal hosted by the National Archives advertises 32 million descriptions of records held by it and more than 2,500 archives across Great Britain. What it captures in breadth, however, it sacrifices in depth. The Loyalist Claims Commission records are broken down by division and subseries but cataloged only at the level of the volume.[35] No names are associated with the record.[36] The Inspection Roll is even more buried. The level of online cataloging in the Sir Guy Carleton Papers offers only a description at the container level.[37] The question becomes: What level of description is the National Archives willing to attach to their records? Would they enter the level of just the memorialist? the witness? or all the mentioned? Can we imagine that the descendant of an enslaved person should have to go through the record of the person who enslaved her ancestor to find her name (and possibly her story)? Or might they be directed over to the Maryland Loyalism Project for that information?

Constructing a digital archive and biographical database provides a certain frame for reconnecting and recovering a fragmented archive. Doing this work can be a form of preservation. Future work with this digital archive and database could go in different directions. Both are built on extensible platforms. The Inspection Roll and Loyalist Claims Commission records of Loyalist refugees from other colonies could be ingested, expanding readers' understanding of the impact of the war on women and men who refused to waver in their allegiance to the Crown. It could also lay the groundwork for a comparative study of Loyalist treatment across the British colonies. Another avenue could involve the ingestion of other types of records, whether they be held by the National Archives, state or local archival repositories, or the oral and family histories of descendants. More details might be discerned for those who were rendered inarticulate by the state, especially people who might restore their thoughts on the revolutionary conflict and the experience of living in colonial Maryland, through their own voices.

NOTES

1. Maryland Loyalism Project, http://loyalismproject.com.
2. The only published history of Loyalism in Maryland, for example, largely ignores the experience of enslaved or free African Americans and focuses mostly on elite white male Loyalists instead. See M. Christopher New, *Maryland Loyalists in the American Revolution* (Centreville, MD: Tidewater, 1996). Scholarship on white Loyalists, often based on LCC records, has been a feature of the historiography of the American Revolution beginning with Lorenzo Sabine's *American Loyalists, or Biographical Sketches of Adherents to the British Crown in the War of the Revolution* (Boston: C. C. Little and J. Brown, 1847). Other prominent studies focusing primarily on white Loyalist exiles include Mary Beth Norton, *The British-Americans: The Loyalist Exiles in England, 1774–1789* (Boston: Little, Brown,

1972); Wallace Brown, *The Good Americans: The Loyalists in the American Revolution* (New York: William Morrow, 1969); Wallace Brown, *The King's Friends: The Composition and Motives of the American Loyalist Claimants* (Providence, RI: Brown University Press, 1965); and Robert M. Calhoon, *The Loyalists in Revolutionary America, 1760–1781* (New York: Harcourt, 1973). Black Loyalism has received less scholarly attention, though the field has expanded greatly since the 1990s. See Cassandra Pybus, *Epic Journeys of Freedom: Runaway Slaves of the American Revolution and Their Global Quest for Liberty* (Boston: Beacon Press, 2006); Graham Russell Hodges, ed., *The Black Loyalist Directory: African Americans in Exile after the American Revolution* (New York: Garland, 1996); James W. St. G. Walker, *The Black Loyalists: The Search for a Promised Land in Nova Scotia and Sierra Leone, 1783–1870* (Toronto: University of Toronto Press, 1992); Mary Beth Norton, "The Fate of Some Black Loyalists on the American Revolution," *Journal of Negro History* 58, no. 4 (1973): 402–26; and Simon Schama, *Rough Crossings: Britain, the Slaves and the American Revolution* (New York: HarperCollins, 2006).

3. Notable exceptions include Maya Jasanoff, *Liberty's Exiles: American Loyalists in the Revolutionary World* (New York: Alfred A. Knopf, 2012); and Jerry Bannister and Liam Riordan, eds., *The Loyal Atlantic: Remaking the British Atlantic in the Revolutionary Era* (Toronto: University of Toronto Press, 2012).

4. The memorial of James Chalmers is the only one from the seventy-three in the five LCC volumes from Maryland published on the MLP that names the people he enslaved. The Claims of James Chalmers, Maryland Loyalism Project, http://ctsdh.org/kroberts/maryland-loyalism-project-redux/the-claims-of-james-chalmers---page-2. Nearly eight hundred other enslaved individuals are represented as numbers rather than people.

5. Jessica Marie Johnson, "Markup Bodies: Black [Life] Studies and Slavery [Death] Studies at the Digital Crossroads," *Social Text* 36, no. 4 (137) (December 1, 2018): 57–79; Britt Rusert, "New World: The Impact of Digitization on the Study of Slavery," *American Literary History* 29, no. 2 (May 24, 2017): 267–86; Daryle Williams, "Digital Approaches to the History of the Atlantic Slave Trade," *Oxford Research Encyclopedia of African History*, November 20, 2018, https://doi.org/10.1093/acrefore/9780190277734.013.121; Jamelle Bouie, "We Still Can't See American Slavery for What It Was," Opinion section, *New York Times*, January 28, 2022, https://www.nytimes.com/2022/01/28/opinion/slavery-voyages-data-sets.html.

6. The syllabus can be found at https://revolutionwillbedigitized.wordpress.com/.

7. Loyalist Claims Commission, Maryland Loyalism Project, http://ctsdh.org/kroberts/maryland-loyalism-project-redux/lcc2?t=1599825861341. See also Norton, *British-Americans*, for a detailed summary of the LCC.

8. There are seventy-three memorials in the five volumes in the LCC series on Maryland (AO12/6, 7, 8, 9, and 60).

9. The last full-length published study of Maryland is New, *Maryland Loyalists in the American Revolution*. More recent unpublished work includes Michelle Fitzgerald, "Confiscating the Castle: The Construction of Loyalist Identity in Governor Robert Eden's Annapolis House" (MA thesis, University of Delaware, 2017); Kimberly Michelle Nath, "Difficulties in Loyalism after Independence: The Treatment of Loyalists and Nonjurors in Maryland, 1777–1784" (MA thesis, University of Maryland, 2009); Kimberly Michelle Nath, "The British Are Coming, Again: Loyalists, Property Confiscation, and Reintegration in the Mid-Atlantic, 1777–1800" (PhD diss., University of Delaware, 2016); Richard Arthur Overfield, "The Loyalists of Maryland during the American Revolution" (PhD diss., University of Maryland, 1968); and Timothy James Wilson, "'Old Offenders': Loyalists in the Lower Delmarva Peninsula, 1775–1800" (PhD diss., University of Toronto, 1998).

10. Historians range widely in their estimates of how many British North American colonists identified as Patriot, Loyalist, or (likely the largest number) were ambivalent

and tried to remain neutral. Upward of as many as three-quarters of a million people (out of a population of 2.5 million) might have supported the British Crown. Schama, *Rough Crossings*, 131.

11. By the end of 1782, for example, between 6,000 and 10,000 Black women and men had left Charleston. Schama, *Rough Crossings*, 135.

12. Most notably this included article 7, which stated that the British would withdraw from the United States "with all convenient Speed and without Causing any destruction or carrying away any Negroes or other Property of the American Inhabitants." Lawrence Hill, "Freedom Bound," *Beaver* 87, no. 1 (February–March 2007): 16–23, 19. The preliminary treaty was signed on November 30, 1782. The final treaty was agreed upon on September 3, 1783. See also Sean Gallagher, "Black Refugees and the Legal Fiction of Military Manumission in the American Revolution," *Slavery & Abolition* 43, no. 1 (2022): 140–59, https://doi.org/10.1080/0144039X.2021.1963192.

13. Hill, "Freedom Bound," 18.

14. There were four inspectors on every dock—two British, two American—who verified both the passenger list from the ledgers and the passengers themselves. Schama says the books could also be used as evidence for the compensation of Patriot owners for their losses. *Rough Crossings*, 146. Library and Archives Canada says "Carleton also appointed three commissioners to superintend all embarkations from New York." Carleton Papers—Book of Negroes, 1783, https://www.bac-lac.gc.ca/eng/discover/military-heritage/loyalists/book-of-negroes/Pages/introduction.aspx.

15. Helpful studies that consider the impact of the histories of remediation on sources include Bonnie Mak, "Archaeology of a Digitization," *Journal of the Association for Information Science & Technology* 65, no. 8 (August 2014): 1515–26; and Molly O'Hagan Hardy and Lindsay DiCuirci, "Critical Cataloging and the Serials Archive: The Digital Making of 'Mill Girls in Nineteenth-Century Print,'" *Archive Journal*, November 2019, https://www.archivejournal.net/essays/critical-cataloging-and-the-serials-archive-the-digital-making-of-mill-girls-in-nineteenth-century-print/.

16. Johnson, "Markup Bodies"; Kim Gallon, "Making a Case for the Black Digital Humanities," *Debates in the Digital Humanities 2016*, https://dhdebates.gc.cuny.edu/read/untitled/section/fa10e2e1-0c3d-4519-a958-d823aac989eb?_ga=2.243199942.5522 52181.1598319113-1065681637.1598319113.

17. Michael J. Kramer, "Going Meta on Metadata," *Journal of Digital Humanities* 3, no. 2 (Summer 2014), http://journalofdigitalhumanities.org/3-2/going-meta-on-metadata/.

18. See Katherine J. Riestenberg, "Meaningful Interaction and Affordances for Language Learning at a Zapotec Revitalisation Program" (paper presented at the Indigenous Studies Seminar, American Philosophical Society, Philadelphia, December 10, 2019).

19. The project team was very much influenced by Digital Paxton, which is built on Scalar, in thinking about what its digital archive might look like. http://digitalpaxton.org/works/digital-paxton/index.

20. *Omeka S User's Manual*, https://omeka.org/s/docs/user-manual/content/vocabularies/.

21. This exercise forced the project team to define the *biographical* fields they had selected, the content they envisioned being recorded for each field, the type of value (text, number, URL) the field would accept, and any notes on how the data should be formatted.

22. Loyalist Migrations, https://loyalistmigrations-westernu.opendata.arcgis.com/; New Brunswick Loyalist Journeys, https://unbgis.maps.arcgis.com/apps/MapSeries/index.html?appid=074bbc635b0b464e94f72ffc2b4bda6a.

23. Gregory Palmer, *Biographical Sketches of Loyalists of the American Revolution* (Westport, CT: Meckler, 1984); Peter Wilson Coldham, *American Migrations, 1765–1799* (Baltimore: Genealogical Publishing, 2000).

24. Many had certificates issued by General Samuel Birch or General Thomas Musgrave that certified their loyalty, that they had "resorted to the British Lines, in consequence of the Proclamations of Sir William Howe, and Sir Henry Clinton, late Commanders in Chief in America," and allowed them to go to Nova Scotia "or wherever else [he or she] may think proper." Yet over eight hundred, according to the historian Simon Schama, "made no pretence of having answered proclamations." Schama, *Rough Crossings*, 151. In addition, parents were given a piece of paper certifying their children were "Born [or Born Free] Behind British Lines." Schama, 155.

25. The developer called the ontology "bon" (for Book of Negroes).

26. The Memorial of James Chalmers, Maryland Loyalism Project, http://ctsdh.org/kroberts/maryland-loyalism-project-redux/the-claims-of-james-chalmers---page-2.

27. The Memorial of Hugh Dean, Maryland Loyalism Project, http://ctsdh.org/kroberts/maryland-loyalism-project-redux/the-claims-of-hugh-dean---page-1.

28. The Memorial of Robert Eden, Maryland Loyalism Project, http://ctsdh.org/kroberts/maryland-loyalism-project-redux/the-memorial-of-robert-eden?path=ao128. "Homing Pot" = "Hominy Pot," the name of Eden's country estate which was run by enslaved people.

29. Anne J. Gilliland-Swetland and Sue McKemmish, "Recordkeeping Metadata, the Archival Multiverse, and Societal Grand Challenges" (paper presented at the DCMI International Conference on Dublin Core and Metadata Applications, Kuching, Sarawak, Malaysia, 2012), section 2.

30. These records are in the State Papers at the Maryland State Archives: S134–1, Commissioners to Preserve Confiscated British Property (Sale Book), 1781–1785; S964, Governor and Council, Confiscated British Property; S168–1, Sale Book 1784; S132–1, Ledger and Journal, 1781–1782; S32–1, Return Book for Reserved Lands; S133–1 Sales Ledger, 1781–85; S123–1, Claims Book.

31. Control of the furnaces had been lost even before 1781. "After 1776 they had no real control what ever over any of their American property. Russell continued to operate the furnaces and forges, and supplied bar-iron in large quantities to the government for public purposes, and balls for the use of the rebel cannon. In 1780 the Maryland General Assembly passed an act to seize and confiscate all British property within the State. This was the end of the British property within the State." Henry Whiteley, "The Principio Company," *Pennsylvania Magazine of History and Biography* 11, no. 3 (October 1887): 63–68, 292.

32. Claim of Principio Company, May 17, 1786, Schedule of Claims, 4, Maryland Loyalism Project, http://ctsdh.org/kroberts/maryland-loyalism-project-redux/the-claims-of-the-principio-company---page-4.

33. Maryland State Archives, State Papers, S134–1, Sale Book, 1781–85, 29.

34. Maryland State Archives, State Papers, S132–1, Ledger and Journal, 1781–1782, 21–22.

35. Discovery, National Archives, Kew, UK, accessed August 17, 2022, https://discovery.nationalarchives.gov.uk/details/r/C3439378.

36. The record for AO12/8, for example, provides information only about the colony ("Description"), date, holding archive, and legal status ("public record").

37. Discovery, National Archives, Kew, UK, accessed August 17, 2022, https://discovery.nationalarchives.gov.uk/details/r/C3070438.

3

DISCOVERING REVOLUTION IN DIGITAL SOURCES
Other[ed] Colonial Voices

Dorothy Berry

Starting in March 2020 and through autumn 2021, access to special collections and archives was seriously curtailed by safety regulations relating to the COVID-19 pandemic. There has been much public discourse around the effects of these restrictions on scholars whose research is centered in archives, mostly from the perspective of researchers who have had to scramble to request digital reference copies of unique materials and in some cases to completely reroute their research journey. At the same time that researchers faced an unprecedented roadblock to access, archivists and special collections librarians were furiously planning ways to keep classes engaged with primary source research, even if those sources were not available in person.

While the utility of digitized access to special collections has become fairly accepted, there remain feelings that the best access is being able to touch and engage with the physical materiality of a source.[1] This feeling was exacerbated throughout 2020 as digital access moved from a choice of convenience to one of necessity. University libraries mobilized quickly to figure out how to provide as many services as possible in the face of disruption. Harvard College Library put out calls for a variety of ad hoc groups approaching the numerous unpredicted issues, including one for Open Educational Resources (OER)/Primary Source Access led by Claire DeMarco, associate director of user experience and discovery. With little direction and no knowledge of how the pandemic would affect our lives, let alone our job responsibilities, the group's members set off on individual pursuits exploring enhanced digital access to library resources.

At the time I finished this chapter, I was the digital collections program manager at Houghton Library, Harvard University's largest individual repository for rare books and manuscripts. I was immediately drawn to figuring out how our already digitized material could be presented most effectively for teaching and learning. The bulk of the public-facing work of a special collections library is inviting students, faculty, and researchers into our physical spaces to conduct classes and discussions dealing with the physical materials we place in front of them.[2] While serving on the OER/Primary Source Access group, I was also serving on a smaller, internal Houghton committee focusing on primary resource teaching and learning in our new environment. Working at home in the face of a terrifying pandemic, I dove into a new topic and new platform hoping to provide an example of what we could provide to teachers facing the same unsurety that we were.

Before I could build a site, I had to find compelling, digitized material with a broad teaching appeal. My background is in African American special collections digitization, and I have long held as a goal the increased discoverability of hidden Black collections.[3] The largest holdings of digitized special collections material from Harvard Library can be found in Colonial North America, a multiyear, cross-repository digitization project that set out to digitize all of Harvard's eighteenth-century manuscript material.[4] I searched through the CNA digital collections for materials by Black creators or featuring Black voices, but as is common in the historical materials collected by prestigious institutions, the majority of what had been preserved from the past and digitized for posterity included Black people only as background figures to white legal and/or economic documentation.

Looking for materials related to slavery, indenture, and other Black labor, I stumbled across the Evert Jansen Wendell Collection of Contracts for the Sale of Slaves, 1796–1829, since renamed the Evert Jansen Wendell Collection of Slavery and Indenture Contracts, 1796–1829.[5] By no means an early Americanist, I was curious about one of the contractors in the collection: Moses Judah. My interest was piqued by this early Jewish settler, and then even more when I learned, in a collection of "contracts for the sale of slaves," that Judah was an early member of the New York Manumission Society who, two years after signing this contract, was elected to the executive committee of that same society.[6] People are complex and often hypocritical, but this contradiction caused me to look more closely at the small collection of three contracts, and to further investigate my suppositions around views about manumission in turn-of-the-nineteenth-century New York. Looking with less-than-expert eyes, I associated manumission directly with abolition—a view that historians might immediately look askance at, but that I fear might be common among a more casual digital-collection browser. Shane

White points out that the New York Manumission Society was more oriented toward a genteel model of working for better treatment for free Black people in New York and the enslaved, to the extent of having a good number of slaveowners in leadership positions. White details that "of the 120 men who had joined the organization by the end of 1790, a minimum of 27 were listed in the 1790 census as owning slaves. . . . Three out of every ten members who had joined the society in the first six years of its existence were listed as slaveowners in either the 1790 or 1800 census."[7]

I quickly learned that many of the assumptions I held after reading the collection title were false. The Evert Jansen Wendell Collection of Slavery and Indenture Contracts, 1796–1829 told a story, a story about marginalized people marginalizing other people, and about the complexities of abolition and emancipation. Researching the archival subjects represented in the manuscripts, I knew this document was perfect for my digital teaching exhibit. With a short turnaround due to the scrambled confusion following campus closures and rapidly changing health advising, I spent two weeks researching and building Other[ed] Colonial Voices: Slavery and Indenture in New York.[8]

The Object

If I, an archivist with a background in African American historical research, brought false assumptions into my reading of the description of the Evert Jansen Wendell Collection of Slavery and Indenture Contracts, it bears spending a brief amount of time on why. The rules that guide archival description are shaped by their own history, which places a high value on collectors. The first words you read in the Evert Jansen Wendell Collection of Slavery and Indenture Contracts, 1796–1829 are Evert Jansen Wendell. Wendell himself had nothing to do with the people or business documented in this set of contracts; in fact, he was not even born until forty years after the oldest was created. He was a Harvard alumnus and a successful man about town.[9] With a deep commitment to Harvard, philanthropy, and collecting rare and unique items, upon his death in 1917 Wendell bequeathed his vast personal collection to Harvard Library.[10] Materials that began in general circulation, over time, can become rare. Houghton Library now contains 362 named Wendell collections and thousands of individual items that list him as a previous owner.

Familiar with the provenance-focused naming structure, what led me most into confusion was the title and brief description. Quickly processed, the three contracts were described with a single phrase "contracts for the sale of slaves and indentured servants." Materials were, historically, often described rather quickly based on the balance of available staff resources, perceived importance, and staff interest/knowledge. A more focused examination of the documents revealed that only one was a contract for the sale of an enslaved person, and that all three were uncommon—not just as eighteenth- and early nineteenth-century manuscripts but as manuscripts representing the intersections of men and women, and Christian, Jewish, Black, and white people.

Each of the contracts has at least three parties, the current enslaver or indenture holder, the new enslaver or indenture holder, and the Black people whose lives and labor were being contracted. The contracts are from the state of New York during flashpoints in the state's path to gradual abolition. Like any primary source, they provide a peek into entire lives, family networks, and economic structures. The figures represented therein are all mostly mysteries, but the information they do provide shines a small, narrow light on the conflicted tiny revolutions that sparked across a populous settling a new nation and dependent on Black labor.

The contracts were presented digitally with discussion prompts, with the idea that lessons for virtual classes could be shaped around the contracts. The prompts are a mixture of factual identification and archival imagination. This section of my chapter will be a mixture of the same. Describing the site, its structure and development, and content is not a particularly complex project, but exploring the possibilities that haunt the dearth of information around any single manuscript is expansive. The hope of sharing primary resources with students is, in part, to spark the realization that historical figures were once as human and material as we are today. Sandwiching the object's possibilities between its technical spaces, both archival and digital, is an active attempt at centering these archival figures. By presenting the facts as we have them, and the possibilities that research illustrates, I will attempt what I desire for the reader of Other[ed] Colonial Voices: to imagine the full humanity of archival subjects listed only by name.

Catharine Bleecker, Daniel Paris, and Tom

> Know all Men, that I, Catharine Bleecker, of the City and County of Albany, & State of New York; for & in consideration of the sum of Fifty pounds to me in hand paid by Daniel Paris of Canajoharie in the county of Montgomery & state aforesaid; the receipt whereof, I do hereby confess & acknowledge: have bargained, sold, and let, and by these presents, do bargain, sell and let unto the said Daniel Paris, his Executors, Administrators, and Assigns for the Term of Eight Years, from the date hereof my Negro man slave Tom, aged about seventeen years. To have and to hold the said Negro man slave Tom, to the said Daniel Paris, his

Executors, Administrators & Assigns for and during the said term of Eight years and no longer.

And I, the said Catharine Bleecker do hereby manumit and set free, my said Negro man slave Tom at the expiration of eight Years from the date of these presents. In witness whereof, I have hereunto set my hand and Seal, this Twenty fifth day of February in the Year of Our Lord, One Thousand, seven hundred & ninety six.[11]

In 1796 Catharine Elmendorf Bleecker was a widowed mother of eleven. Very little is known about Bleecker herself apart from a brief biographical sketch from the New York State Museum and a portrait of a five-year-old in an ochre gown with a lace stomacher, red carnation in hand, held at the Metropolitan Museum of Art.[12] Most information about Bleecker and her sale and enslavement of seventeen-year-old Tom must be surmised from other, more documented women and their legal lives.

The discussion questions around this first contract ask about Bleecker's legal position as a widow, about the reasons behind the manumission clause, and what we can learn about her society's ideas from those two discussions. In 1790, six years before this contract was signed, between one-in-three and one-in-five households around Albany held at least one enslaved laborer.[13] While many of those households held only a single man or woman as a household laborer, it is important to note that the idea of slavery in the North as more genteel or harmless than plantation slavery in the South is a false one. It is not known how long Tom lived and worked for the Bleecker family or whether he had a local family of his own. What is known is that Tom was legally the property of Catharine Bleecker, who as a widow was free to sign a contract selling Tom to Daniel Paris.

The year 1796 was three years before New York State would pass An Act for the Gradual Abolition of Slavery, which stated that children born to an enslaved woman after July 4, 1799, would be legally free when male children had turned twenty-eight and females twenty-five, though they were required to remain with their previous enslaver as an indentured servant—possibly with very little material effect on their actual lives. Enslaved people born before July 4 were redefined as indentured servants. Enslavers who did not want to deal with these civilly liminal children could turn them over to the care of the overseer of the poor who would bound them out until the age of twenty-one. Completely abandoned formerly enslaved children would be supported by the state.[14]

Bleecker's contract called for her own small gradual manumission, specifying that Tom was less "sold" to Daniel Paris and more let out for a term of eight years after which he was to be manumitted at age twenty-five. Whether this illustrates simply a practical desire to make immediate cash or pay off debts through the transfer of forced labor, a desperate need to remove Tom from the home, a widow's struggle to get by with eleven children and limited income is entirely unknowable. The manumission clause plays into ideas of more empathetic enslavement in the North, but extrapolating emotions from a single commercial document is beyond the bounds of archival imagination.

Moses Judah, John Oakley, and Lewis

> Know all men by these Presents that I John Oakley of Jamaica, in the county of Queens and state of New York for and in consideration of the sum of one hundred and seventy five dollars to me in hand paid by Moses Judah of the City of New York Merchant the receipt whereof I do hereby acknowledge, I have granted bargained and sold assigned transferred and set over and by these presents doth grant bargain and sell, assign, transfer, and set over unto the said Moses Judah his executors, administrators or assigns a negro boy named Lewis aged ten-years or thereabouts. To have and to hold the said negro boy unto the said Moses Judah his heirs and assigns for an during the term of twenty five years next ensuing from the day of the date of these presents. And I the said John Oakley do hereby covenant and agree to and with the said Moses Judah his heirs and assigns that I the said John Oakley now am the lawful and absolute owner of the said negro boy and have good right and authority to sell and dispose of the said negro boy in manner hereinbefore mentioned. And that I the said John Oakley my heirs executors and administrators shall and will warrant and defend the said negro boy to the said Moses Judah his executors administrators and assigns against the lawful claims and demands of all persons whomsoever and during the said term of twenty five years. In witness whereof I have hereunto set my hand and do seal in the City of New York this twenty third day of June in the year of our God one thousand Eight hundred and four.[15]

In 1804 Moses Judah had been a free man for thirty-six years.[16] He had gained the legal right to take part in retail trade and had become a fairly successful merchant. Judah has primarily been remembered in US history for his membership in the New York Society for Promoting the Manumission of Slaves.[17] That the few references to Judah in the archival record are his long-term membership in New York's most prominent manumission society, and this contract for the indenture of ten-year-old Lewis points again to how much interiority is missing when looking through archives.

Here, students are asked to think about Judah's experiences as a Jewish immigrant, how that status placed him in different social and civil roles than immigrants classed as white or nonforeign, and perhaps most compellingly, how they understand Judah's active membership in a manumission society and his indenture of a child. Judah was part of a small Jewish community in New York, migrating before the larger waves of Ashkenazi Jews starting around the 1820s. He was a member of New York City's first synagogue, the Sephardi Congregation Shearith Israel, and was successful enough not only eventually to serve on the New York Manumission Society's standing committee but also to have the equivalent of almost $4,000 to purchase Lewis's labor for the next twenty-five years of his life.[18] It is unclear whether Judah enslaved any people outright. If he did, however, he would not necessarily have stood out among the society's membership. Perhaps the most famous member, the chairman John Jay, enslaved people while in a leadership position.[19]

The year 1804 had seen multiple modifications to the Act for the Gradual Abolition of Slavery. In 1802 the amount of and timeline for state care of abandoned Black children were reduced and in 1804 eliminated altogether. That same 1804 law stated that former enslavers holding children transferred into indenture were required to provide biblical literacy education before the age of twenty-one. If it could be proven that this clause was ignored, four to seven years could be subtracted from the term of indenture. Judah signed this contract at a point when indenture was commonplace for servants of any race, and may not have felt any cognitive dissonance in the purchase of a boy's childhood into young adulthood and the active legal fight for abolition.[20]

Peter Conover, Samuel Fleet, and Cynthia Haycorn

> This indenture witnesseth, that Cynthia Haycorn a coloured girl—now aged six years by and with the consent of Peter Conover and Joseph Herbert overseers of the poor of the town of Brooklyn hath put herself, and by these Presents doth voluntarily, and of her own free will and accord put herself with Samuel Fleet of the Town aforesaid to learn house and kitchen work and after the manner of an Apprentice, to serve from the day of the date hereof, for and during, and until the full end and term of Twelve years or till she arrives at the age of Eighteen years—next ensuing; during all which time, the said apprentice his master faithfully shall serve, his secrets keep, his lawful commands every where readily obey; he shall do no damage to his said master, nor see it done by others, without letting or giving notice thereof to his said master: he shall not waste his said master's goods, nor lend them unlawfully to any: he shall not contract matrimony within the said term: at cards, dice, or any unlawful game he shall not play, whereby his said master may have damage: with his own goods, nor the goods of others, without

license from his said master, he shall neither buy nor sell; he shall not absent himself day nor not from his master's service, without his leave; nor haunt ale-houses, taverns dance-houses, or play-houses; but in all things behave himself as a faithful apprentice ought to do during the said term. And the said master shall use the utmost of his endeavour to teach, or cause to be taught or instructed, the said apprentice in the trade of mystery of house and kitchen work and procure and provide for her sufficient meat, drink, washing lodging mending and clothing fitting for an apprentice during the said term and to instruct or cause the apprentice to be instructed to read and write and at the expiration of the said time to give her a new Bible.

And for the true performance of all and singular the covenant and agreements aforesaid, the said parties bind themselves, each unto the other, firmly by these presents.

IN WITNESS WHEREOF, the parties to these Presents have hereunto set their hands and seals the 27th day of July in the year of Lord one thousand eight hundred and twenty nine.

Sealed and Delivered in the Presence of
Cynthia Haycorn
Her X mark
Samuel Fleet
Joseph Herbert
Peter Conover Overseer of the Poor[21]

In 1829 Cynthia Haycorn was a ward of a state. An Act Relative to Slaves and Servants was passed on March 31, 1817. This act was another slow trickle of legal freedom ending with complete emancipation in 1827.[22] Haycorn was born four years before that total emancipation and, at some point, was placed under the guidance of the Overseer of the Poor. Peter Conover and Joseph Herbert, city officials, scratched out irrelevant or inaccurate portions of an apprentice template to produce legal documentation that Haycorn "of his her own free will and accord" agreed to indentured servitude until she reached the age of eighteen. She signed with an X.

Haycorn's contract is the only one in this collection that was signed after New York State abolition and the first that details the labor and training required for her indenture. Students were invited to look not only at Haycorn's life but the world she lived in: What was Black community like in early Brooklyn? Why have a clause in the law and therefore the contract requiring scriptural literacy? What, if anything, makes this contract different from Tom's and Lewis's? Cynthia Haycorn was a child in the town of Brooklyn, which was decades away from becoming a borough of New York City. Black communities were growing across the area, including in a Carnasie community known as the "Colored Colony."[23] Brooklyn was farmland and, in the decades before Haycorn's birth, Dutch farm owners enslaved the same proportion of Black laborers as Virginians did. Many of those enslaved laborers remained on as farmworkers postemancipation, leading to the potential for a local Black community. By 1830 the New York area had around fourteen thousand free Black Americans, who formed their own social clubs, salons, newspapers, and churches.[24] Haycorn's contract was signed just a few years before the Black community in Brooklyn would begin flourishing, with Black land speculators purchasing lots in the Ninth Ward as early as 1832. As the decade moved on, Haycorn may have been aware of the aspirational Weeksville community, where formerly enslaved people and Free People of Color had purchased plots of land and begun to establish an independent community that would progress to high levels of political and economic success by the middle of the century.[25]

Despite these legal freedoms, Haycorn was also growing up in a time of white tension over the increasing population of free Black Americans. They began passing laws that limited social access and, most threateningly, supported the rights of southerners hunting formerly enslaved fugitives to kidnap as they saw fit from the city's streets. The sizable free community fought back, most famously with the 1835 founding of David Ruggle's New York Committee of Vigilance, which argued for jury trials for recaptured fugitives and hired lawyers for their defense. Haycorn's contract offers just enough information to imagine the life of a six-year-old being trained for service. If Samuel Fleet was kind, the eighteen-year-old Cynthia Haycorn might have been ready to seek employment with good references and the ability to read and write. If he was less so, or even just inattentive, Haycorn was in a very dangerous situation.

The Site

While the Other[ed] Colonial Voices site is designed to invite this style of deeper imagination for those so inspired, it is purposefully as simple as possible. The primary focus is on the object itself. This choice was shaped by the truncated timeline and my admitted lack of scholarly expertise in early American slavery in New York State but most heavily by the need to replicate the sorts of teaching styles that often take place in special collections classes but were clearly going to be impossible for the foreseeable future.

The collection has been digitally accessible for a few years, but each form of access requires a level of research effort that felt like an overload for faculty reconfiguring their semesters on the fly. All public digitized archival materials at Harvard University are discoverable through the general catalog, Hollis; the finding aid directory, Hollis for Archival Discovery; and the digital collections platform, Harvard Digital Collection. The Evert Jansen Wendell Collection of Slavery and Indenture Contracts, 1796–1829 is also discoverable through Colonial North America's digital platform. As discussed earlier, however, the unique details of this collection did not make it into the description. The site is designed to bridge the gap between the broad catalog description and the sorts of deep knowledge that only expert researchers would have.

There are many methodologies for inviting students to consider primary sources that I have learned mainly from experienced archivist and special collections librarians, rather than the literature from the field.[26] My goal was to provide a site that faculty could plug into larger units around related topics using the provided discussion questions, or that they could use as an in-class activity along the lines of "See, Think, Wonder."[27]

The site's structure mimics the See, Think, Wonder exercise by first centering images of each contract itself, then a transcription of the contract text, followed by the brief historical context. By highlighting the unique representations of identity in each contract, the design also allows for any individual item to stand alone for classroom integration. Courses specifically about the long history of Jewish identity in the Americas could focus on Moses Judah's contract; those on Black childhood could focus on Cynthia Haycorn. As part of the experimental process for spinning out primary resource learning tools in rapid response, the combination of broad appeal across subject specializations and incredibly simple design became key.

Other[ed] Colonial Voices is built using Scalar, an open source publishing platform from the Alliance for Networking Visual Culture. In its most fully realized forms, Scalar can be used to create data-rich sites with built-in visualizations and a complex custom design.[28] The librarian for collections and digital scholarship Carol Chiodo has provided multiple workshops for Harvard librarians and archivists to begin implementing Scalar in simpler classroom integrations. Having attended one of these workshops before the pandemic, I knew that Harvard staff had free access to Scalar hosting and that the linear progressions or "paths" Scalar is based on would work well to mirror special collections teaching environments.

The site opens with a basic introduction to the Colonial North America project and the archival object. The context and provenance of a primary source are foundational in special collections, but fewer classes across universities focus on that particular area, so the section is brief and high level. The first content-rich page—Slavery, Indenture, and Freedom in New York—provides a simple framework: an introduction to the object, a glossary disambiguating the key terms "enslavement," "indenture," "emancipation," and "manumission"; an interactive timeline chronicling slavery in New York from Dutch colonialism through gradual abolition; and summaries of contract law and the New York Manumission Society. I tried to follow the advice I personally give students who are writing on primary sources: "Researchers have dedicated years of their lives to creating entire books on each of these topics—don't worry about covering everything." My interest in this section, as throughout the site, was to intrigue readers enough for them to want to move forward on the site and to give just enough information to make sense of the sources they would encounter.

The site moves forward mirroring the experience of looking at the materials laid out in their acid-free folders in a special collections reading room. On the first page, there is a full screen image of a single contract. The second page is embedded with an International Image Interoperability Framework (IIIF) viewer presenting the contract in a more explorable manner, followed by full transcriptions and short summaries of the figures and identities represented therein: "White Women in Colonial New York," "Jewish People in Colonial New York," "Black Life in Colonial New York." Readers can move through the materials chronologically, ending with a discussion guide for each page, but can also navigate from a front-page table of contents or a drop-down menu from the top of every page.

Undeniably, the most difficult part of building this site was doing enough research to present the topic in a manner that would be useful to expert instructors and selecting sources to illustrate and support each section. Like the teaching faculty I was hoping to serve, I too was on lockdown without access to nonelectronic resources. I work for an institution that provides access to a wealth of electronic databases and e-books, but designing a site for the general public meant that I purposefully excluded most material from those sites. At the time, JSTOR was providing free access in response to pandemic-based library shutdowns, so I included some articles from that database but primarily stuck to Wikidata, open access material from museums and historical societies, and other publicly and freely available content.[29]

Most of those resources are not linked directly in the brief articles, in hopes of keeping them succinct and fitting more into the imaginative possibilities of a See, Think, Wonder framework. Instead, I created a Further Learning page with blogs, digital humanities projects, teaching guides, and videos. The material on this page is directed less at the student reader, though it is certainly accessible, and more toward the transitioning instructor in need of fast access to digital resources to fill in the gaps in pandemic-based library service interruptions.

The overall mission to experiment in creating a close-read primary source opportunity for newly remote learning environments was achieved with minimal time investment by accepting the reality that the best I could do was a simple introduction. By expanding my own imagination around the collection, I was able to strip the text down to the simplest form, leaving just the framing pieces for a modular implementation into scrambling courses in transition.

Response

Before bringing the site back to the OER/Primary Source Access group, I first presented it to Houghton's curator for early modern books and manuscripts, John Overholt. He helpfully added some corrections to my transcriptions, but neither of us are specialists in this area. That was the extent of his editorial comments, though he was very supportive of the project. The OER/Primary Source Access group was receptive to the new site, not necessarily for the content itself but as an example of how our library staff could quickly spin out a new site using existing resources. Other[ed] Colonial Voices was shared out locally to Houghton and more broadly at Harvard Library with a hope for future iterations on other special collections materials.

Remote classroom integration was a primary objective of this experiment and was successfully achieved in Leah Whittington and Ann Blair's Texts in Transition. Texts in Transition—a general education course with forty-four students from across the four years of Harvard undergraduate education—focused on the transmission of books and manuscripts across history and institutions. I was invited to give a guest lecture during the week of the course focusing on African American history in rare books and manuscripts and assigned Other[ed] Colonial Voices as the class reading. I hoped that this would provide a simple framework for exploring how materials come to a place like Houghton Library and how Black lives are often documented primarily as secondary characters.

In advance of the class, Whittington and Blair collected questions from the students, which completely blew away my expectations—and my ability to answer! While the basic goal of the site was an introduction to the manuscripts and the repository, the student questions illustrated that the simple text on the site would invite deeper questions and considerations of life in late eighteenth- and early nineteenth-century New York. Of the eighteen questions I received, twelve were directly about the site's content. A small portion asked questions about the material itself: "How frequently do documents like slave contracts survive?" "How common was the manipulation of already-written documents essentially disguising domestic servitude as an apprenticeship?" "Are there extant sources written by enslaved people during their enslavement? I know most were not literate, but is there any sort of primary source that directly records their experiences?" Most of the questions, however, reflected a deep engagement with imagining the lives of the archival subjects. One student wrote

> One question I had while reading the contract between Catharine Bleecker and Daniel Paris was: to what extent were these honored? Did Catharine ever check back with Paris to make sure Tom was freed? I have heard in some history classes that slaves who were supposed to be free ended up not being manumitted, and I am wondering if you have more information on these contracts and if they were upheld or not?

Another asked

> Was the story of the life of Cynthia Haycorn a common one? And do we know if she ever was taught to read and write; are there any more accounts from or about her?

The "wonder" stage of See, Think, Wonder was clearly sparked in the students without any specific instruction. There were detailed historical questions, but the presentation of the material directed the students toward human-centered inquiries. They wanted to know what happened to Tom and Cynthia Haycorn. They wanted the exact details that are almost never available, the gaps in the historical record where Black life happened.

Giordana Mecagni, the head of archives and special collections at Northeastern University, once quipped that "the internet is littered with the shipwrecked hulls of abandoned digital humanities projects."[30] That quote stuck with me, and has generally made me wary of building one-off project sites. Other[ed] Colonial Voices served an experimental goal in providing a new form of access to primary source materials, but it also served a larger, less-expected role. The internal response to the site made it clear that Harvard Library had, up to that point, done a poor job of providing access to our rich holdings related to African American history in the eighteenth and nineteenth centuries. Those student questions, specifically about Tom or Cynthia Haycorn, might not have been answerable but our stacks did have examples of early American Black print culture from figures like Absalom Jones, Prince Hall, and Jupiter Hammon. I felt the urgent need to expand on that access, beyond a single Scalar site.

Realizing the gap in our digital access, I worked remotely with my colleague, the metadata librarian Vernica Downey, to pull a series of records from our catalog. As we discovered more and more rare and relevant African American history materials hidden in the stacks, the idea for a much more ambitious digital access point emerged. A new project began to follow the smaller-scale realization from

Other[ed] Colonial Voices. Weeks after the completion of the site, I proposed Slavery, Abolition, Emancipation, and Freedom: Primary Sources from Houghton Library.[31] The new project exponentially multiplies the scale by adding a curated group of over fifteen hundred manuscripts, rare books, and ephemera to our digital collections, this time with interpretative text written not by me but by a group of student scholars. Other[ed] Colonial Voices showed us that simple, focused inquiry was as useful as heady and complex research from subject area experts. Slavery, Abolition, Emancipation, and Freedom: Primary Sources from Houghton Library will take that further by centering the student See, Think, Wonder experience as the core interpretative device.

NOTES

1. Alexandra Chassanoff, "Historians and the Use of Primary Source Materials in the Digital Age," *American Archivist* 76, no. 2 (September 2013): 458–80, 463.

2. Emily Kader, "Object Lessons: An Assessment of Special Collections Pedagogy" (MA thesis, University of North Carolina at Chapel Hill, 2013), 6.

3. Dorothy Berry, "Digitizing and Enhancing Description across Collections to Make African American Materials More Discoverable on Umbra Search African American History," The Design for Diversity Toolkit, Northeastern University, August 2, 2018, https://des4div.library.northeastern.edu/digitizing-and-enhancing-description-across-collections-to-make-african-american-materials-more-discoverable-on-umbra-search-african-american-history/.

4. "About," Colonial North America at Harvard Library, Harvard University, accessed June 29, 2021, https://colonialnorthamerica.library.harvard.edu/spotlight/cna/about/about.

5. Evert Jansen Wendell Collection of Contracts for the Sale of Slaves, Hollis for Archival Discovery, accessed May 20, 2020, https://hollisarchives.lib.harvard.edu/repositories/24/resources/1442; Processing Information, Hollis for Archival Discovery, accessed June 29, 2020, https://hollisarchives.lib.harvard.edu/repositories/24/resources/1442.

6. Howard B. Rock, *Haven of Liberty: New York Jews in the New World, 1654–1865* (New York: New York University Press, 2012), 97.

7. Shane White, *Somewhat More Independent: The End of Slavery in New York, 1770–1810* (Athens: University of Georgia Press, 1991), 79.

8. Dorothy Berry, Other[ed] Colonial Voices, Harvard University, June 3, 2020, https://scalar.fas.harvard.edu/hou-colonial-voices/index.

9. "E. J. Wendell Died in France," *Harvard Crimson*, September 21, 1917, accessed June 25, 2021, https://www.thecrimson.com/article/1917/9/21/e-j-wendell-died-in-france/.

10. "Wendell Library Goes to Harvard," *New York Times*, October 12, 1919, accessed June 25, 2021, https://timesmachine.nytimes.com/timesmachine/1919/10/12/97981974.html?pageNumber=26.

11. "Bleecker, Catharine. Contract with Daniel Paris of Canajoharie, New York: AD, Albany, New York, 1796 Feb. 25," Harvard Library Digital Collections, accessed June 29, 2021, https://digitalcollections.library.harvard.edu/catalog/hou00197c00001.

12. Stefan Bielinski, "Catherine Elmendorf Bleecker," Colonial Albany Social History Project, New York State Museum, March, 14, 2017, https://exhibitions.nysm.nysed.gov/albany/bios/e/caelmendorf.html#portrait; unknown artist, *Catherina Elmendorf*, Metropolitan Museum of Art, accessed June 24, 2021, https://www.metmuseum.org/art/collection/search/19022.

13. Shane White, "Slavery in New York State in the Early Republic," *Australasian Journal of American Studies* 14, no. 2 (1995): 1–29, 3.

14. "An Act for the Gradual Abolition of Slavery, 1799," New York State Archives, accessed June 24, 2021, http://www.archives.nysed.gov/education/act-gradual-abolition-slavery-1799.

15. "Oakley, John. Contract with Moses Judah of New York City: ADS, Jamaica, Queens County, New York, 1804 June 23," Harvard Library Digital Collections, accessed June 29, 2021, https://digitalcollections.library.harvard.edu/catalog/hou00197c00002.

16. Max J. Kohler, "Civil Status of the Jews in Colonial New York," *Publications of the American Jewish Historical Society*, no. 6 (1897): 81–106, 103.

17. Jerrold Nadler, "In Honor of the 350th Anniversary of the Congregation Shearith Israel," *Congressional Record* 150, no. 126 (October 7, 2004), https://www.govinfo.gov/content/pkg/CREC-2004-10-07/html/CREC-2004-10-07-pt2-PgE1830.htm.

18. "Value of $175 from 1804 to 2021," CPI Inflation Calculator, accessed June 29, 2021, https://www.officialdata.org/us/inflation/1804?amount=175.

19. Ron Chernow, *Alexander Hamilton* (New York: Penguin, 2005), 214.

20. John McNelis O'Keefe, *Stranger Citizens: Migrant Influence and National Power in the Early American Republic* (Ithaca, NY: Cornell University Press, 2021), 138.

21. "Haycorn, Cynthia. Indenture contract as apprentice to Samuel Fleet of Brooklyn, New York: DS, 1829 July 27," Harvard Library Digital Collections, accessed June 29, 2021, https://digitalcollections.library.harvard.edu/catalog/hou00197c00003.

22. "An Act Relative to Slaves and Servants, 1817," New York State Archives, accessed June 30, 2021, http://www.archives.nysed.gov/education/act-relative-slaves-and-servants-1817.

23. "Black Canarsie: A History," Brooklyn Public Library, accessed June 29, 2021, https://www.bklynlibrary.org/locations/jamaica-bay/black-canarsie-history.

24. Eric Foner, *Gateway to Freedom: The Hidden History of America's Fugitive Slaves* (New York: Oxford University Press, 2015), 46.

25. Judith Wellman, *Brooklyn's Promised Land: The Free Black Community of Weeksville, New York* (New York: New York University Press, 2014), 13–17.

26. Christine Woyshner, "Inquiry Teaching with Primary Source Documents: An Iterative Approach," *Social Studies Research and Practice* (Board of Trustees of the University of Alabama) 5, no. 3 (2010): 36–45, 38.

27. "See, Think, Wonder," Facing History & Ourselves, accessed June 26, 2021, https://www.facinghistory.org/resource-library/teaching-strategies/see-think-wonder.

28. "About Scalar," Alliance for Networking Visual Culture, accessed June 26, 2021, https://scalar.me/anvc/scalar/features/.

29. "Expanded Access to JSTOR during COVID-19 Crisis," JSTOR, April 15, 2020, https://about.jstor.org/news/expanded-access-to-jstor-during-covid-19-crisis/.

30. Giordana Mecagni, "Where Do We Go from Here? Applications and Next Steps," Design for Diversity closing forum, Northeastern University, Boston, MA, August 23, 2018.

31. Anna Burgess, "This Year, a Single Digitization Focus at Houghton," *Harvard Gazette*, July 30, 2020, https://news.harvard.edu/gazette/story/2020/07/houghtons-2020-21-digitization-focus-black-american-history/.

4

BUILDING A RELATIONAL DATABASE TO EXPLORE ENSLAVED MIDWIVES' WORK IN EARLY AMERICA

Sara Collini

Few are the names of enslaved women recognized and remembered during the era of the American Revolution. More familiar are the names of their enslavers, men like George Washington, George Mason, and Thomas Jefferson, who founded a new republic and set the stage for revolutionary movements around the Atlantic. While those men fought battles against British tyranny, extended diplomacy abroad, and debated new constitutions, the women they enslaved at home enacted their own important changes. They became midwives on southern plantations.

The work of enslaved midwives formed part of the rhythms of motherhood and the traumas of plantation life. Kate at George Washington's Mount Vernon provided essential health care services for women and children in her community. Nell and Nan at George Mason's Gunston Hall sustained networks of care between plantation landscapes. Rachael at Thomas Jefferson's Monticello connected families across generations, assisting mothers and their daughters. Enslaved women also advocated for themselves as midwives with valuable skills in medicine and family support, engaging in complicated economic relationships with those who enslaved them. Enslaved midwives often earned compensation for each child they safely delivered into a mother's arms and into the brutal system of racial slavery that their enslavers had built and maintained.

The relationships enslaved midwives developed with mothers, families, and enslavers reverberated beyond the perimeters of the plantation and formed part of the discords of revolutionary change in the eighteenth century. Slavery was pervasive throughout the British North American colonies and was used to fuel and expand colonial economic horizons. Those in power during the colonial era, through the legal doctrine of *partus sequitur ventrem*, had ensured that their legacies of white stability and prosperity depended on the forced labor of Afro-descended people and heritable enslavement of their children.[1] While important outcomes of the revolutionary moment for enslaved people included differing forms of freedom in some northern and mid-Atlantic states, slavery expanded in the South and eventually in the West.[2] The birth of enslaved children on American soil and the midwives who aided in those births became increasingly important to buttressing the political economy of the young United States.

Reconstructing the history of enslaved midwives provides an important new lens through which to view the revolutionary era, yet studies of their lives and work are missing from the canon of American revolutionary history. This historical silencing descends in part from the violence of slavery's archive, comprising disconnected fragments written by enslavers and those in positions of social and political power. For decades innovative historians and genealogists have worked with incomplete evidence, deconstructing slavery's archive from the perspective of its authors to bring forward groundbreaking histories of enslaved women and families around the Atlantic.[3]

Along with the decolonizing methodologies of slavery studies, there is the newer field of digital humanities and, more specifically, Black digital humanities. Black digital humanities calls for critical investigations into digital tools and frameworks themselves, as they are often born from racialized systems that have marginalized and oppressed. As Kim Gallon explains, this methodology often involves the "technology of recovery," in which scholars seek to bring histories of subjugated peoples forward, as well as the critical evaluation of those very tools which may still reinforce that subjugation.[4] The violence of slavery's archive and its historiographies are inescapable, and as scholars like Jessica Marie Johnson point out, "there is no bloodless data in slavery's archive," in the print or digital form. The tenets of "black digital practice" call for practitioners, especially those engaging in slavery studies, to "feel this pain and infuse their work with a methodology and praxis that centers the descendants of the enslaved, grapples with the uncomfortable, messy, and unquantifiable, and in doing so, refuses disposability."[5] It is imperative for people engaging in this work, such as a white female historian like myself, to understand the responsibility and ethics involved and to continuously reflect on biases, choices, and ramifications of that digital practice.

This project thus attempts to bring the lives and work of enslaved midwives during the revolutionary period forward from slavery's archive and from historical erasure, intermixing methodologies from slavery studies and digital humanities. It aggregates fragmentary sources from the archives of powerful men like George Washington and Thomas Jefferson and reframes and highlights the histories of enslaved women and families through a relational database model. Studying women like Kate, Nan, Nell, Rachael, and other women enslaved on plantations in the Upper South, reveals the intricacies of their agency as midwives working within the world of revolutionary America as that world increasingly intertwined freedom with racial slavery and childbirth. This project, which is still a work in progress, begins to show how enslaved women's work in women's and children's health during the revolutionary period supported generations of enslaved families, paradoxically underpinned white stability and freedom, and contributed to the expansion of a new nation.

The Sources

The sources that scholars have long studied to understand and interpret histories of enslaved peoples largely include plantation documents such as enslavers' diaries, correspondence, account books, farm productivity reports, tax lists, and property inventories. These collections are often contained in well-funded private and public libraries across the nation, scattered themselves over time and place. Jefferson, Washington, and other slaveowners whose names title and organize these collections recorded parts of enslaved people's lives with the incisional pen marks of capital gain and the confidence of social power that ensured the archives' posterity.

Within this archive enslaved people were mutated into data. Those engaged in the business of slavery transformed African people and their descendants into economic data for private profit and white generational wealth.[6] Required to pay taxes on enslaved people of certain ages, enslavers generated annual lists of enslaved people's names, farm locations, and ages. Wanting to keep track of capital gains, enslavers kept generational lists of the birth of enslaved children, documenting children's names and genders, mothers' names, birth dates, birth locations, death dates, and sell dates. Slave traders advertised enslaved people for sale in newspapers. Slaveowners kept daily cash accounts and financial ledgers to track local and global business dealings, which included tracking enslaved people's productivity in the field, in the market, and in the womb.

It is within these sources where the lives, work, and agency of enslaved midwives is recorded. Midwifery was a valuable service in a society built on racial slavery, and it was a service that often resulted in financial exchange. In cash accounts, enslavers often documented the cash and goods they paid to enslaved midwives for their services. Some slaveowners documented this health work in more formalized individual accounts within plantation ledgers created for enslaved midwives with women's names recorded as the account holder. For

CHAPTER 4

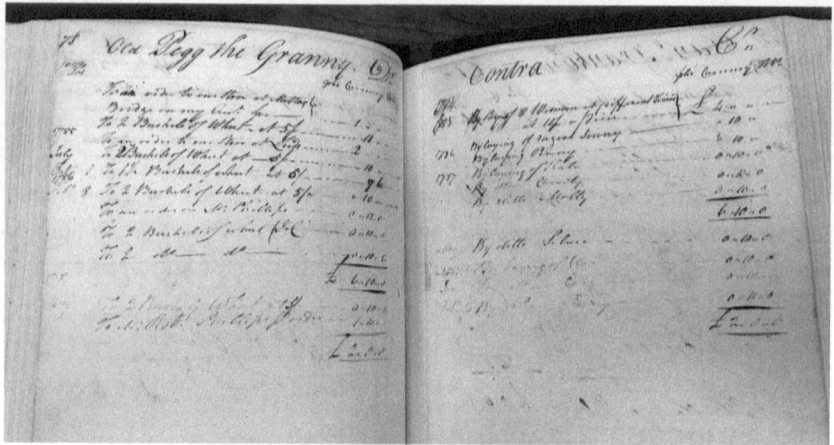

FIGURE 4.1. "Account with Old Pegg the Granny," 1784–1787, William Augustine Washington Account Book, 1776–1792, Library of Congress.

example, George Washington's nephew, William Augustine Washington, maintained an account with a midwife named Pegg, who was most likely enslaved, in Westmoreland County, Virginia, following the Revolutionary War.[7]

The right side of the account made up the credit side, in which William Washington recorded the names of mothers Pegg had aided in childbirth, including Jenny, Penny, Kate, Charity, Molly, Silvia, and Cloe. The debit side of the account, recorded on the left side of the page, is where he documented the valuations of midwifery services and recorded when compensation—in the form of cash, goods, or the extension of store credit—was provided to Pegg. The pages of plantation cash books, ledgers, birth lists, newspapers, and property inventories reveal the complex relationships enslaved midwives engaged in with both families and enslavers during the revolutionary period, and these are the records of slavery's archive that must be reframed.

The Digital Methodology

The power of digital tools makes possible this refocusing and aggregation of fragmentary evidence. Information from primary sources can be collected and interpreted with a data model. Information is filled into columns and rows, creating records that are contained in tables. These tables can then be linked to one another, undergirding the digital form known as the database.[8] As defined by the media studies scholar Lev Manovich, a database is a "structured collection of data," usually organized into tables that are connected and can be searched to recover information.⁹

The data model for this project includes the creation of four interrelated tables: people, events (including births and payments to midwives), locations, and sources. This organization makes up the basic structure of a relational database model, which links separate tables of information together through key identifiers. It follows the model set forth by the Database of Mount Vernon's Enslaved Community and the Jesuit Plantation Project.[10] The person table includes biographical and relationship information on midwives, children, family members, and enslavers. The events table includes a record of enslaved midwives' work and birth records of children. The location table documents the places where people lived, labored, and moved between and records enslaved midwives' movements and parameters of practice. Finally, the source table records basic information about each primary source in slavery's archive. Databases are arguments, and this structure strives to best answer the historical questions I pose.

The next step, and the most important, is metadata creation. Metadata is the information or set of terms that catalog and describe documents, artifacts, and other entities of study in a standardized format. Specifically, descriptive metadata involves the creation of labels or descriptors for each column in the tables that make up the database so that each record can be organized and explored. In the context of digital archives, metadata can "[bring] to the forefront opportunities for critical reflection," which this project aims to continuously do.[11] As Michael Kramer explains, historians often add interpretation to existing metadata created by archivists and librarians, or "add *meta*-metadata to the archival database." In the digital realm, the work of archivists and historians comes together through digital archives and can provide a "new kind of useful fluidity . . . among linked open source archives and scholarship using the materials in those archives."[12]

Metadata creation does not come without problems. There are serious potential issues of imposing digital violence on historically marginalized groups through this process. Creating metadata runs the risk of turning exploited peoples into objective numbers, just as their exploiters did in the past, to repeatedly splinter their lives. A dataset of enslaved people's lives uncomfortably resembles the organization of ledgers and property inventories that enslavers used to transfigure them centuries ago. As Jessica Marie Johnson warns, "the legacy of commodifying black bodies and truncating black life infuses and informs digital design and execution."[13] This project uses the model of a relational database to center the lives and relationships of enslaved midwives and the families they supported, reinterpreting the sources in slavery's archive written by enslavers and providing as much context as possible.

70   CHAPTER 4

Approaching metadata creation with this critical lens, the next step is to translate and reframe information in slavery's archive into these tables with the appropriate metadata. The first table I started to fill out was the person table, which includes biographical and family relationship information on midwives, parents, children, and enslavers. Cash accounts and property lists, such as those recorded by William Washington, are a useful set of sources for doing this work. Thomas Jefferson also recorded the work of midwives at Monticello and surrounding farms in his Memorandum Books. Jefferson tracked daily payments made to merchants, farmers, tavern keepers, blacksmiths, instructors, enslaved men, and enslaved women for their services and goods. Every time he or an intermediary—such as a daughter or overseer—paid someone, that financial exchange was recorded in the cash account. For example, in March 1813, a few years after he retired from the presidency, Jefferson recorded this line in his cash ledger: "Pd. [paid] the midwife (Rachael) 6.D. [dollars] for attending Edy, Moses's Mary and Esther."[14] The midwife Rachael, mothers Edy, Mary, and Esther, Mary's husband Moses, the newborn children, and Thomas Jefferson are all included in the person table.

Each person is documented with as much biographical information as can be found from cross-referencing other plantation records, such as Jefferson's Farm Book, and other digital records, such as the Monticello Plantation Database. From filling out the person table, details of enslaved people's lives can be highlighted, and families can be connected together, information that is otherwise obscured and fragmented in slavery's archive.

TABLE 4.1 Transcription of Thomas Jefferson's memorandum book, March 1813, Founders Online

Mar.	9.	Pd. for butter 1.50.
	10.	Hhd. exp. 3.D.
	12.	Do. 1D.
	14.	Do. 562.
	17.	Promised Barnaby to give him one barrel out of every 31. he sends to the mill.
	20.	Pd. Salmons for his work at the corner of the toll mill 1.D.
	21.	Hhd. exp. 50.22. Do. hhd. exp. 2.D.
	23.	Do. 5.D.
	24.	Recd. of Saml. J. Harrison an order on Gibson & Jefferson for 1000.D. at 10. days sight, on acct. of my tobo. which I inclosed to Mr. Gibson.
	27.	Pd. Wm. Johnson for oysters on acct. 5.D.
	28.	Pd. Ned &c. for sewers 1.D. hhd. exp. 1.5.
	29.	Pd. the midwife (Rachael) 6.D. for attending Edy, Moses's Mary and Esther.
	31.	Pd. Mclure for Mary C. Oglesby for weaving 3.50.
		Pd. Cooley on acct. blowing 5.D. for 6. days in the Canal.
		Gave order on Gibson & Jefferson for 40.D. in favr. Dunlap McKinney, substitute for E. Bacon, which I make no charge for, being a gratuity to E. Bacon.

The events table is the major digital record that builds off the person table. It enables the births recorded in enslavers' account books and property lists to be refocused on families and the midwives who supported them. Jefferson's record that reads "Pd. [paid] the midwife (Rachael) 6.D. [dollars] for attending Edy, Moses's Mary and Esther" can be reframed into a payment record to Rachael for midwifery work and three individual records of children's births: one each for Edy's daughter Betsy-Ann, Mary's child, and Esther's child Lindsay. Rachael earned two dollars for aiding in each of these births, and the digital records reflect this complicated financial agency. The information in the events table includes source text, title and type of event, description, midwife, mother, child, location, date, payment value, payment method, payment mediator, source references, and event key identifier.

Through this digital reframing of slavery's archive, it is clear that Rachael worked as the primary midwife at Monticello from at least 1810 to the year Jefferson died in 1826, taking on the role after several white women had been delivering enslaved children on the estate since 1774. Rachael likely started working as a midwife as early as 1808. She maintained relationships with several families enslaved at Monticello, including the Hemings, the Fossetts, the Herns, and the Grangers. Rachael also worked as a midwife across generations for mothers and daughters within the same families. She earned two dollars for each of these births.[15]

Other enslavers in the Upper South recorded enslaved women's midwifery work and the birth of children in account books. After the Revolutionary War, Richard Tilghman, a wealthy slaveowner on the Maryland Eastern Shore, maintained economic relationships with two midwives, Peg and Lucy, who were

TABLE 4.2 Abbreviated snapshot of payments to Rachael for midwifery work at Monticello

SOURCE TEXT	EVENT TYPE	RECIPIENT OF PAYMENT	DATE	PAYMENT VALUE	PAYMENT METHOD
"Pd. [paid] the midwife (Rachael) 8.D. [dollars] for Rachael Bedf. [Bedford], Scilla, Cretia and Ursula"	Payment	Rachael	1814-01-10	$8	Cash
"Pd. [paid] Rachael midwife for Lazaria, Virginia, Lucy & Fanny 8 D. [dollars]"	Payment	Rachael	1814-11-23	$8	Cash
"Pd. [paid] Rachael the midwife 10 D. [dollars] to wit for Edy, Virginia, Ursula & Mrs. Marks's Sally, the 2 D. [dollars] overpd. on account"	Payment	Rachael	1816-01-22	$10	Cash
"Midwife Rachael for Rachael & Scilla (Lego) & Fanny 6 D. [dollars]	Payment	Rachael	1816-09-03	$6	Cash

enslaved by different female slaveowners in the community. These accounts were similar to the one William Washington maintained with Pegg in Westmoreland County, Virginia. In Tilghman's ledger, Peg and Lucy each held their own independent accounts, and they worked concurrently between 1791 and 1803. Lucy, who was enslaved by Margaret Gardiner, delivered children for several enslaved women on Tilghman's plantation. Peg, a woman enslaved by the widowed Mary Gordon, delivered children for Priss, Phillis, Poll, Kate, Memory, and Chloe, and earned approximately 12 pounds.[16]

A specific example from this source speaks to the power of digital records to reframe the datafication of people of African descent in history and in the archive. In 1797 Richard Tilghman recorded in the account that Peg assisted an enslaved woman named Kate in childbirth: "Mar. 9th By attending Kate I believe nothing £0.0.0." Within the same ledger, Tilghman kept a list of enslaved children's births to track his family's capital assets for tax purposes, and for the year 1797, he wrote "Kate a son - - [born] March 1797 - - [named] Richmond - died Mar 29th."[17] By making this connection, it is evident that Peg helped Kate deliver her son Richmond who lived a short time before passing away. In addition to aiding Kate through childbirth, Peg likely supported Kate during the loss of her child, a grievous moment that a mother carries for a lifetime. To Richard Tilghman, Richmond's death signified a loss in potential profit, as he wrote "I believe nothing" for the child's death in his account book, a contraction of human life. Connections made through digital records reframe Tilghman's dehumanization of Kate and her son into an emotional narrative of the loss of a child while enduring the traumas of enslavement.[18]

Not all enslavers paid the women they enslaved for midwifery services, of course. Some women in Virginia and Maryland were not paid for this work. There is also a case where an enslaver used the labor of an enslaved midwife *as payment* for other services. In Accomack County, Virginia, Edward Taylor used the midwifery work of Sarah to pay for other services in the local community in the 1780s. Taylor used Sarah's labors as payment for a physician's services, schooling for his children, and accounts with innkeepers, recording payments as "To Negro Sarah Delivering your whench."[19] Taylor maintained business relationships within the community that were on occasion tied together with Sarah's labor as a midwife. Studying these accounts is another example of how the creation of data from slavery studies—even as that process strives to dismantle slavery's archive—reimposes digital forms of dehumanization. Within this database, Sarah is documented in both the "midwife" column and the "payment method" column. While enslaved women exercised social and medical agency with midwifery work, enslavers realized the value of that work and began to turn midwives themselves into financial tools for increased profit.

Initial Patterns

The records in this database reveal individuals, families, and relationships that have been previously overlooked in slavery's archive. The project also reveals important insights and patterns of enslaved women's midwifery work in the context of the revolutionary period in which they lived. The first important pattern that begins to emerge is that enslaved women began regularly surfacing as midwives in slavery's archive during the late 1760s through the 1780s. In my research, one of the first mentions of an enslaved midwife in colonial newspapers appeared in a slave sale advertisement in 1768.[20] These records point to the historical moments in which enslavers paid more attention to enslaved midwives and acknowledged their skills in written and printed sources.

The late 1760s was the exact moment when many colonies banned the importation of "foreign slaves" from Africa and the West Indies as part of the nonimportation resolutions. Responding to the passage of the Townshend Acts in 1767 by the British Parliament, colonial merchants in Virginia and South Carolina especially prohibited the importing, buying, or selling of any enslaved people from Africa and the West Indies. Colonists boycotted the importation of wine, tea, and other British goods, including human beings; engaged in nonconsumption agreements; and promoted North American manufacturing like homespun.[21] The British Parliament eventually repealed the Townshend Acts in 1770, except for the import tax on tea. However, after the imposition of the Coercive Acts, colonists formed the First Continental Congress and reimposed the nonimportation resolutions with the Continental Association. The ban on the importation of foreign slaves was reinstated across the colonies in 1774, and it continued through the end of the Revolutionary War.[22] The passage of the nonimportation resolutions—an important economic moment of the Revolution that encouraged the patriotic American production of goods—catalyzed a reliance on the domestic reproduction of slaves and pushed enslaved midwives into the nexus of economic tension, revolutionary change, and the expansion of racial slavery in the South.

Relatedly, another pattern emerges from these digital records. Along with the natalist shift in American slavery, enslaved women started moving the role of midwife on plantations away from white midwives, advocating for families across plantation landscapes and across generations. This pattern is especially apparent in spaces where men like Thomas Jefferson and George Washington enslaved hundreds of people over decades. At Monticello, records from the person and event tables show that at least nine white women acted as midwives for enslaved women from 1773 to 1808, the year Rachael most likely started to appear as the primary midwife.[23]

A similar pattern occurred on plantations in Fairfax County, Virginia. At Mount Vernon, George Washington employed several white midwives to deliver children for the women he and the Custis family enslaved from the 1750s through the Revolutionary War. However, enslaved women began to appear as midwives in the archive of Mount Vernon in the telling year of 1776. On February 20, 1776, Washington's ledgers document a payment of 10 shillings made to Jane, recorded as "Mrs. Frenches Jane," for delivering a child for another woman named Jane at Dogue Run Farm. This was most likely the first record of payment made to an enslaved woman for midwifery work at Mount Vernon. During the 1790s (and likely starting from 1785), enslaved women regularly delivered babies for other women enslaved on Washington's Mount Vernon plantation, including Kate, who petitioned Washington to become a midwife and to be paid for that work in 1794, and Nell, who was a midwife enslaved by the Mason family at Gunston Hall three miles south on the Potomac River.[24] During this period, enslaved women became the primary midwives on plantations, where members of their own communities and families provided maternal, nutritional, and emotional support. Enslaved midwives also helped deliver their own grandchildren and passed their knowledge down to their female family members, friends, and members of their communities.

The third major finding revealed by these relationship tables is how common it was for enslaved women in the Upper South to earn compensation for their midwifery skills, especially after the Revolutionary War. Enslaved midwives like Pegg, Rachael, Peg, Lucy, and Nan earned regular payments in cash, goods, and even store credit during a time of expanding political freedom for white men and the development of early American capitalism.[25] For the most part, enslavers who compensated enslaved midwives did so at the same rates as white female midwives. These financial exchanges reveal the complexity of enslaved women's roles in commercial networks of the revolutionary period. Although they found financial and social agency within the confines of the plantation environment, the mechanics of racial slavery ensured that enslaved women working as midwives expanded the very system that oppressed them and the families they supported.

While outcomes of the American Revolution included an increase in antislavery sentiments and gradual emancipation for some enslaved peoples, women and families in the South endured an expansion of the system that enslaved them. The bans on foreign slave importation from the 1760s precipitated a reliance on domestic reproductive slavery. Coupled with the ratification of the Constitution—a document that espoused republican ideals yet gave southern slaveowners tremendous political leverage through the Three-Fifths Compromise—slavery and childbirth became further enmeshed in the infrastructure of the republic.[26] The natalist shift in slavery culminated in 1808, when Congress banned the importation of slaves on a federal level, after a twenty-year delay written into the Constitution. Yet again, the country did not ban slavery as an institution. What resulted was the expansion of slavery into the South and West, especially following the cotton boom.[27] Enslavers and other capitalists forced enslaved people into these new areas through interstate human trafficking and increased the importance of childbirth, which included the work of midwives, to the success of the new American nation. Enslaved women, the children they bore, and the midwives who aided them became important financial and political capital in securing the prosperity of the American political economy.

Next Steps

This project on enslaved women's midwifery work will eventually be integrated into the content management system Omeka S with linked open data. Linked data enables records to be connected to other databases and datasets on the web that use the same ontologies, or classification systems. It employs the principle of a semantic web of data, in which data is undergirded with custom vocabularies that enable it to be accumulated and queried systematically across the internet.[28] Omeka S was built with linked open data in mind, enabling the integration of shared ontologies and the use of URIs, or Uniform Resource Identifiers, like a URL, to connect information across the web.[29]

This project will incorporate the vocabularies created specifically to help document enslaved people's lives through Enslaved.org, which revolves around "inclusive and reparative scholarship about historical slavery and responsible stewardship of historical data about enslaved people in digital spaces."[30] Previously, there was no linked data model specific to slavery studies, as Sharon Leon notes.[31] For example, within the available ontologies, there was a term for "employer" but not "enslaver," which is the exact problem of metadata creation that Black digital humanities brings forward. The Enslaved.org project pluralized the "Archival Multiverse," which "challenges archivists and recordkeepers to use metadata to address the experiences, needs and aspirations of marginalized and under-represented groups as well as addressing the wider social imperative to ensure that recordkeeping can help to document, empower and enfranchise."[32] By connecting this project to Enslaved.org—which aggregates projects that center the lives of enslaved and freed peoples in an ethically conscious digital realm—the names, lives, and work of midwives and families brought forward from slavery's archive will be responsibly recovered and made fully accessible.

The intermixing of methodologies in slavery studies and digital humanities opens up the possibilities for context-rich records that forward histories of enslaved women, men, and children, providing new bridges across the fragmented sources of slavery's archive. As the digital records of this project show, these methods also provide new perspectives through which to understand the history of the American Revolution and how white freedom was intertwined with racial slavery and childbirth. Ultimately, enslaved midwives advocated for and secured change in maternal and infant health on southern plantations, becoming caretakers for generations of enslaved families. Increasingly after the Revolution, those engaged in the business of slavery used enslaved midwives as financial tools to help secure white generational wealth and the vitality of the American political economy. The lives and work of enslaved midwives were thus paradoxically and inextricably linked to the freedoms won by their enslavers.

NOTES

1. Philip D. Morgan, *Slave Counterpoint: Black Culture in the Eighteenth-Century Chesapeake and Lowcountry* (Chapel Hill: published by the Omohundro Institute of Early American History and Culture and the University of North Carolina Press, 1998); Kathleen M. Brown, *Good Wives, Nasty Wenches, and Anxious Patriarchs: Gender, Race, and Power in Colonial Virginia* (Chapel Hill: published by the Omohundro Institute of Early American History and Culture and the University of North Carolina Press, 1996); Wendy Warren, *New England Bound: Slavery and Colonization in Early America* (New York: Liveright, 2016); Jennifer L. Morgan, *Laboring Women: Gender and Reproduction in New World Slavery* (Philadelphia: University of Pennsylvania Press, 2004).

2. Sylvia R. Frey, *Water from the Rock: Black Resistance in a Revolutionary Age* (Princeton, NJ: Princeton University Press, 1991); Joanne Pope Melish, *Disowning Slavery: Gradual Emancipation and "Race" in New England, 1780–1860* (Ithaca, NY: Cornell University Press, 1998); Eva Sheppard Wolf, *Race and Liberty in the New Nation: Emancipation in Virginia from the Revolution to Nat Turner's Rebellion* (Baton Rouge: Louisiana State University Press, 2006); Adam Rothman, *Slave Country: American Expansion and the Origins of the Deep South* (Cambridge, MA: Harvard University Press, 2005).

3. Herbert G. Gutman, *The Black Family in Slavery and Freedom, 1750–1925* (New York: Pantheon Books, 1976); Deborah Gray White, *Ar'n't I a Woman? Female Slaves in the Plantation South* (New York: W. W. Norton, 1985); Wendy Anne Warren, "'The Cause of Her Grief': The Rape of a Slave in New England," *Journal of American History* 93, no. 4 (March 2007): 1031–49; Annette Gordon-Reed, *The Hemingses of Monticello: An American Family* (New York: W. W. Norton, 2008); Jessica Millward, *Finding Charity's Folk: Enslaved and Free Black Women in Maryland* (Athens: University of Georgia Press, 2015); Marisa J. Fuentes, *Dispossessed Lives: Enslaved Women, Violence, and the Archive* (Philadelphia: University of Pennsylvania Press, 2016); Michel-Rolph Trouillot, *Silencing the Past: Power and the Production of History* (Boston: Beacon Press, 1995); Sasha Turner, *Contested Bodies: Pregnancy, Childrearing, and Slavery in Jamaica* (Philadelphia: University of Pennsylvania Press, 2017); Katherine Paugh, *The Politics of Reproduction: Race, Medicine, and Fertility in the Age of Abolition* (Oxford: Oxford University Press, 2017); Erica Armstrong Dunbar, *Never Caught: The Washingtons' Relentless Pursuit of Their Runaway Slave, Ona Judge* (New York: Simon and Schuster, 2017); AfriGeneas: African Ancestored Genealogy, https://www.afrigeneas.com; BlackPast, blackpast.org.

4. Kim Gallon, "Making a Case for the Black Digital Humanities," in *Debates in the Digital Humanities*, ed. Matthew K. Gold and Lauren Klein (Minneapolis: University of

Minnesota Press, 2016), 42–49. See also Safiya Umoja Noble, "Toward a Critical Black Digital Humanities," in *Debates in the Digital Humanities*, ed. Matthew K. Gold and Lauren Klein (Minneapolis: University of Minnesota Press, 2019), 27–35.

    5. Jessica Marie Johnson, "Markup Bodies: Black [Life] Studies and Slavery [Death] Studies at the Digital Crossroads," *Social Text* 36, no. 4 (December 2018): 57–79, 71, 70. See the Early Caribbean Digital Archive as an example of a digital project that decolonizes slavery's archive: Nicole Aljoe and Elizabeth Maddock Dillon, Early Caribbean Digital Archive, Northwestern University, 2017.

    6. Morgan, *Laboring Women*; Stephanie E. Smallwood, *Saltwater Slavery: A Middle Passage from Africa to American Diaspora* (Cambridge, MA: Harvard University Press, 2007); Sowande' M. Mustakeem, *Slavery at Sea: Terror, Sex, and Sickness in the Middle Passage* (Champaign: University of Illinois Press, 2016); Daina Ramey Berry, *The Price for Their Pound of Flesh: The Value of the Enslaved, from Womb to Grave, in the Building of a Nation* (Boston: Beacon Press, 2017); Edward Baptist, "Toward a Political Economy of Slave Labor: Hands, Whipping-Machines, and Modern Power," in *Slavery's Capitalism: A New History of American Economic Development*, ed. Sven Beckert and Seth Rockman (Philadelphia: University of Pennsylvania Press, 2016), 31–61; Edward Baptist, *The Half Has Never Been Told: Slavery and the Making of American Capitalism* (New York: Basic Books, 2014); Caitlin Rosenthal, *Accounting for Slavery: Masters and Management* (Cambridge, MA: Harvard University Press, 2018); Nora Doyle, *Maternal Bodies: Redefining Motherhood in Early America* (Chapel Hill: University of North Carolina Press, 2018).

    7. "Account with Old Pegg the Granny," 1784–1787, in William Augustine Washington Account Book, 1776–1792, Manuscript Division, Library of Congress; Morgan, *Slave Counterpoint*, 325.

    8. Johanna Drucker, "Data and Databases," in *Intro to Digital Humanities: Concepts, Methods, and Tutorials for Students and Instructors*, UCLA Center for Digital Humanities, 2014, 28–33.

    9. Lev Manovich, "Database as a Genre of New Media," *AI and Society* 14, no. 2 (June 2000): 176–83.

    10. Mount Vernon Ladies' Association, Database of Mount Vernon's Enslaved Community, 2016, https://www.mountvernon.org/george-washington/slavery/slavery-database/; Sharon M. Leon, Jesuit Plantation Project, https://jesuitplantationproject.org. This project also draws from the Monticello Plantation Database, 2008, http://plantationdb.monticello.org/.

    11. Itza A. Carbajal and Michelle Caswell, "Critical Digital Archives: A Review from Archival Studies," *American Historical Review* 126, no. 3 (2021): 1102–20. This article offers a guide for all scholars, especially historians, engaging with digital archives and metadata.

    12. Michael J. Kramer, "Going Meta on Metadata," *Journal of Digital Humanities* 3, no. 2 (Summer 2014), http://journalofdigitalhumanities.org/3-2/going-meta-on-metadata/.

    13. Johnson, "Markup Bodies," 60; Sharon Leon, "The Peril and Promise of Historians as Data Creators: Perspective, Structure, and the Problem of Representation," *[bracket]*, November 24, 2019, https://www.6floors.org/bracket/2019/11/24/the-peril-and-promise-of-historians-as-data-creators-perspective-structure-and-the-problem-of-representation/.

    14. Thomas Jefferson, entry for March 29, 1813, Jefferson's Memorandum Books, vol. 2, in *The Papers of Thomas Jefferson Digital Edition*, 2nd ser., ed. James P. McClure and J. Jefferson Looney (Charlottesville: University of Virginia Press, Rotunda, 2008–2019), 1287; "Memorandum Books, 1813," Founders Online, National Archives, https://founders.archives.gov/documents/Jefferson/02-02-02-0023.

    15. Data compiled from Jefferson's Memorandum Books, vols. 1 and 2, in *The Papers of Thomas Jefferson Digital Edition: Jefferson's Memorandum Books, Volumes 1 and 2: Accounts,*

with *Legal Records and Miscellany, 1767–1826*, ed. James A. Bear and Lucia C. Stanton (Princeton, NJ: Princeton University Press, 1997), 1247–1413; Thomas Jefferson, Farm Book, 1774–1824, in *Thomas Jefferson Papers: An Electronic Archive*, Massachusetts Historical Society; Monticello Plantation Database.

16. "Account with Old Peg a Negroe belonging to Mrs. Gordon," 1791–1803, p. 38 of digitized microfilm, and "Account with Old Lucy a Negroe belonging to Mrs. Gardiner," 1794–1796, p. 48 of digitized microfilm, both in Richard Tilghman, Account Book, 1787–1807, Tilghman Collection, Microfilm SCM 1346, Maryland State Archives.

17. Tilghman, Account Book, 1787–1807.

18. Sasha Turner, "The Nameless and the Forgotten: Maternal Grief, Sacred Protection, and the Archive of Slavery," *Slavery and Abolition* 38, no. 2 (April 2017): 232–50.

19. Edward Taylor, Ledger, 1687–1805, miscellaneous reel 532, Library of Virginia. The bulk of this ledger covers 1775 to 1789.

20. David Fogartie, "To Be Sold," *South Carolina Gazette*, August 9, 1768, America's Historical Newspapers, Readex.

21. "Resolutions," *South Carolina Gazette*, July 27, 1769, America's Historical Newspapers, Readex; George Mason, Virginia Nonimportation Resolutions, May 17, 1769, Founders Online, National Archives; Glenn Curtis Smith, "An Era of Non-Importation Associations, 1768–73," in *William and Mary Quarterly* 20, no. 1 (January 1940): 84–98; John P. Kaminski, ed., *A Necessary Evil? Slavery and the Debate Over the Constitution* (Madison, WI: Madison House, 1995). Merchants in New York City and Philadelphia had passed earlier nonimportation resolutions in response to the Stamp Act of 1765, but those did not prohibit the importation of enslaved people.

22. Journals of the Continental Congress, 1774–1789, October 20, 1774, in *A Century of Lawmaking for a New Nation: U.S. Congressional Documents and Debates, 1774–1875*, 75–81, American Memory, Library of Congress; Kaminski, *Necessary Evil?*, 1–4.

23. Data pulled from Thomas Jefferson's Memorandum Books, 1774–1826, *The Papers of Thomas Jefferson Digital Edition*, 2nd ser., ed. James P. McClure and J. Jefferson Looney (Charlottesville: University of Virginia Press, Rotunda, 2008–2019); Ann Carey Randolph, Household Accounts, 1805–1808, in vol. 1, Household Accounts and Notes of Virginia Court Legal Cases, Series 7: Miscellaneous Bound Volumes, Thomas Jefferson Papers, Library of Congress.

24. Entry for February 20, 1776, George Washington Ledger B, 1772–1793; Mount Vernon Ladies' Association, Database of Mount Vernon's Enslaved Community, 2016; George Washington to William Pearce, August 17, 1794, in Presidential Series (24 September 1788–30 September 1794), vol. 16 (1 May–30 September 1794), in *The Papers of George Washington Digital Edition*, ed. Theodore J. Crackel (Charlottesville: University of Virginia Press, Rotunda, 2007–), 574; entry for "Kate B" on August 17, 1794, Mount Vernon Ladies' Association, Database of Mount Vernon's Enslaved Community; James Anderson, "Manager Ledger," October 11, 1798, 187, accessed through Database of Mount Vernon's Enslaved Community.

25. Seth Rockman, *Scraping By: Wage Labor, Slavery and Survival in Early Baltimore* (Baltimore: Johns Hopkins University Press, 2009); Ellen Hartigan-O'Connor, *The Ties That Buy: Women and Commerce in Revolutionary America* (Philadelphia: University of Pennsylvania Press, 2009); Paul A. Gilje, ed., *Wages of Independence: Capitalism in the Early American Republic* (Madison: Madison House, 1997). Male physicians occasionally delivered babies during this period; they were usually paid twice as much as female midwives.

26. David Waldstreicher, *Slavery's Constitution: From Revolution to Ratification* (New York: Hill and Wang, 2009).

27. Steven Deyle, "An 'Abominable' New Trade: The Closing of the African Slave Trade and the Changing Patterns of U.S. Political Power, 1808–60," in *William and Mary Quarterly*, 3rd ser., 66, no. 4 (October 2009): 833–50; Rothman, *Slave Country*.

28. Jonathan Blaney, "Introduction to the Principles of Linked Open Data," *Programming Historian* 6 (2017), https://doi.org/10.46430/phen0068.

29. Corporation for Digital Scholarship and the Roy Rosenzweig Center for History and New Media, Omeka S, https://omeka.org/s/.

30. Statement of Ethics, *Enslaved: Peoples of the Historical Slave Trade*, https://enslaved.org/statementofEthics.

31. Sharon M. Leon, "Thinking with Linked Data; Representing History," *[bracket]*, November 1, 2018, https://www.6floors.org/bracket/2018/11/01/thinking-with-linked-data-representing-history/; Sharon M. Leon, "About," Jesuit Plantation Project, https://jesuitplantationproject.org/s/jpp/page/About.

32. Anne J. Gilliland and Sue McKemmish, "Recordkeeping Metadata, the Archival Multiverse, and Societal Grand Challenges," DC-2012: Proceedings of the 12th International Conference on Dublin Core and Metadata Applications, 2012, 109; Carbajal and Caswell, "Critical Digital Archives," 1109–11.

Part II
SPATIAL REVOLUTIONS

5

GEOGRAPHIES OF EMANCIPATION

Geospatial Technology in Mapping Black Thought in the Age of Revolutions

Jessica M. Parr

In his preface to the 1818 translation of *The Haytian Papers*, the Black attorney and abolitionist Prince Saunders sought to set before the British "more correct information with respect to the enlightened systems of policy, the pacific spirits, the altogether domestic views; and the liberal principals of the Government," of the young Black republic.[1] Saunders, a formerly enslaved and Dartmouth-educated man, sought to reframe how white observers conceived of a Black-governed political space, which Saunders called a "new and truly interesting empire."[2] His writing exemplified one of the ways Black abolitionists sought to center Black liberation on Black political spaces. In addressing white audiences, Saunders demonstrates the ways discussions of Black spaces and slavery were entangled with white, European cartographies of power during the Age of Revolutions.

It is impossible to eliminate problems of power and erasure that create difficulties for scholars working to map Black experiences within the geographies they traveled. However, writings by Saunders and other Black intellectuals focus on Black understandings of place and space. Techniques from data visualization, text mining, and spatial analysis offer opportunities to recover these experiences. Centering these techniques on Black thought and Black concepts of space and place necessitates thinking outside of the frequently white-centric methodologies that frequently dominate the digital humanities.[3] Black DH methodologies—described by Kim Gallon as "technologies of recovery" that connected the cartographies to the humanity—provide particular promise for scholarship on Black lives on Blackness.[4]

Much of Western cartography has its roots in the Enlightenment and in imperial policies. As Martin Brückner observed, maps were less a practice of representing terrain than an expression of possession and power over colonized lands. Where maps were functions of empire and imperial policy, they have created exclusionary spaces that tend to override, if not erase, Indigenous African concepts of African geographies. But, as Brückner argues, maps have a social and political life as well that, when combined with digital techniques, can go beyond carto-coded or georectified representations of place, space, and human experience.[5] Within the paradigm of Black DH, techniques from data visualization and radical cartography, viewed through the lens of "entanglements," can be utilized to better account for the power dynamics of the Age of Revolutions.[6] Entanglements acknowledge that there is a power imbalance yet allows for Black experiences and understandings of place and space as a valid coauthor of these human geographies.

Reassessing the ways scholars can use digital techniques to allow for complex geographies entails an exploration of some of the limitations of mainstream approaches to mapping. Consider this 1789 map of Africa from the David Rumsey digital collection by the English cartographers Robert Sayer and Thomas Jefferys and the French cartographer Jean Baptiste Bourguinon. The map's authors acknowledge the land's settlement by Maghrenbine Berbers but use the term "Moors," which is not a term that any African ethnicity used for self-identification. And the map's focus is on European political and economic ambitions on a continent that, by 1789, had been stripped of significant numbers of its people for over two centuries.

The cartouches in these maps are reflective of Enlightenment-era fascinations with ancient Egypt. Playing on European tropes about Africa, with Eurocentric terminology like "Moors," they are a form of intellectual colonizing and cartographical appropriation of Africa.[7] These cartographical colonialisms worked in tandem with popular travel narratives like *Leo Africanus*, to erase the sociocultural and political complexities of a vast and diverse continent.[8] This map's centeredness on spaces where the dominant population consisted of Afro-Arab Muslims—who tended to have higher literacy rates than people in other parts of Africa and were among the economic elites—reinforced European biases around textual systems of knowledge over the oral traditions in other parts of Indigenous Africa. Despite some acknowledgment of historical land claims that predate European colonization, the map is nonetheless an expression of empire rather than one that offers an African insight into the land. Strict adherence to political borders shaped by colonization can therefore be limiting. But thinking more specifically about what those boundaries—which were often metaphysical boundaries—meant to the enslaved and free people of African descent, in

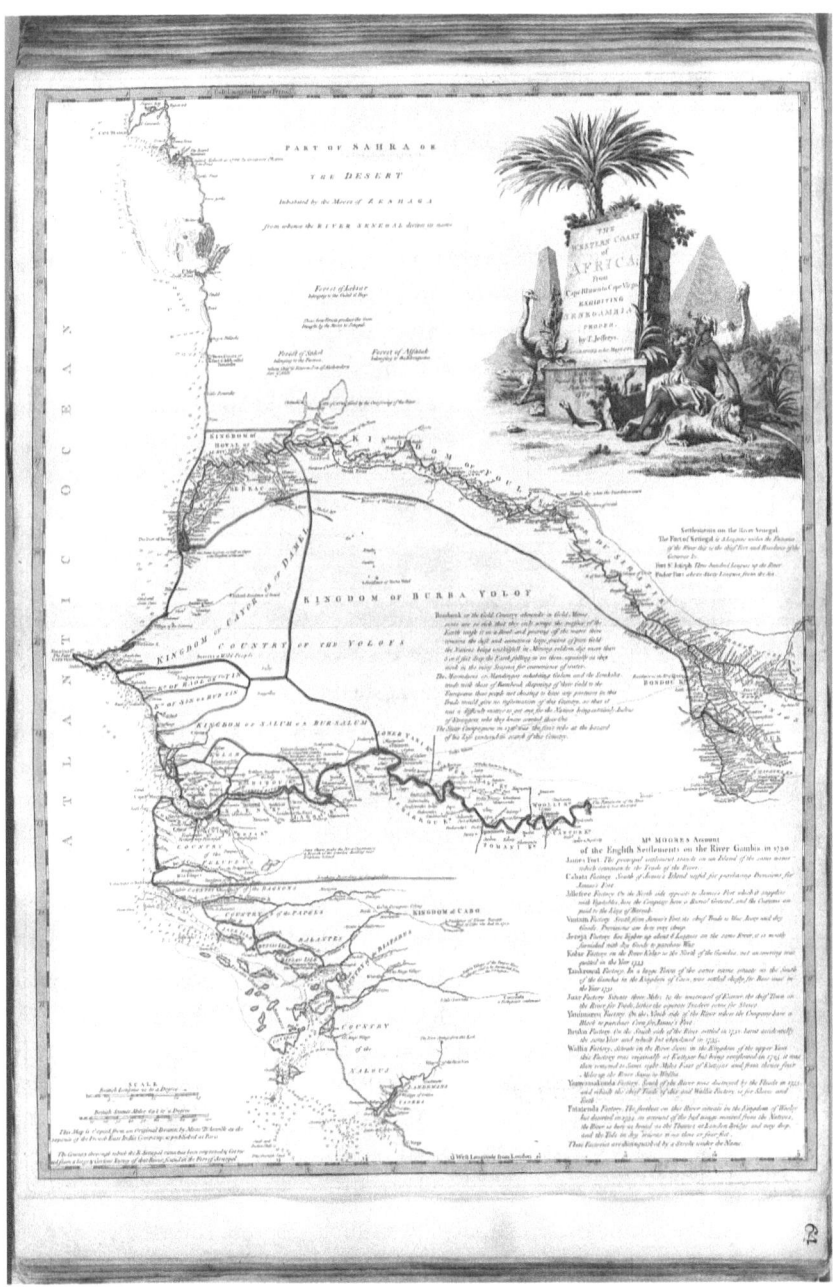

FIGURE 5.1. The Western Coast of Africa . . . By T. Jefferys, Geographer to his Majesty. London, Printed by Robert Sayer, No. 53 Fleet Street, 1789. David Rumsey Map Collection.

combination with digital humanities techniques, offers new opportunities for layering Black political and social experiences on these spaces, albeit potentially with some limitations that privilege text sources over oral histories unless the practitioner deliberately seeks out oral sources and treats them as text.[9] As will be described later in this chapter, there are ways of using archival fragments and writing against the archive that can mitigate but not eliminate this problem entirely. Aside from questions of literacy, conditions like fugitivity are powerful motives for enslaved individuals to try to make themselves geographically absent or inconspicuous, evading both the archive and mapping.

To address the problems in Eurocentric representation of places on maps we can start by thinking about the geography they are intended to describe in different ways. Specifically, radical countermappings allow us to address the histories of dispossession and colonization.[10] For example, as part of a sociological study of the status of Black Americans, W. E. B. Du Bois and a team of sociologists produced a series of data visualizations for the 1900 World's Fair that was intended to show the progress African Americans had made in the decades since the end of slavery. Du Bois's aim was to enhance a collection of photographs that was intended to undercut racist stereotypes of their African American subjects. Yet the photographs themselves did not tell the whole story that Du Bois wanted to tell. So he set about creating sixty data visualizations that drew on empirical sociological data to demonstrate the economic and other ongoing barriers to African American progress.[11]

Among the many striking visualizations produced by Du Bois was a series of maps that represented a cartography of population demographics in the decades leading up to the end of slavery, during reconstruction, and into the early years of Jim Crow. This visualization, titled "Proportion of Negroes in the total population of the United States," shows a set of four simple maps of the United States between 1800 and 1890. The total population of the United States is outlined in red. Smaller black maps of the United States show the relative percentage of the Black population relative to the white population, with the size of both maps growing to illustrate population growth in the nation as a whole. The maps are simply labeled with titles indicating the proportions "one-fifth," "one-sixth," "one-seventh," and "one-eighth."[12] His aim was to show that the Black population of the United States had shrunk during the nineteenth century.

The political boundaries are limited to a simple disproportionately drawn set of maps, because Du Bois's intentions did not require precision mapmaking or careful attention to cartographic practice. Rather, his emphasis was on representing the relative shrinking of the Black population to make the larger point that African Americans were becoming a smaller and smaller minority group, and that this was an obstacle to progress. The maps also demonstrate that the end of

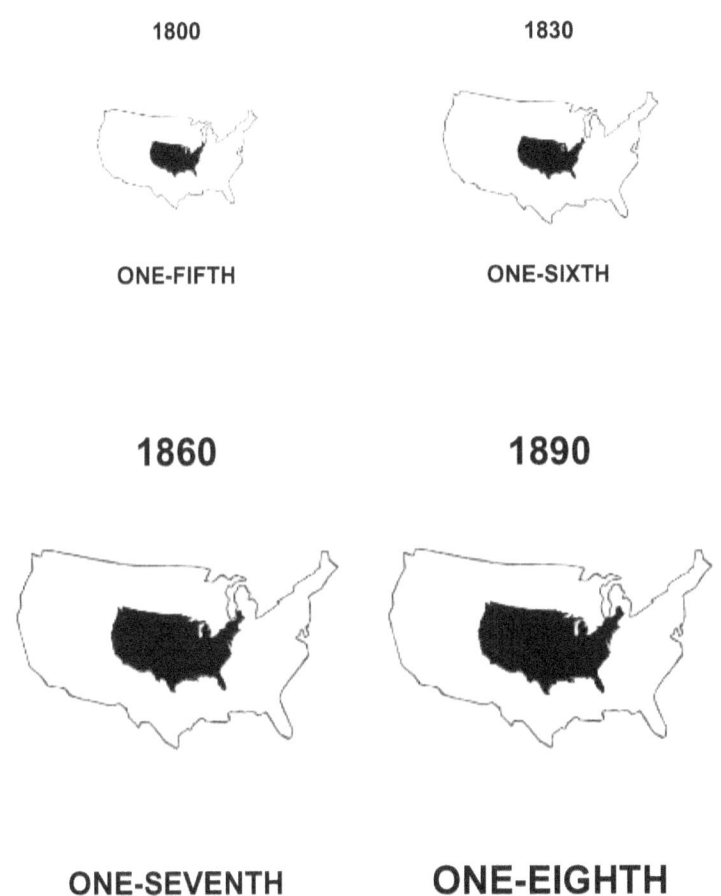

FIGURE 5.2. W. E. B. Du Bois Data Portraits of "Proportion of Negroes in the total population of the United States." Courtesy of the Library of Congress.

slavery and the start of the Second Industrial Revolution did little in the United States to curtail a steady shrinking of the African American population. A simple graphic conveyed a critical bit of information about the Black experience.

Contemporary approaches to radical mapping that recenter the human experiences include Vincent Brown's cartographic narrative of slave revolts during Jamaica's First Maroon War. The First Maroon War (1728–1740) was, in many respects, a revolutionary conflict between the Windward and Leeward Maroons (self-liberated societies of Africans) and the British Empire. The war, which was waged intermittently over twelve years, was part of an effort by the Maroons to dislodge British control over land in Jamaica. Brown's digital project draws from data that aligns the movements of the Maroon rebels with those of the British forces and planters they opposed.[13] It particularly illuminates the care taken in their military strategies and careful positioning of themselves within Jamaica's mountainous terrain, showing that it was anything but happenstance, although many of his carefully curated sources were from colonial archives. Similarly, Gergely Baics and Leah Meisterlin's work on urban planning and radical mapping reassesses some of the problems in traditional geospatial analysis, particularly its tendency to omit the human topographies of freedom and inequity.[14] Jeffrey Kok Hui Chan and Ye Zhang write of the need to account for what they call the "missing dimension" of "sharing spaces," which is a methodology for looking at different types of spaces and places that might be created by the human experience.[15] Not all histories of the enslaved and unfree were urban stories, but the notion of entangled shared spaces can be used to describe the ways that Black intellectuals shaped both intimate and metaphysical geographies that existed both alongside and outside of white notions of place and space.

In the Age of Revolutions and beyond, one of the major obstacles to Black freedom was an uneven network of state, national, and imperial laws governing slavery. As Katherine McKittrick, Elizabeth Stodeur Pryor, and Martha Jones demonstrate, this uneven legal terrain meant that Black mobility became part of how African Americans understood their relationships to place.[16] McKittrick, in particular, observes that diasporan geography does not inherently follow Western cartographical traditions. It may be forced to consider settler colonialist political boundaries as well as the wages of whiteness, but it creates imagined geographies that differ from African and, more broadly, diasporan concepts of space. Also important are explorations of freedom and unfreedom in what she calls the "cartography of struggle."[17] This cartography of struggle involves evaluating landscapes for the degrees of freedom and mobility they afforded enslaved and free Africans. Not only did the denizens of the Black Atlantic negotiate their own diasporic cartographies, they also had to navigate an uneven political and legal geography that determined their status and constricted their movements.

Uneven and changing legal geographies made a tremendous difference in the Black experience, to the point where abolitionist newspapers reported on legal cases involving emancipation or writs of habeas corpus.[18] In his attempts to petition the Massachusetts legislature for relief from the kidnapping of Black Bostonians, Prince Hall ran into a problem caused by a weak and ineffectual national government that lacked the legislative authority to address a problem tied to laws around slavery, there being no meaningful mechanism for interstate commerce, and a federal government that was in large part deeply unmotivated to intervene in any way that might undermine the business of slavery. There was no political, legal, or diplomatic remedy to address the concerns laid out by Hall and the eight other African American Masons.[19]

African American attorneys like Robert Morris, who specialized in antislavery cases, collected cases involving Black freedom.[20] Others collected and distributed pamphlets that codified and advocated for the validity of Black political spaces—physical Black-controlled political spaces, such as postrevolutionary Haiti, as Prince Saunders described in his preface to *The Haytian Papers*.[21] Part of their emancipatory geography meant mounting what were initially local legal challenges to slavery and other unfreedoms. Drawing from Du Bois's approach to geographies to convey population shrinkage, it is also possible to convey unevenness in the legal landscape of early America to help illustrate obstacles to emancipation that African Americans confronted. Because structures like the law could produce vastly different experiences, even in localities that were adjacent to one another, it is also less important to accurately portray state boundaries to convey the broader points about geography and law.

This rendering of the legal geography of slavery shows the states of most of the major Eastern Seaboard that was once part of the British Empire, a confederacy of slave societies from the American Revolution onward.[22] The states are rendered in polygons and are not georectified to a conventional map because the

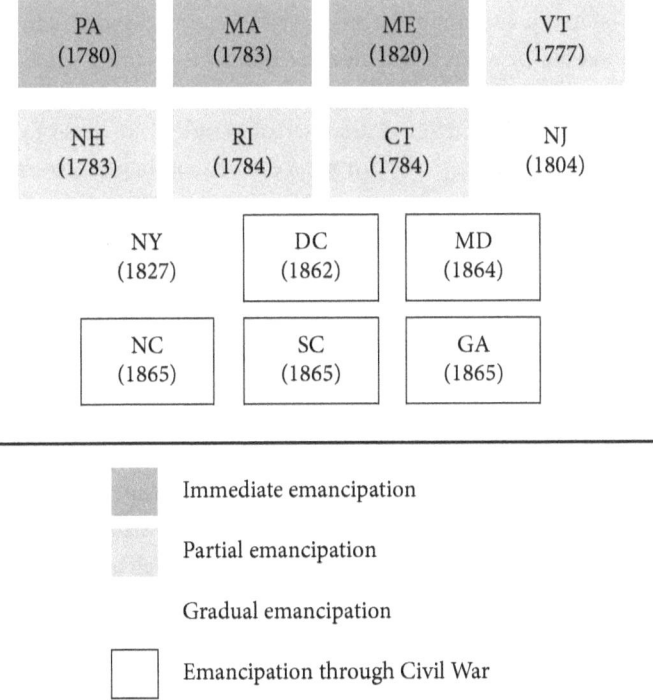

FIGURE 5.3. A Confederacy of Slave Societies. Visualization by the author.

focus of this representation is on comparing the legal status of the enslaved from state to state. Each state is illustrated with a grayscale gradient to offer a visualization of simplified data on when and how emancipation occurred—whether it was immediate or gradual.[23] It illustrates what Van Gosse describes as "Racial Orders [and Disorder]," which meant that the sometimes rapidly changing law created shifts in racial orders within these geographies.[24] In the case of some states, there was a mixture of emancipatory moments; those states are represented with a gradation in their grayscale rather than a solid color. And each state had its own racial order. What this approach offers is not so much a nuanced analysis of the law but a snapshot of how the African American experience was entangled with white legal structures and how that experience differed from place to place. A more nuanced detail that expands on change across time might entail creating an interactive or animated graphic in R-Studio or Python, akin to Lincoln Mullen's Spread of Slavery visualization but without the necessity of shapefiles.[25]

This approach can also be used to consider the forces—voluntary and involuntary—that continued to shape Black cartographies of struggle within an Anglo-American paradigm from the Age of Revolutions and beyond. In this graphic, the gradients represent spaces that were colonized by Britain and the United States. Africa and other European spaces are rendered in white, but the visualization could be altered with further gradients to bring the Black experiences of migration, slavery, and freedoms in other colonized spaces into the analysis. Migrations that involved a subset of people who chose to move are depicted as solid lines, though the historical events that shaped these migrations also produced involuntary movements. One of the limitations of visualizations is that, particularly when they are static rather than interactive, they do not satisfactorily capture all nuances—for example, the colonization of Sierra Leone and Liberia, which included a mixture of formerly enslaved Africans who chose to resettle in West Africa; and forced migrations of populations like the Jamaican Maroons, who were moved from Nova Scotia to Sierra Leone. Involuntary migrations—such as the Maroons who were transported out of Jamaica or the enslaved Africans who were brought by their enslavers to the mainland United States during the Haitian Revolution—are depicted as dashes. Labels note major organizations and individuals involved in Black freedom struggles.

In total, this visualization shows the rather complex web of Black movements that resulted from the second Jamaican Maroon War, the Haitian Revolution, the start of colonization movements in the United States and Britain, and "antislavery without abolition."[26] There is acknowledgment of Black agency, though the power dynamics behind all the migrations are uneven. As with the Du Bois models, the visualizations represent both progress and obstacles to progress.

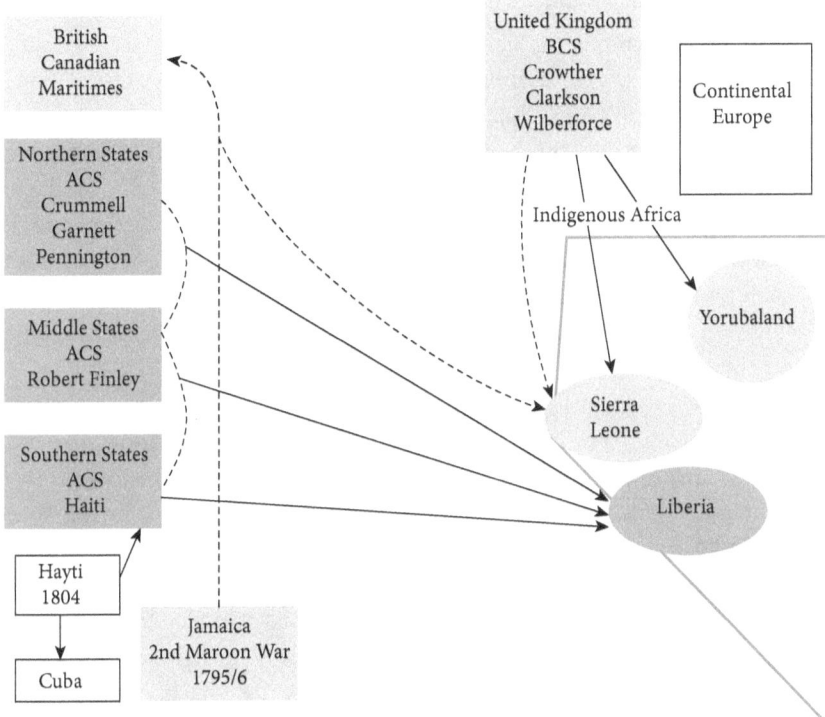

FIGURE 5.4. British Atlantic geographies and the Black experience. Visualization by the Author.

And the models illustrate spaces where race was not only performed, but the experiences ascribed to racial identity were defined by these shifts in law and geography.[27]

Finally, these visualization techniques can also be used to explore the ways that Black abolitionists in the United States and Britain came to project their emancipatory geographical imaginations on Africa, and the ways these metaphysical cartographies were entangled with those of white British and American geographical imaginations.[28] For example, in the 1840s, the African American missionary Alexander Crummell ruminated on the meaning of the Liberian Republic for freed slaves. He saw it as proof that people of African descent could self-govern, although his vision was colored by colonization and respectability politics. Among his numerous pamphlets was one that touted English, "the speech of Chaucer and Shakespeare" as a "gift of the almighty." It portrayed a Black-controlled space that was centered on Africa yet dismissive of Indigenous African languages and customs.[29]

Crummell was part of a network of Black abolitionists that included H. H. Garnett, J. W. C. Pennington, Samuel Crowther, and abolitionist and missionary organizations that saw themselves as the saviors of Africa from the degradations of slavery. Crummell's Africa is just one of a number of metaphorical spaces that appear in diasporan literature, but it demonstrates that, to be truly effective, these geospatial approaches need to be more than quantitative. More than numbers, the datasets need to be designed to recognize multiple themes in Black thought, to draw on multiple corpora, and to allow for new datasets as more data becomes available and the questions change and evolve. The potential corpus is vast and, like the Transatlantic Slave Voyage database, ripe for data collection from a community of scholars.

This visualization in figure 5.5 accounts for entanglements between white Americans and Britons, Black Americans and Britons, organizations (treated as transnational), and Indigenous Africans. Lines are used to represent projections of geographic imagination (dotted line), known relationships between people and organizations (solid line), and arrows to represent power dynamics. Political geographies are represented more vaguely since geographic imaginaries have little to do with defined kingdoms or nation-states; rather, they are about aspiration—in this case, both spiritual and economic. They represent a way that competing forces projected their aspirations onto Africa, which, in the coming decades, resulted in further dispossession and dislocation of Indigenous Africans as Africa became more and more colonized by the United States and European powers through missionary work and economic ventures. Even Crummell, who saw himself as a liberator of Africa, contributed to a Liberia that was ruled by African American elites and squeezed out those who descended from people whose physical ties to Africa had remained unbroken by slavery.[30]

But these approaches do not require quantitative data the way that other approaches to considering Black emancipatory geographies do. In spatial studies of Black human geographies, the problems of erasure and exclusion are compounded by archival silences in the records that form the corpora.[31] The survivors of the Middle Passage and their descendants came from societies steeped in oral tradition and/or were specifically prohibited from learning to read. As a result, mentions of Blackness in the early modern Atlantic are all too frequently filtered through a white lens. Moreover, there are structural issues in the collection and practices of archives; as Marisa Fuentes's recent landmark study of Black women and the archives demonstrates, scholars of early Black thought face several challenges.

The first challenge is within the archive itself. Aside from high-profile writers like Phillis Wheatley and Olaudah Equiano, the sources tend to be what Fuentes calls "fragmentary bodies," which are scattered around the archives.[32] References

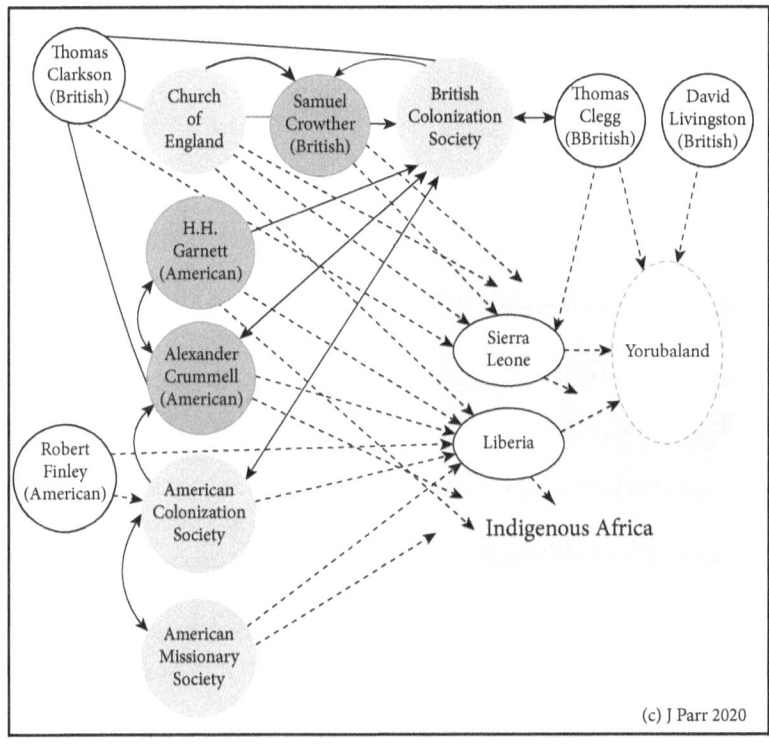

Key

---- Geographic imaginary (economic, religious, etc) imposed

— Connection between people, organizations, etc.

→ Indicates the direction of a power relationship

○ White British or American individual

● Black British or American individual

Organization

indigenous Africa and African peoples

FIGURE 5.5. Africa: Competing geographical imaginations. Visualization by the author.

to lower-profile individuals frequently appear as brief mentions in other sources, and often as what Nicole Aljoe and John Blassingame have described as "testimonials." Their use has invoked some criticism because of the difficulties of working with multiauthored sources that are filtered through white lenses and white print networks.[33] Yet dismissing them erases any agency that Black writers and speakers might well have had in creating them. Even so, historians as well as literary studies, religious studies, and African American studies scholars, and others who work on these topics, may struggle with problems of representativeness.[34] Fuentes's work brilliantly demonstrates how some of these source limitations can be overcome by using interdisciplinary methods from history, linguistics, literary analysis, and gender studies. The challenge then becomes a matter of making the geography align with cartographic practices that not only did not consider but often actively erased populations of people who were either Indigenous to the colonized lands or had been forcibly migrated to those lands via slavery or other colonizing mechanisms. Drawing together research methodologies that take into account different types of texts along with Fuentes's "fragments," offers a decolonized approach to the archive and to the treatment of sources that can be deployed by digital humanists in the construction of their corpora and their datasets.

Textual mining and geospatial techniques can also be used to better illustrate the Black experience, disparity, and archival silences. As scholars of Black studies are aware, the cartographies of dispossessed people are necessarily different from conventional academic cartographies in that they must explicitly account for the relationship between space and power. Accounting for dispossessed pasts is particularly challenging. Critical cartography, deep mapping, and geospatial studies involving displaced peoples are among the spatial methodologies developed over the past fifty years to address geographic disparities. For example, the geographer Anne Knowles noted the use of mapping as way of "ascribing meaning according to hierarchies that structure social relationships by regulating who is allowed or denied access to particular places."[35] Knowles does not explicitly define her work as critical cartography, but her work is applicable to exploring the spatialities of other dispossessed pasts: in particular, the cartographies of the Holocaust. For example, Knowles, Tim Cole, and Alberto Giordano have convincingly argued that the Holocaust had multiple geographies, created by camps, by Nazi policies and propaganda, by trains, and by other circumstances of the Holocaust. In this historical context, they are also the geographies of the "places and spaces that people created, occupied, passed through, and endured."[36] "The material landscapes," they continue, "were essential to the implementation of the Holocaust and people's experience of it."[37] Knowles and her colleagues' "material landscapes" share similarities with the "mental maps" described by the geographers Eric Boschmann and Emily Cubbon.[38]

The geographies in the cases Knowles describes were not only physical but also metaphysical: geographies of human experience. Geographies can be about how a given location affected the human experience but can also include geographical imagination—both the physical and the human geographies. Boschmann and Cubbon describe it as a cartography "distort[ed] to represent human perceptions, beliefs, and imaginations."[39] These "distortions" were tied to mental images carried by the individuals who traveled these spaces.[40] In the case of the diaspora, these mental images are contained in slave narratives, pamphlets, interviews, and other sources.

In histories of dispossession (and eradication), the researcher is tasked with considering the dynamics of space and power.[41] "The emphasis on the historical conditions" became part of the cartographical analysis.[42] It is analysis in which maps are acknowledged as having the capacity to produce knowledge and inscribe power.[43] While the historical contexts of the Holocaust and the slave trade are very different, the transatlantic slave trade produced multiple geographies that are also rooted in the dynamics of human geography, space, and power. Some of the digital projects on the Atlantic slave trade have focused on exploring the movement of people. The Transatlantic Slave Voyages Database project draws on a corpus of ships' manifests; the data has subsequently been turned into an animated geospatial rendering of the forced migration of the Middle Passage across time and place.[44]

Deep mapping techniques that use this data offer promise for building a tool that can be used to incorporate experience. Ethnographers have wrestled with this problem of making maps representational of the narrative. Deep maps are open-ended explorations of space.[45] They allow for "interwoven paths" and "trajectories" that Mia Ridge, Don Lafreniere, and Scott Nesbit argue offer scholars the means to account for the exploration of "innumerate questions," dependent only on the availability of "historical data."[46] They contend that "humanistic interpretation" requires "situation in archives" that offer "myriad traces of evidence about a site."[47] Evidence can include mentions of towns, cities, city squares, and buildings. In some cases, there may not be a precise street address, but the spaces can be represented either by bounding boxes or by geospatial points approximated through historical research. The techniques recently used by the PLACE Project to determine the geospatial coordinates of locations described in the New England Intercollegiate Geospatial Conference field guides might prove a fruitful methodology for approximating geospatial data when more specific locations are not available.[48]

As noted earlier, there are many digital projects that piece together fragmented, dispossessed pasts; however, with the exception of McKittrick's and Nelson's work, many studies focus primarily on a single corpus. Black thought was disseminated through correspondence, poetry, slave narratives, sermons, antislavery tracts, pamphlets, and by the 1820s, Black-owned newspapers.[49] While each corpus may not be equal in prominence across all time periods, gathering data from multiple corpora provides a richer picture of Black thought and allows researchers to ask better questions.

As Ridge, Lafreniere, and Nesbit note, in drawing from multiple corpora, researchers must assess the "richness and paucity" of particular sources "within their timeframe."[50] For a corpus where the research questions might change across time and space, the inclusion of chronological data as part of the dataset is essential. The American Panorama's Forced Migration of Enslaved People project provides a useful model.[51] It combines state-level census data with excerpts from slave narratives and the interface allows users to select by location and time. Additional datasets that consider questions like where and when these slave narratives were sold might help enhance the user's ability to explore broader patterns in the dissemination of Black writing. Other datasets might consider the locations within the travels described by early Black writers. The location and temporal data might need to be approximated in at least some of these cases but would allow users to find Black writers who were in close proximity to each other. And the addition of datasets from more fragmented sources, like correspondence, could help scholars "see" less visible members of the Black Atlantic who interacted with the broader Black print network.

A survey of the literature shows that, while there were some Black writings that predate 1760, the texts of only a few of those writings appear to have survived to the present day. The year 1860 was selected as the current stop date for two reasons. The first is that the Civil War led to the abolition of slavery, which marks a shift in Black thought that still contends with slavery and the effects of slavery but should be distinguished from the colonial and antebellum periods. Second, though the design of the tool will allow the addition of data from the postbellum period, there are far more written sources from the late nineteenth century onward as well as digital sources like the Library of Congress's Works Progress Administration (WPA) slave narratives.

Many North American slave narratives have been digitized and encoded by the Documenting the American South project at the University of North Carolina.[52] The University of Detroit at Mercy has a large, digitized corpus of Black-authored abolitionist literature.[53] The Internet Archives, Digital Public Library of America, and American Antiquarian Society collectively offer large, digitized collections. And some church archives, like the Congregational Library, have been expanding their digitized collections to include church records that document Black conversion narratives. Most newspapers remain in proprietary databases or are poorly indexed by Google's now-defunct Google newspapers, but there are growing open source corpora available through projects like the Black Press Research Collective, the Library of Congress's Chronicling America project, Colored Conventions, and the Caribbean Newspaper Digital Library.[54]

Imprint data from these digital collections offers a means to take quantitative data from across archives and types of sources to create a spatial analysis of the movement of Black thought across geography. Figure 5.6 shows a sample from a dataset of approximately four hundred Black publications that were published between 1760 and 1860. Facets include the name and gender of the author, the date(s) of printing, and the location(s) of printing—all cleaned in OpenRefine to standardize capitalization, spelling, and other facets of data as well as checking for holes within the dataset and reassessing the criteria for data collection where needed.

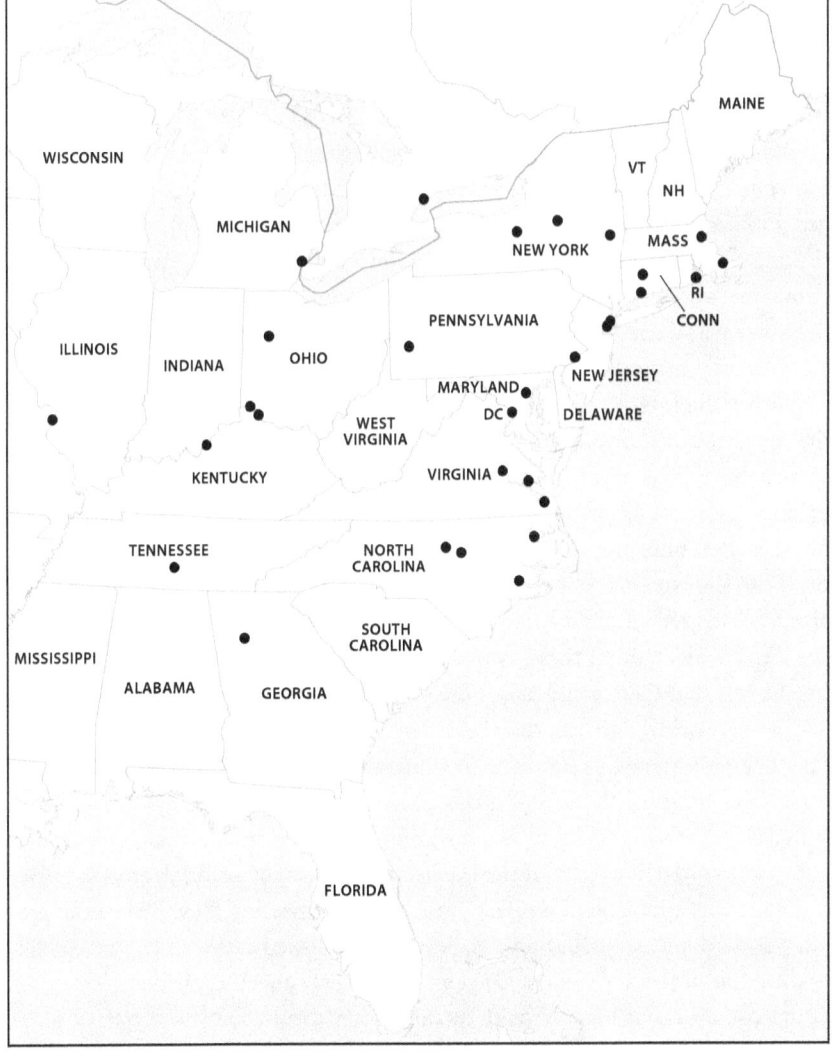

FIGURE 5.6. Map of Black correspondence, 1734–1825. Map by the author.

Geospatial data has been appended to each string. Since, in most cases, it is not possible to trace the addresses of specific print shops, location has been approximated with the coordinates for the center of each city. In the case of archival sources where there is no precise location, but specific geographical features are described, a center is calculated from the center of a radius that encompasses that location.

Using Palladio, it is possible to produce, first, a quantitative measure of Black writings from the period 1760–1860, where we can (not surprisingly) see spikes in Black writing that are contemporary with David Walker, Nat Turner, Maria Stewart, and other major figures of Black abolition and then another large increase in the decade leading up to the Civil War. A geospatial rendering of this same data produces an analysis of the intellectual hotspots produced by correspondence within the Black liberation movement between 1734 and 1825 (see figure 5.6).

While the data on women is sparse due to the limitations of the archive, we can show that the Black women who did produce writing appear in the same geographies as their male counterparts. Using Vanessa Holden's exploration of the role of women in the aftermath of the Nat Turner Rebellion, and Julius Scott's rumor networks during the Age of Revolutions, it is possible to read against the archive and argue for a greater presence of Black women activists than the archives might suggest.[55]

Within the interactive map, specific data is preserved for each person within the dataset, but to avoid the cliometrics historiography that once defined the scholarship of slavery, maps appear alongside narration of the people within the primary sources that produced the corpus for the mapping. This type of approach helps us overcome problems of representation, fragmentation, and their reliance on white sponsorship and white printing networks. Some of these writings are also reported secondhand and transcribed by white writers, though Nicole Aljoe has argued for treating these sources as testimonials.[56] Aljoe's approach allows for the fact that these writings are the product of entanglements between Black and white worlds, particularly where they are being used as data points rather than being dismissed as products of white abolitionists entirely. After all, in the Age of Revolutions, abolitionism was generally not segregated by race. As such, it makes sense to treat these sources as an entangled intellectual contribution and acknowledge it as part of the intellectual geography of Black emancipatory thought. This approach also offers promise for connecting these texts more precisely to slave uprisings and revolts, beyond established cases—like David Walker's *An Appeal to the Coloured Citizens of the World* and Nat Turner's Rebellion. Using visualizations together with quantitatively driven spatial analysis of Black emancipatory geographies can also allow the researcher to consider more localized examinations or analysis across larger swaths of time and space.

GEOGRAPHIES OF EMANCIPATION   99

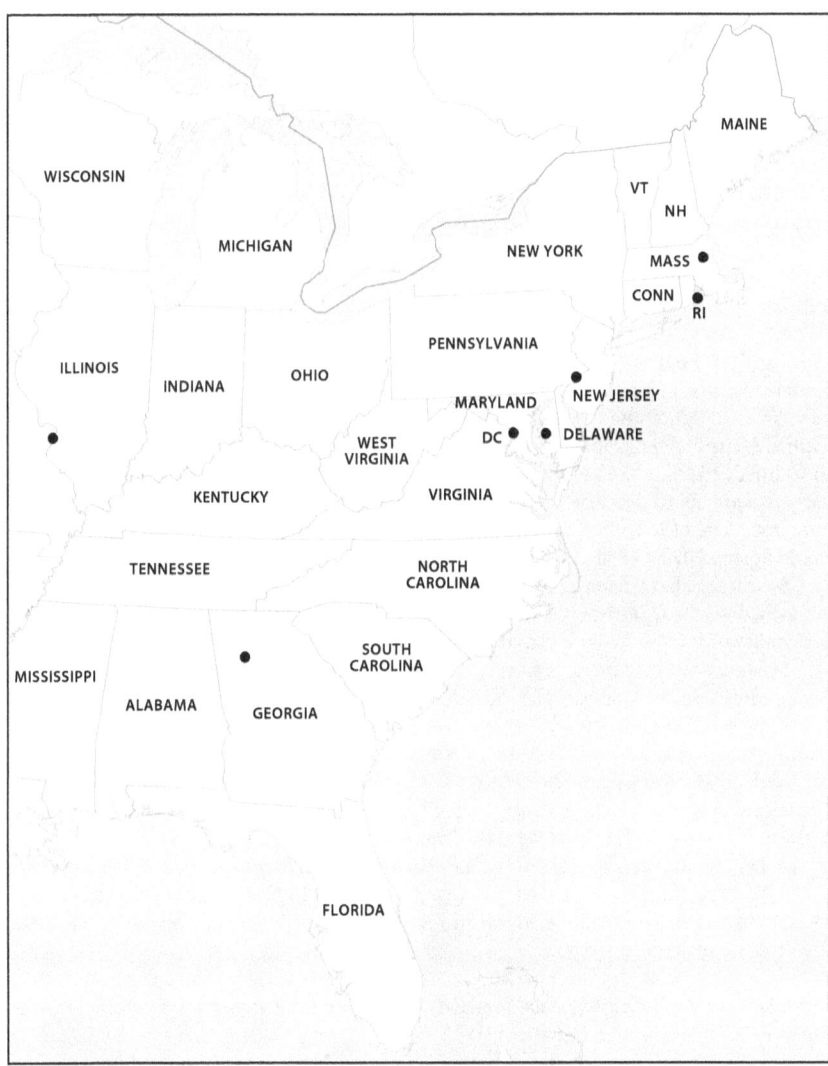

FIGURE 5.7. Women writers in early America, 1734–1825. Map by the author.

NOTES

1. Prince Saunders, ed., *Haytian Papers: A Collection of the Very Interesting Proclamations and Other Official Documents; Together with Some Account of the Rise, Progress, and Present State of the Kingdom of Hayti* (London: printed for W. Reed, 1816), 13. On Haiti, the United States, and Black humanism, see Julia Gaffield, *Haitian Connections in the Atlantic World: Recognition after Revolution* (Chapel Hill: University of North Carolina Press, 2015); Julian S. Scott, *The Common Wind: Afro-American Currents in the Age of Haiti* (London: Verso, 2018); and Marlene Daut, *Baron de Vastey and the Origins of Black Atlantic Humanism* (New York: Palgrave Macmillan, 2017).

2. Saunders, *Haytian Papers*, 13.

3. Roopika Risam, *Decolonizing the Digital Humanities in Theory and Practice* (New York: Routledge, 2018), 7.

4. Kim Gallon, "Making the Case for Black Digital Humanities," in *Debates in the Digital Humanities*, ed. Mathew K. Gold and Lauren F. Klein, open access edition (Minneapolis: University of Minnesota Press, 2016).

5. Martin Brückner, *The Social Life of Maps in America, 1750–1860* (Chapel Hill: University of North Carolina Press, 2017), 1–15.

6. Trevor J. Barnes and Erick Sheppard, eds., *Spatial Histories of Radical Geography: North America and Beyond* (New York: Wiley and Sons, 2019); Whitney Battle-Baptiste and Britt Rusert, *W. E. B. Du Bois's Data Portraits: Visualizing Black America* (New York: Princeton Architectural Press, 2018): Eliga H. Gould, "Entangled Histories, Entangled Worlds: The English-Speaking Atlantic as a Spanish Periphery," *American Historical Review* 112, no. 3 (June 2007): 764–86.

7. On Enlightenment-era fascinations with Egypt, see Jason Thompson, *Wonderful Things: A History of Egyptology*, vol. 1, *From Antiquity to 1881* (Cairo: American University in Cairo, 2015). Van Gosse notes that the term "Moors" is part of the language used by early Americans to describe racial orders. Van Gosse, "Patchwork Nation: Racial Orders and Disorder in the United States, 1790–1860," *Journal of the Early American Republic* 40, no. 1 (Spring 2020): 45–81, 47–48.

8. Scholars have been skeptical about Leo Africanus's claim to have visited all the lands he described. See Natalie Zemon Davis, *Trickster Tales: A Sixteenth-Century Muslim between Worlds* (New York: Hill and Wang, 2007).

9. Saidiya V. Hartman, *Scenes of Subjugation: Terror, Slavery, and Self-Making in Nineteenth-Century America* (New York: Oxford University Press, 1997), 1–3.

10. Trevor J. Barnes and Eric Sheppard, eds., *Spatial Histories of Radical Geography: North America and Beyond* (New York: Wiley and Sons, 2019); Antipode Collective, eds., *Keywords in Radical Geography: The Antipode at 50* (New York: Wiley and Sons, 2019); Yomaira C. Figuera-Vasquez, *Decolonizing Diasporas: Radical Mappings of Afro-Atlantic Literature* (Evanston, IL: Northwestern University Press, 2020).

11. Battle-Baptiste and Rusert, *W. E. B. Du Bois's Data Portraits*.

12. W. E. B. Du Bois, "The Georgia Negro: A Study" (1900). Library of Congress.

13. Vincent Brown, *A Slave Revolt in Jamaica, 1760–1761: A Cartographic Narrative*, revolt.axismaps.com/project.

14. Gergely Baics and Leah Meisterlin, "Zoning before Zoning: Land Use and Density in Mid-Nineteenth-Century New York City," *Annals of the American Association of Geographers* 106, no. 2 (September 2016): 1152–75.

15. Jeffrey Kok Hui Chan and Ye Zhang, "Sharing Space: Urban Sharing, Sharing a Living Space, and Shared Social Spaces," *Space and Culture* 24, no. 1 (2021): 157–69.

16. Katherine McKittrick, *Demonic Grounds: Black Women and the Cartographies of Struggle* (Minneapolis: University of Minnesota Press, 2008); Elizabeth Stodeur Pryor, *Colored Travelers: Mobility and the Fight for Citizenship before the Civil War* (Chapel Hill: University of North Carolina Press, 2016); Martha S. Jones, *Birthright Citizens: A History of Race and Rights in Antebellum America* (New York: Cambridge University Press, 2019).

17. McKittrick, introduction to *Demonic Grounds*, xi–xxxi.

18. On slavery and the law, see especially Alejandro de la Fuente and Ariela J. Gross, *Becoming Free, Becoming Black: Race, Freedom, and the Law in Cuba, Virginia, and Louisiana* (New York: Cambridge University Press, 2020); R. J. M. Blackett, *Making Freedom: The Underground Railroad and the Politics of Slavery* (Chapel Hill: University of North Carolina Press, 2017); and Paul Finkelman, ed., *Slavery and the Law* (New York: Rowan Little, 1998).

19. Prince Hall, Petition to the Massachusetts Legislature (1777), in *Proceedings of the One Hundredth Anniversary of the Granting of Warrant 459 to African Lodge* (Boston: Franklin Press, 1885), 12–13.

20. Jessica M. Parr, "Black Mobility, Law, and Freedom," *Black Perspectives* (blog), African American Intellectual History Society, October 27, 2017.

21. Saunders, preface to *Haytian Papers*.

22. On "slave societies," see Ira Berlin, *Many Thousands Gone: The First Two Centuries of Slavery in North America* (Cambridge, MA: Belknap Press of Harvard University Press, 2000): 109–216.

23. Particularly useful here are Edward R. Tufte, *Envisioning Information* (Cheshire, CT: Graphics Press, 1990); and Edward R. Tufte, *Beautiful Evidence* (Cheshire, CT: Graphics Press, 2006).

24. Gosse, "Patchwork Nation," 45–81.

25. Lincoln Mullen, "The Spread of U.S. Slavery, 1790–1860," interactive map, https://lincolnmullen.com/projects/slavery/; Minnesota Population Center, "National Historical Geographic Information System: Version 2.0," (Minneapolis: University of Minnesota, 2011), http://www.nhgis.org.

26. On the complexities of British versus American abolitionism, see Christopher L. Brown, *Moral Capital: Foundations of British Abolitionism* (Chapel Hill: University of North Carolina Press, 2006), 33–105, 209–332.

27. Gosse, "Patchwork Nation," 47.

28. On "geographic imaginaries," see especially Edward W. Said, *Culture and Imperialism* (London: Chatto, 1993).

29. Alexander Crummell, *The English Language in Liberia: The Annual Address before the Citizens of Maryland County, Cape Palmas, Liberia* (1860), 5.

30. On Black Christian imperialism, see Sylvester A. Johnson, *African American Religions, 1500–2000* (New York: Cambridge University Press, 2015); and Emily Conroy-Krutz, *Christian Imperialism: Converting the World in Early America* (Ithaca, NY: Cornell University Press, 2018).

31. David Thomas, Simon Fowler, and Valerie Johnson, *The Silence of the Archive* (London: Facet, 2017).

32. Marisa R. Fuentes, *Displaced Lives: Enslaved Women, Violence, and the Archives* (Philadelphia: University of Pennsylvania Press, 2016), 12.

33. Nicole Aljoe, *Creole Testimonies: Slave Narratives from the British West Indies, 1709–1838* (New York: Palgrave Macmillan, 2012), 27–56; John Blassingame, introduction to *Slave Testimonies: Two Centuries of Letters, Speeches, Interviews, and Geographies* (Baton Rouge: Louisiana State University Press, 1977) xvii.

34. Smadar Lavie and Ted Swedenburg, eds., *Displacement, Diaspora, and the Geographies of Identity* (Durham, NC: Duke University Press, 1996).

35. Anne Kelly Knowles, "GIS and History," in *Placing History: How Maps, Spatial Data, and GIS Are Changing Historical Scholarship*, ed. Anne Kelly Knowles and Amy Hillier (Redlands, CA: ESRI Press, 2008): 1–25, 5.

36. Anne Kelly Knowles, Tim Cole, and Alberto Giordano, eds., *Geographies of the Holocaust* (Bloomington: Indiana University Press, 2014), 2.

37. Knowles, "GIS and History," 2.

38. Knowles, "GIS and History," 2; E. Eric Boschmann and Emily Cubbon, "Sketch Maps and Qualitative GIS: Using Cartographies of Individual Spatial Narratives in Geographic Research," *Professional Geographer* 66, no. 2 (April 2014): 236.

39. Boschmann and Cubbon, "Sketch Maps," 236–38.

40. Boschmann and Cubbon, "Sketch Maps," 237.

41. Paul Jakot, Anne Kelly Knowles, and Chester Harvey, "Visualizing the Archive: Building Auschwitz as a Geographic Problem," in Knowles and Hillier, *Placing History*, 163.

42. Jeremy W. Crampton and John Krygier, "An Introduction to Critical Cartography," *ACME: An International E-Journal* 4, no. 1 (2005): 15.

43. Crampton and Krygier, "Introduction," 16.

44. David Eltis and others, *Trans-Atlantic Slave Trade—Database* (2007), accessed October 24, 2017, https://www.slavevoyages.org/voyage/database; Andrew Kahn and Jamelle Bouie, "The Atlantic Slave Trade in Two Minutes," *Slate*, June 25, 2015, http://www.slate.com/articles/life/the_history_of_american_slavery/2015/06/animated_interactive_of_the_history_of_the_atlantic_slave_trade.html.

45. Mia Ridge, Don Lafreniere, and Scott Nesbit, "Creating Deep Maps and Spatial Narratives through Design," *International Journal of Humanities and Arts Computing: A Journal of Digital Humanities* 7, no. 1–2 (2013): 176.

46. Ridge, Lafreniere, and Nesbit, "Creating Deep Maps," 176–78.

47. Ridge, Lafreniere, and Nesbit, "Creating Deep Maps," 177.

48. PLACE is an IMLS-funded project to build a geoportal to make geospatial-ready digital collections searchable by location. This technique used Google Earth. A description will be available at http://docs.sr.place.unh.edu once the administrator makes the page public.

49. See Patrick S. Washburn, *The African American Newspaper: Voice of Freedom* (Evanston, IL: Northwestern University Press, 2007).

50. Ridge, Lafreniere, and Nesbit, "Creating Deep Maps," 179.

51. "Forced Migration of Enslaved People, 1810–1860," American Panorama, Digital Scholarship Lab, University of Richmond, https://dsl.richmond.edu/panorama/.

52. North American Slave Narratives, Documenting the American South, Libraries of the University of North Carolina; P. Gabrielle Foreman, Jim Casey, and Sarah Lynn Patterson, eds., *The Colored Conventions Movement: Black Organizing in the Nineteenth Century* (Chapel Hill: University of North Carolina Press, 2021).

53. Black Abolitionist Archives, University of Mercy, Detroit, MI.

54. Black Press Research Collection; Chronicling America, Library of Congress; Colored Conventions; and Caribbean Newspaper Digital Library, Digital Library of the Caribbean.

55. Vanessa M. Holden, *Surviving Southampton: African American Women and Resistance in Nat Turner's Community* (Urbana: University of Illinois Press, 2021); Julius S. Scott, *The Common Wind: Afro-American Currents in the Age of the Haitian Revolution* (New York: Verso, 2019).

56. Nicole Aljoe, *Creole Testimonies: Slave Narratives from the British West Indies, 1709–1838* (New York: Palgrave Macmillan, 2012).

6

VISUALIZING CITY-SPACES DURING THE AGE OF REVOLUTIONS

Molly Nebiolo

In 1744 Dr. Alexander Hamilton, a Scottish immigrant who shared his name with a Founding Father, was sick of his cough. This ailment, a remnant of a near-death illness of "fevers and spitting blood," stemmed from "seasoning" after his transatlantic voyage from Scotland to Maryland.[1] He and his circle of doctor-friends had concluded that the way to improve from such a long bout of sickness was to travel up the coast of the Atlantic to a cooler, drier climate. Besides regaining his health, the trip was also prescribed to Hamilton as "an antidote to personal indisposition and the torrid strife of Maryland politics as well as relief from ennui and the opportunity to see the colonies."[2] A trip through the colonies would also be an opportunity to escape the travails of Annapolis to leisurely explore the colonies for the first time. The travel-narrative format of the next four months of Hamilton's life became one of the most detailed accounts of early colonial cities for historians of eighteenth-century America. Through Hamilton's diary, we can see colonial America from the eyes of a "truthful traveler" and European immigrant.[3]

Journals like Hamilton's offer a rare and colorful glimpse at living in a past century. Through diaries, we piece together what life was like visiting or living in cities that continue to exist but are completely transformed into something unfathomable to an eighteenth-century inhabitant. For historians of early American cities, these sources are rich in detail but have their limitations in scope or perspective to get a comprehensive look at an early Anglo-American city. Particularly throughout the eighteenth century, cities were complex places; historians still do not have a good handle on visualizing them. Maps of colonies and fragments of city plans offer a close rendering of historical space but in very one-sided, two-dimensional ways that often perpetuate colonial narratives of space.[4] Computational tools and digital history methods are a way to lessen the limitations of these sources by examining their contents in a new, spatially accountable way. This chapter uncovers some of the ways historical records, like diaries, can be used to conceptualize space and demonstrates how 3D and VR tools further our examination of how historical spaces were occupied and used in the eighteenth century.

During the Age of Revolutions, cities were the backdrop to the events and actions between British and American soldiers and politicians.[5] Cities, however, were more than just landscapes for activities. If we look at the way space was described and written about by people living in the eighteenth century, we can imagine their subjective experiences. 3D representations of colonial and revolutionary spaces can collate these subjective experiences and merge them with contemporary representations of cities on maps and blueprints of the period in order to give a larger picture of what spaces of the past might have looked like, a process historians and digital humanities have not been able to do up until the advent of 3D and VR programs. These renderings of historical spaces, in the way that they are immersive, lend themselves as sources to interpret the diverse experiences of the urban past.

Early Americanists have begun to harness the resources of 3D and VR programs using video games to visualize early American spaces with historical narratives found in archives.[6] One of the larger projects that uses virtual reality to step closer into the history of revolutionary America is Witness to the American Revolution.[7] In this video game, the user pieces together the events of the Boston Massacre as the news spreads by word of mouth immediately after March 5, 1770. The sources for this game come from historical narratives of the event, and the makeup of Boston was re-created from maps and images depicting the city days and weeks after the bloody event.[8] On another scale, historical re-creation of revolutionary spaces is done with more creative license, like with the Assassin's Creed video games.[9] The franchise has been lauded for its pseudohistorical accuracy of bygone places, from ancient cities to revolutionary Paris.[10] In 2019 the company that makes the Assassin's Creed video games, Ubisoft, donated their virtual plans of Notre Dame to the Parisian government; it was one of the most detailed and accurate renderings of the building that existed before a fire destroyed much of the antiquated church.[11] When this donation occurred, it reinforced the idea that 3D models and virtual reality are additional ways to become informed about historical spaces, not just synonymous interpretations of 2D plans. Paris had blueprints of Notre Dame they could use to rebuild the city, but the 3D visualization of the church gave spatial assistance to its accurate reconstruction. While video games are a truly engaging way to be immersed in a historical space, the gaming platform is not the only way 3D modeling and virtual reality can be used to make and study city-spaces of the past. The simple act of mapping and modeling, without any user-interfaced gaming aspect, provides as much of an impact on historical inquiry as these games.

Philadelphia is a central location to the events of early America and the history of this country, but a 3D rendering of the city during the revolutionary period had not been attempted before the beginning of my project, Visualizing Colonial Philadelphia.[12] A robust collection of primary sources must be available to make a visually and historically accurate rendering of eighteenth-century space. 3D modeling and then the re-creation of virtual cities is time consuming and complex, and the process of unearthing the representational space of Philadelphia, or the subjective perspective of the city, is equally important to the first stages of modeling as the architectural details for virtually constructing Philadelphia's colonial buildings. The *Itinerarium* of Alexander Hamilton helps in this regard, as do other diverse perspectives, like the diary of Elizabeth Drinker.[13] Drinker was a native Philadelphian who experienced its importance during the Age of Revolutions, its centrality as a national capital in the 1780s, and its role as the epicenter to the yellow fever pandemic of 1793. Her perspective on the city, detailed in her diaries, offers a unique glimpse into the past through the eyes of a Philadelphian, a Quaker, and a woman. Both journals aided in the conceptualization and rendering of early Philadelphia's city-space.

Through the spatial analysis of two rich journals on these locations—that of Dr. Alexander Hamilton and the popularly analyzed diaries of Elizabeth Drinker—we can see how cities have been depicted in primary sources, and how this information can be transformed into a closer 3D depiction of early colonial space. Hamilton holds a unique authority in his descriptions of early American cities as he grew up and moved through Edinburgh, London, and the other metropoles of Europe before moving to America. His thoughts and descriptions of what were the most urbanized locales in the American colonies helps us grasp what might have defined a city in a larger global context of the period. Elizabeth Drinker—as a literate female and inhabitant of Philadelphia during its peak as both a colonial hub and a national capital—provides a rare look at the aspects and boundaries of city-spaces from the viewpoint of a different gender. The breadth and digital accessibility to these two journals also makes them ideal sources for highlighting what "urban" might have meant in the eighteenth century.

Journals like those by Hamilton and Drinker help us reconstruct the lived experiences of an urbanizing early Philadelphia. Computational tools help us reconstruct the absolute space for early American cities, and maps sit between these ideas of city-spaces and absolute spaces of a city. As Richard White has discussed when talking about the benefits of using digital tools to understand historical spaces, there is a "giveness" to the boundaries depicted on maps that can be better studied and represented through digital tools and, I believe, through centering the personal writings of these spaces in digital projects.[14] Virtual reality moves us closer to a fuller visualization of historic spaces which allows us to examine the importance of space during the War for Independence and the formation of the United States, something that other platforms cannot do.

By examining Philadelphia, we see how a city during the revolutionary period was both a rural place and a cityscape to its inhabitants.[15] By better visualizing early revolutionary space, we can see the early makings of suburbs forming around cities and how certain infrastructures appeared and remained during and after a period of chaos. Philadelphia is also a unique example because of its prominence as one of the oldest planned cities in America and its centrality to the British Empire and later the United States. As is seen with the history of Philadelphia, the concept of the city was more ambiguous in the eighteenth century; people would often call the same place a city in one document and a town in another. In this period, they were interchangeable and meant the same thing to those who lived in these spaces. The concept of "urban," however, did not exist in its current definition and should be questioned when it is used to discuss and re-create early colonial and revolutionary city-spaces.[16]

The word "urban," as it was meant in the early modern period, defined society rather than a city. In seventeenth-century dictionaries, urban meant gentlemanly or well behaved.[17] It did not mean a densely populated area or define a city-space. While early modern cities were often the locations of people with the best manners (over time "urban" evolved into "citylike"), the term still described people not spaces. When studying urban space in the eighteenth century, then, we first need to ask ourselves if "urban space" is the best term to use to study that type of space in that period. I argue that, broadly speaking, "city-spaces" is a better term to describe the areas that constitute early American urban history in order to account for the discrepancy between our interpretation of urban and the historical use of the term.

Confronting the term "urban cities" in the eighteenth-century aids in the process of recentering the field of urban history to better incorporate the early modern period. Urbanizing—as it has been studied in the early American context—is a better way to conceptualize city-spaces as they grew and expanded, which is a label for the process rather than the space of the city. Digital tools help with the endeavor to highlight the significance of early American spaces in the history of urban space because 3D modeling and VR move us toward properly doing the interdisciplinary work that is needed to accurately provide a comprehensive notion of early American cities and the lives that were lived there.

Hamilton's and Drinker's Philadelphia

In his journal, Alexander Hamilton depicted spaces in the eighteenth century through the relationships he describes at length. The hundreds of pages of his *Itinerarium* are composed of recollections of the conversations he had while traveling on horseback and the types of people he saw at different stops and cities along his route. He used these connections to posit depictions of the regions, even if these folks were travelers themselves. For example, while in Philadelphia, Hamilton indirectly noted an uptick in the number of Barbadians when he was in Philadelphia. "After dinner, Mr. V[ena]bles, a Barbadian gentleman, came in who, when we casually had mentioned the free masons, began to rail bitterly against that society as impudent, assuming, and vain cabal pretending to be wiser than all mankind besides, an imperium in imperio, and therefore justly to be discouraged and suppressed as they had lately been in some foreign countries."[18] Not only do the topics found in this quotation open a window into eighteenth-century news and politics; they also elaborate on an undescribed point that Hamilton made in his sections on cities. Particularly in Philadelphia, the city was a locale for people to convene from across the European empires to interact and exchange news. This point also shows the education level or worldliness of the people who frequented the city, and the types of discussions that were commonly overheard in town.

In the countryside, on the road between cities, many of the conversations Hamilton noted were simply about roads, the weather, or his religious practices. He also had lively conversations with travelers going from one town to the next. "I put up this night at one Miller's att the Sign of Admiral Vernon [in Brunswick, New Jersey] and supped with some Dutchmen and a mixed company of others.... Our conversation at supper was such a confused medley that I could make nothing of it."[19] Brunswick, to Hamilton, was a "neat small city" in which he came across a variety of people, including a Dutchman, who may have been traveling.[20] To Hamilton, this resulted in a cacophony of social interactions that ranged so much he was too confused to recall them; a reflection of the way he felt about Brunswick as a more rural space but also a small city. To Hamilton, the types of conversations he had with people during his travels reveal the "giveness" of the boundaries of what he defined to be cities. For New York, Newport, Boston, and in a way Brunswick, these places were also considered cities to Hamilton because one could easily hold educated discussions with people who traveled from all over the British Empire. It was not just affluent men who unconsciously defined city-space as a location to find educated discussion. The interactions Elizabeth Drinker had with people she found in Philadelphia also stand out as experiences unique to the city.

Social interactions were central to Elizabeth Drinker's experience of city-space in Philadelphia. Her diary entries are painstaking lists of the people she spoke with throughout the day. "[June 1764] 19. Sister. self, John, and Sally, went this Morning, after breakfast, to see our little dear, call'd at Abels, came home a little after 11; I went this Afternoon to see Neighr. Levy -Sister went down Town with Sally -Betsy, and Richd. Waln here this Evening."[21] Here, we see how Drinker's diary differs from Hamilton's. She does not include many anecdotes, opinions, or colorful asides throughout her diary, but instead, we can interpret her understanding of city-space in the presence of activities she listed in her diary; social interactions being one of them. "[January 1772] 2. I went this Morning to see my F[rien]d. peggy who still continues in the same Malancholy situation: -Sister went there after dinner, Aunt Jervis and the Girles spent this Afternoon with me -Sister came home in the Evening Peggy still the same -Judah Foulk came after 10 o'clock, to desire sister to go to Wm. Parrs, as the Doctor Expected a change e'er long in Peggy; she went accordingly with him stayd all Night."[22] During Drinker's life, her family owned a home outside of what is considered to be within Philadelphia's boundaries, in Frankford, but her depiction of her days did not differ much from her times in the city. "July 8, 1772 HD. on Horse back MS. ED. and Sally D in the Chaise, Harry, with us -went to Frankford after Dinner to our place: Sammy and Hannah Sansom and their daughter Sally, Richd. and Betsy Waln, and John Glover, drank tea with us there, return'd in the Evening."[23] Not only were her social connections unbroken by her movement from Philadelphia to Frankford but she was also able to travel back and forth within a day from her house in the country to her home in Philadelphia. The continuation of movement and sociality between these two locations reveals how far-reaching her city network was, and how blurred the boundaries were between what was considered within Philadelphia and the way inhabitants moved about and treated the spaces in and around the city. The city-space of the city was ever-changing based on the types of interactions and conversations Drinker and Hamilton had while visiting the area.

Hamilton's and Drinker's recordings of their acquaintances are not unique to the personal writings of the period. Many of the diaries from the seventeenth, eighteenth, and nineteenth centuries are often recollections of the people the writer encountered each day rather than the introspective style of diary writing that is depicted in the twentieth- and twenty-first centuries. Alexander Hamilton and Elizabeth Drinker fit the stereotype of the removed observer whose entries are often lists of names or account-book style recording of day-to-day banalities. However, reading these diaries through a spatial lens helps us decipher what constituted eighteenth-century city-space. Digital humanities provides a context for many of the details, such as social interactions, that could be overlooked in sources like Hamilton's and Drinker's diaries

If we look at a map of Philadelphia from when it was first settled, the presence of people through the labeling of land parcels illustrates how important population was to the actuality of city-space. The location of the people and their land on these maps also shows where the representational boundaries of Philadelphia lay as compared with the actual boundaries of its constructed city. Using ArcGIS, I was able to georectify a seventeenth-century map of Philadelphia onto contemporary coordinates to study and highlight where people's lands were found and where they lay with respect to the city in its historic and contemporary boundaries. The images in figures 6.1 and 6.2 are a simplified representation of this work.

As you can see, their lands lay outside of the boundaries of what we may assume to be the city—the grid of streets and blocks—but in representational space, might not be the boundaries or makeup of the city as exemplified by the movement and social interactions that Drinker had nearly a century later in the same region. Indeed, Hamilton seems to have illustrated Philadelphia's city-space before he set foot on the grid of streets shown on this map. "The country round the city of Philadelphia is level and pleasant, having a prospect of the large river

FIGURE 6.1. A georectified map of Thomas Holme's 1681 map of the province of Pennsylvania to contemporary coordinates of the Philadelphia region. Courtesy of Molly Nebiolo.

FIGURE 6.2. A close-up of 1681 mapped city-space of Philadelphia over a contemporary map of the city. The parcels of land that surround the city (the grid of blocks at the bottom of the page) emulate a sense of population density. Maps like this one reveal some of the "giveness" of what might have constituted Philadelphia's city-space. Courtesy of Molly Nebiolo.

of Delaware and the province of East Jersey upon the other side. You have an agreeable view of this river for most of the way betwixt Philadelphia and Newcastle."[24] Interactions between people were highly regarded in early American city-space. Rural diaries, like the well-known diary of the midwife Martha Ballard, show that those living in country locales were also social but not at the level that Hamilton or Drinker noted in their journals, unless they had a profession like Ballard's.[25] While social interactions existed outside of cities, the sheer density of interactions and, to Hamilton, the level of educated discussion defined a city in the eighteenth century.

Access to comforts, particularly social drinks like tea, coffee, and beer were experiences Hamilton and Drinker denoted to show they were in a populated city or town in revolutionary America. Tea was central in the lives of colonists. People sat down to drink tea throughout the day for breaks and as hospitable acts when visitors came to call. In his *Itinerarium*, Hamilton mentions drinking coffee twenty-seven times, all of which took place in major towns or cities.

However, the presence of drinking, particularly tea and coffee, reveals how cities, to Hamilton, represented places where these comforts were easily accessible and central to the culture of city living. Coffee as a commodity illustrates some of the spatial makeup of revolutionary Philadelphia, as we can trace where coffee might have come from and the spaces it occupied once it came into the city. VR and 3D modeling provide a creative outlet that can help historians reimagine how coffee was present in the cities. Its proximity to major roads might have led inhabitants to smell it as they walked to and from the market, or the movement of coffee might have been an exciting site to see for someone walking by as containers were carried through winding streets from the wharves. The experience of these commodities comes alive through VR in a way that is hard to replicate in other media.

Indeed, when Hamilton writes about his experiences drinking coffee, the experiences are all in early American cities: Philadelphia, New York, Albany, Newport, and Boston. The twenty-four instances where Hamilton visits or notes the presence of a coffee shop are all in towns and cities. To a European immigrant, coffeehouses had been in existence since the 1650s.[26] In America, the presence of a coffee shop reinforced the transatlantic significance of a place—like Philadelphia or Boston—since it had easier access to the coffee trade and there was a need for such a metropolitan space for conversation and the sharing of news.

Alexander Hamilton used coffeehouses as places to examine the inhabitants of the town and to find comfort in the cities he visited. "In the afternoon, I went to the coffee house where I was introduced by Dr. Thomas Bond to serval gentlemen of the place, where the ceremony of shaking of hands, an old custom peculiar to the English, was performed with great gravity and the usuall compliments."[27] Hamilton gravitated toward coffeehouses when he found himself in city-space because he knew that any eighteenth-century city would have this amenity. In the coffeehouses of New York, Hamilton found comfort in games and afternoon pastimes that fit with his status as an educated and wealthy man from Europe. "After dinner the doctor [Dr. Colchoun] and I went to the coffee-house and took a hit att backgammon. He beat me two games. At 5 in the afternoon I drank tea with Mrs. Boswall and went to the house again, where I looked on while they played at chess."[28] The presence of games at coffeehouses was an important point for Hamilton, who noted in his *Itinerarium* the presence or lack of leisure in the cities he visited. When describing Philadelphia, he noted, "They apply themselves strenuously to business, having little or no turn towards gaity (and I know not indeed how they should since there are few people here of independent fortunes or of high luxurious taste)."[29] Here, Hamilton ruminated on an aspect of Philadelphia that might have been one reason why the city was often thought of as both rural and a bustling city: the simplicity and austerity that often came with Quaker culture was found in the city. While he might have been particularly sensitive to the opportunity to luxuriate because part of his trip was for relaxation, Hamilton also seemed to be noting this fact as it was a necessity in his expectation of a city-space to have access to fun and luxury items.

The presence of taverns played a similar role for Hamilton in the way he measured up the opportunity for socializing and discussing politics. City-spaces held bustling taverns and beer has always been a great convener of conversation. In early modern Europe and in European colonies, the tavern held a central role for both travelers and inhabitants of the towns.[30] It was the place to go to find housing if you were on the road, but it also was the place to visit away from work to socialize, get food, and hear the news. Taverns were present in nearly all populated areas of the Anglo-American colonies, but their density and diversity were particularly unique in city-spaces.[31] Hamilton found this to be the case in most of the cities he visited, particularly Philadelphia. "I dined att a tavern with a very mixed company of different nations and religions. There were Scots, English, Dutch, Germans, and Irish; there were Roman Catholicks, Church men, Presbyterians, Quakers, Newlightmen, Methodists, Seventh day men, Moravians, Anabaptists, and one Jew."[32] The centrality of the tavern is also important to our more objective understanding of city-space, as Hamilton took the time to detail the physical description of some taverns. He noted in Philadelphia, "The whole company consisted of 25 planted round an oblong table in a great hall well stoked with Flys. The company divided into committees in conversation; the prevailing topick was politicks and the conjectures of a French war."[33] Through Hamilton's pen, we can imagine what a revolutionary tavern must have looked like in colonial Philadelphia. To Hamilton, the social spaces and comforts offered by coffeehouses and taverns were necessities in what defined a city-space. Indeed, for Elizabeth Drinker, similar spaces were necessary to her spatial understanding of Philadelphia and her access to tea.

Elizabeth Drinker, as a Quaker woman, did not frequent taverns and coffeehouses at the rate Hamilton did during his travels, but tea played a major role in her access to comforts while living in Philadelphia. The Quaker religion frowned on the overconsumption of drink and the showiness of luxury goods like coffee and beer, yet men and women drank them through the eighteenth century. In the latter half of the century, however, taverns were under tighter restrictions and a segmentation of taverns happened. Once a place for all walks of life, now there were specific taverns for specific classes of inhabitants. Women were less likely to visit taverns for pleasure by the time Elizabeth Drinker was an adult in the 1760s and 1770s.

Tea, then, was an accessible commodity for Quaker women living in Philadelphia. Many of the entries in Drinker's diaries note the taking of tea with friends or family, whether at her house or when she was out visiting others in town. Of the 569 pages that made up her lifetime of diary keeping, there are over one thousand instances where Drinker notes she had tea. "[July 1764] 4. took a ride this Afternoon to see our little Dear, -drank Tea at A James."[34] "[May 1779] 23 First Day: S Swett din'd -A Parish came from meeting with me and drank tea -Wm. Miller and Jemmy Emlen, Owen Jones Senr., Hannah Catheral D. Drinker, call'd -my little Molly took a dose of Phisick to day: -Docr. call'd -warm to day -Abel James junr. din'd with us -heard cannon fire."[35] "[May 23 1795] 23. Cloudy with rain, wind Westerly. Jacob Downing call'd -Benjn. Swett din'd with us HD. &c. went this Afternoon to the President on Indian Affairs -Betsy Jervis took tea with us, she went with Sister to Jacob Downings -Dr. Rush here this evening."[36] All three quotes pulled from various parts of Drinker's life show that the social comfort of taking tea was central to her experience of life in Philadelphia. While tea drinking was ubiquitous across much of colonial America, tea's availability, and the frequency with which folks could consume it based on their interactions with one another, was much higher in city-spaces.[37] For Drinker, this was definitely true.

The presence of luxury comforts—often tea, coffee, and beer—were a through line in defining a city-space in eighteenth-century America, and their locations and densities assist in the re-creation of colonial city-space. A simple map of an American city in the eighteenth century may denote the presence of the major taverns or coffeehouses in town, but it is the description of them and their proximity to other locations—affluent homes, the wharves, merchants' businesses—that almost come alive with 3D modeling and virtual reality. Some revolutionary bars still exist in some capacity, like the recently shuttered City Tavern in Philadelphia or the Green Dragon in Boston, but through architectural blueprints and descriptive diaries like that of Alexander Hamilton, we can envisage these spaces in more detail than is offered from the bird's-eye view of a city map. With information like Drinker's, the numerous occasions of her tea drinking could be mapped and the walking paths from her house to others' can be virtually walked. The importance of these places within the visualization of the larger city-space is that they form networks for the way people used these places and how they fit into their daily lives as they moved through space.

With just two of the diaries that remain from the eighteenth century, we can still draw robust conclusions about the makeup of city-spaces during the Age of Revolutions. Besides the presence of buildings and wharves, the writings of Dr. Alexander Hamilton and Elizabeth Drinker—resources that have been used for historical inquiry for decades—can be reexamined, refashioned, and brought back to life with 3D and VR technologies.

Using the Digital to See the Past

Why do we need to rethink the concept of city-space when we have maps from the revolutionary period? Maps from the colonial and revolutionary periods give us an idea of what the bare structure of a city may have looked like around the time the maps were created. However, they should not be trusted so readily as representations of early American city-space. As Brückner notes in his writing, maps are often exemplary of the way people of power saw space in a period of time as unbiased depictions of a city.[38] While they could be good starting points for conceptualizing early American city-space, or how one portion of a population described their surroundings, computational tools and modeling historic spaces close the gaps on what is left out if we just look at maps as an absolute framework of early spaces. One of the projects that is attempting to move discussions of historical American city-spaces in the direction of inclusivity is Visualizing Colonial Philadelphia.

Visualizing Colonial Philadelphia, or VCP, approaches the idea of conceptualizing spaces of the past by incorporating the representational spaces of a city, like Hamilton's and Drinker's experiences of Philadelphia, and the absolute boundaries that are often associated with a place. From its inception, VCP has tried to investigate how we can understand early urban space without applying contemporary ideas of "urban" to early American cities. While metropoles like London and other European Continental cities existed as extreme examples of what "urban" might have been like in the early modern period, re-creating an early city requires the historical context and the subjective, representational understanding of space to be the backbone of the process. Otherwise, modeling projects like VCP would end up as inaccurate historical models that fit a teleologic time line of city development. Cities in early America had similarities, but their constructs differed due to their histories and locations in different colonies. The acceptance of this fact is necessary to better visualize city-space in the Age of Revolutions.

Additionally, 3D modeling and VR replications of early American city-space provide opportunities to merge absolute and representational spaces in a visually enticing and creative form. VCP uses the historical vignettes of the city to re-create parts of its infrastructure as well as the architectural information of revolutionary-style buildings. Hamilton described what Philadelphia looked like physically from afar, which is one example of how a vignette can provide both representational and absolute understandings of city-space. "Att my entering the city, I observed the regularity of the streets, but at the same time the majority of the houses mean and low and much decayed, the streets in generall not paved, very dirty, and obstructed with rubbish and lumber, but their frequent building excuses that."[39] Quotations like these give more meaning to the physical infrastructure of the city that is pulled from maps, blueprints, and bird's-eye views of Philadelphia. However, representational space as described in Hamilton's journal provides new understanding and context when it is used to render a 3D model of the city. For example, historians have long called Philadelphia a very rural town, but what does that mean? Hamilton also hints at this when he writes that he can easily walk to the countryside in an afternoon for some fresh air before walking back for dinner.[40] Trees are within view, even from where Hamilton stays in town, which was so close to the city action that he was frequently awoken at 5 a.m. because his bedroom window faced Market Street. In VR we can see how Hamilton may have seen trees and where he might have walked out from the city to the countryside. In 3D a contemporary user might begin to see how the vast networks between Philadelphia proper and its resources just outside of the constructed limits might still be considered part of the city. These brief asides about the city provide more context and add to the visualization of the city as it is built in VR.

VR and 3D modeling also break down the binary concept of urban and rural during the Age of Revolutions, especially for Philadelphia. These tools, along with the help of the firsthand accounts of these locations, reinforce the idea that cities did not have set boundaries and that the immediate surroundings played a role in the rest of the city. For Philadelphia, that included the Delaware River in its role of providing safe passage for boats between the city and the rest of the Atlantic; the immediate countryside with the roads and path into and out of the city to water, food, and trade; and the smaller taverns and inns that lay closest to the city center that travelers would frequent as people are moving inside and outside of city-space. Trying to visualize, let alone define, what was rural and what was part of the city reveals the inherent messiness of this question. There were no specific boundaries between the cityscape of Philadelphia and the countryside that surrounded it.[41]

Virtual reality can also help break down the notion that Anglo cities during the Age of Revolutions were fundamentally white spaces. The creators of historically accurate VR spaces need to account for the diversity of the people, and the events and experiences that took place within the ambiguous boundaries of city-space. While the spottiness of the historical record makes it a challenge to get a full picture of how various groups experienced and moved about cities like Philadelphia, the gaps are more noticeable in virtual space than they sometimes are in written narratives. The interdisciplinarity needed to accomplish 3D and VR models aids in the process of producing an inclusive history of the past.

ArcGIS, 3D modeling, and virtual reality show that city-spaces during the Age of Revolutions were complex and very experiential; often the boundaries between the city and the countryside were blurred or considered part of a broader understanding of revolutionary city-space. Pedagogically, VR and 3D modeling are novel ways to engage students to critically analyze sources and experiences of the past, but the research implications of these tools are vast. Digital historians who are familiar with these tools are aware of how useful they can be to help contextualize the past, but the presence of virtual historic spaces can be used by others, as well. These projects do not need to be complete to be useful or to help in producing research questions that would fill in some of the gaps in our understanding of historical spaces.

3D modeling tools and virtual reality platforms open a unique door into the realm of understanding spaces of the past. During the Age of Revolutions, American cities were expanding as colonization was growing. Transatlantic trade catalyzed the centrality of goods and resources to these various places and the creation of a new style of city—one that was both rural and heavily populated. The primary sources from this period of history give readers a glimpse into the world of visualizing the past with our memories, but we have more opportunities to achieve a thorough understanding of historical space with the computational tools found in the digital present. With 3D modeling and virtual reality platforms, researchers and educators can attempt to reconstruct the past in a way that is both enticing for the viewer and scholastically engaging for the researcher.

Visualizing Colonial Philadelphia, even in its early stages, aims to be an example for how to deploy 3D/VR in a smaller form for academic and pedagogical use. Its existence invites historians, educators, novices, and experts to contemplate how to better articulate early American space. The project's purpose is to attempt to reconstruct Philadelphia's Market Street as it was mapped in 1776, but the hope is that the process of reconstructing historical space is an intervention in the way historians engage with and conceptualize what "urban" really meant in the eighteenth century and how city-space shaped the larger understanding of American spaces during the Age of Revolutions. The journals I analyze in this chapter are only two of the many perspectives of the people who had visited Philadelphia or had called it their home. Yet, even with this limited scope, this chapter illustrates some of the ways people of the eighteenth century experienced city-spaces and differentiated them from other places in the American colonies, and it exemplifies some of the ways historians of many chronologies can interrogate the complex evolution of the concept of "urban."

NOTES

1. Based on the side effects, historians of medicine had concluded that Hamilton was suffering from tuberculosis, as noted in the introduction to *Gentleman's Progress*.

2. Alexander Hamilton, introduction to *Gentleman's Progress: The Itinerarium of Dr. Alexander Hamilton, 1744*, ed. Carl Bridenbaugh (Chapel Hill: University of North Carolina Press, 1946), xi–xiv, xiv.

3. Hamilton, *Gentleman's Progress*, xi.

4. Martin Brückner, *The Social Life of Maps in America, 1750–1860* (Chapel Hill: University of North Carolina Press, 2017), 5.

5. The historiography of cities during the revolutionary period is rich. For foundational discussions on the role of cities, broadly, in the American revolutionary period, see Carl Bridenbaugh, *Cities in Revolt: Urban Life in America, 1743–1776* (New York: Alfred K. Knopf, 1955); Benjamin Carp, *Rebels Rising: Cities and the American Revolution* (New York: Oxford University Press, 2007); and Gary Nash, *The Urban Crucible: The Northern Seaports and the Origins of the American Revolution* (Cambridge, MA: Harvard University Press, 1986). For the role of a specific city during the revolutionary period and more nuanced arguments around the importance of the city as a setting for revolution, see Serena Zabin's *The Boston Massacre: A Family History* (New York: Houghton Mifflin Harcourt, 2020); Jennifer L. Goloboy, *Charleston and the Emergence of Middle-Class Culture in the Revolutionary Era* (Athens: University of Georgia Press, 2016); Joseph S. Tiedemann, *Reluctant Revolutionaries: New York City and the Road to Independence, 1763–1776* (Ithaca, NY: Cornell University Press, 1997); Richard Godbeer, *World of Trouble: A Philadelphia Quaker Family's Journey through the American Revolution* (New Haven, CT: Yale University Press, 2019) among many others. For books on Caribbean and global cities that were influenced by the American Revolution, see Vincent Brown's *Tacky's Revolt: The Story of an Atlantic Slave War* (Cambridge, MA: Harvard University Press, 2020); Gerald Horne, *The Counter-Revolution of 1776: Slave Resistance and the Origins of the United States of America* (New York: New York University Press, 2014); and Michael Rapport, *The Unruly City: Paris, London, and New York in the Age of Revolutions* (New York: Basic Books, 2017).

6. For examples of how video games present and examine history, see Sarah Juliet Lauro, "Digital Saint-Domingue: Playing Haiti in Videogames," *archipelagos*, 2 (July 2017): 1–21. For further analysis on the historical accuracy and the imaginary found in historical video games, read Andrew Denning's "Deep Play? Video Games and the Historical Imaginary," *American Historical Review* 126, no. 1 (March 2021): 180–98.

7. Serena Zabin and Austin Mason, Witness to the Revolution: The Boston Massacre in 3D (under development, Carlton College), video game, https://bostonmassacre3d.amason.sites.carleton.edu/.

8. Molly Nebiolo, "Recreating Revolutionary Cities: An Interview with Serena Zabin," *Age of Revolutions Online Journal*, June 29, 2020, https://ageofrevolutions.com/2020/06/29/recreating-revolutionary-cities-an-interview-with-serena-zabin/.

9. Assassin's Creed Franchise, Ubisoft, https://www.ubisoft.com/en-us/game/assassins-creed.

10. Aris Politopoulos, Angus A. A. Mol, Krijn H. J. Boom, and Csilla E. Ariese, "'History Is Our Playground': Action and Authenticity in Assassin's Creed; Odyssey," *Advances in Archaeological Practice* 7, no. 3 (2019): 317–23.

11. Mike Snider, "Ubisoft Pledges Monetary, Tech Assistance for Notre Dame Cathedral Restoration," *USA Today*, April 17, 2019.

12. Molly Nebiolo, "Visualizing Colonial Philadelphia," https://mollynebiolo.com/digital-humanities/visualizing-colonial-philadelphia/.

13. Elaine Forman Crane, ed., *The Diary of Elizabeth Drinker* (Boston: Northeastern University Press, 1991).

14. Richard White, "What Is Spatial History?," Spatial History Lab, February 1, 2010, https://web.stanford.edu/group/spatialhistory/media/images/publication/what%20is%20spatial%20history%20pub%20020110.pdf.

15. Sylvia Doughty Fries, *The Urban Idea in Colonial America* (Philadelphia: Temple University Press, 1977).

16. The work from this chapter contributes to larger arguments around "the urban" in early America discussed by Jessica Roney in *Governed by a Spirit of Opposition: The Origins of American Political Practice in Colonial Philadelphia* (Baltimore: Johns Hopkins University Press, 2014); and Paul Musslewhite, *Urban Dreams, Rural Commonwealth: The Rise of Plantation Society in the Chesapeake* (Chicago: University of Chicago Press, 2018).

17. "Urban" would appear in early modern dictionaries under the adjective "urbane," until "urban" appears in Norton's 1735 dictionary, which still defined urban as "a proper name of man." Any other time "urban" appeared in early modern dictionaries, the reference was to Pope Urban. The dictionary searches for this chapter used the Lexicon of Early Modern English repository of early modern dictionaries, through the University of Toronto Libraries. Definitions were pulled from Thomas Blount, "Vrbane or Vrbanical vrbanus vrbanicus," *Glossographia or a Dictionary* (London, 1656); Benjamin Norton Defoe, "Urban," *A New English Dictionary* (London, 1735); and John Garfield, "Urbane," *A Physical Dictionary* (London, 1657).

18. Hamilton, *Gentleman's Progress*, 19.
19. Hamilton, *Gentleman's Progress*, 37.
20. Hamilton, *Gentleman's Progress*, 37.
21. Crane, *Diary of Elizabeth Drinker*, 187.
22. Crane, *Diary of Elizabeth Drinker*, 305.
23. Crane, *Diary of Elizabeth Drinker*, 318.
24. Hamilton, *Gentleman's Progress*, 18.
25. Laurel Thatcher Ulrich, *A Midwife's Tale: The Life of Martha Ballard, Based on Her Diary, 1785–1812* (New York: Alfred A. Knopf, 1990).
26. Jonathan Morris, *Coffee: A Global History* (London: Reaktion Books, 2019), 70.
27. Hamilton, *Gentleman's Progress*, 57.
28. Hamilton, *Gentleman's Progress*, 85.
29. Hamilton, *Gentleman's Progress*, 239.
30. Peter Thompson, *Rum Punch and Revolution: Taverngoing and Public Life in Eighteenth-Century Philadelphia* (Philadelphia: University of Pennsylvania Press, 1999).
31. Thompson, *Rum Punch*, 27.
32. Hamilton, *Gentleman's Progress*, 58.
33. Hamilton, *Gentleman's Progress*, 58.
34. Crane, *Diary of Elizabeth Drinker*, 186.
35. Crane, *Diary of Elizabeth Drinker*, 606.
36. Crane, *Diary of Elizabeth Drinker*, 1123.
37. For more on the role of tea consumption, politics, and early America, see Jane T. Merrit, *The Trouble with Tea: The Politics of Consumption in the Eighteenth-Century Global Economy* (Baltimore: Johns Hopkins University Press, 2017); and T. H. Breen, *The Marketplace of Revolution: How Consumer Politics Shaped American Independence* (New York: Oxford University Press, 2004).
38. Brückner, *Social Life of Maps in America*, 4.
39. Hamilton, *Gentleman's Progress*, 56.
40. Hamilton, *Gentleman's Progress*, 237.
41. For more historical work questioning the urban-rural divide, see William Cronon, *Nature's Metropolis* (New York: W. W. Norton, 1991); and Sam Warner Jr., *The Private City: Philadelphia in Three Periods of Its Growth* (Philadelphia: University of Pennsylvania Press, 1968).

7

RETHINKING ENSLAVED CONTAINMENT AND MOBILITY IN NORTH CAROLINA'S 1821 INSURRECTIONARY SCARE

Christy Hyman

A universal panic pervaded Onslow County, North Carolina, in August 1821. "The most daring, cunning, and despicable slaves, who well armed and accoutred [sic]," recounted Colonel William Hill, "committed many felonious acts." Hill alleged that the enslaved insurgents had engaged in "breaking open stores, burning houses and opening gunfire." Another witness claimed that "nothing but the protection of God prevented an attempt being made on his life." The insurrectionary scare compelled white residents of the county to flee, carrying news of the crisis and alerting local authorities in the neighboring counties to call out their militia. For the next three months, six hundred militiamen wandered the swamps and riverbanks of eastern North Carolina in search of enslaved insurgents. But the militia's search proved futile. The armed fugitives were "highly mobile" and evaded capture. The crisis ended not with a bang but a whimper. The Craven County Court was only able to attempt to convict one free person of color, Henry Black, for firing on and raiding the home of a white resident of nearby Jones County. Henry Black was later acquitted.

The 1821 insurrection of Onslow County, North Carolina, was written into the public record not for the actions of the enslaved insurgents but because of the militia's blundered defense of Samuel Street's Bridge. Captain John Rhem led a militia patrol that had been dispatched to guard the bridge. It was believed at the time that the insurgents would need to cross the bridge at some point. Shortly after midnight, Captain Rhem and eight of his men arrived at Street's Bridge. In the darkness, they made out five armed men standing on the other side of the river. Captain Rhem called out for the men to identify themselves. They answered with gunfire. Captain Rhem was shot through the lung, and his arm was shattered. Another of his troops was seriously injured, and others sustained injuries. Rhem's men returned fire. Then both sides retreated until daylight, each holding their side of the bridge. When dawn broke, Rhem's group discovered that their foes had not been armed insurgents at all but another militia company dispatched after word got out that the insurgents were marching on the bridge.[1] The *Fayetteville Observer* reported "no little slaughter on both sides" and that "each captain was dangerously wounded along with five to six privates on each side, also badly wounded."

In North America's history of slave rebellions, the Onslow County insurrection is usually relegated to a footnote or passing mention.[2] The only scholarly work devoted entirely to the 1821 Slave Conspiracy is a master's thesis that centers the militiamen's attempt to obtain monetary restitution from the state for the injuries suffered from the bungled defense of Street's Bridge.[3] Sylviane Diouf devotes some attention to the insurrectionary scare, including a surveying of the evidence presented by militia bands that scoured the region for months but could not find the Maroons. She ultimately concludes that, "when all was said and done, despite the white population's fears there was no 1821 conspiracy, no uprising, and, contrary to some scholars' assertions, no maroon rebellion either."[4]

How could a whole rebellion disappear? However overblown were white fears in North Carolina, the more telling fact is that white militia members were unable to locate the hiding places, trails, and passages through the swamps that were known to enslaved and free Black people. There was a hidden geography of enslaved life in North Carolina. This chapter will revisit the 1821 Insurrectionary Scare of eastern North Carolina utilizing critical readings of transport geography to illuminate the multiplicity of factors that influenced enslaved containment and movement. John James Kaiser's work emphasizes how "militia and patrols worked together in a flexible system designed to suppress both real and potential slave unrest." Here, I will show how the emphasis on internal improvements to increase commerce and navigation in eastern North Carolina created pathways to liberation for enslaved people tasked with their construction. Additionally, the "flexible system" of the domestic slave market resulted in the private revolution of one enslaved man, Manuel, rumored to have taken part in the 1821 insurrection, who eventually gained a measure of freedom not only for himself but also—using his knowledge of navigation—for Eve, his enslaved wife.

My intention, here, is not to advance a particular position on the veracity of the reports concerning the 1821 Insurrectionary Scare nor to speculate on the possible magnitude of it. I contend that the 1821 Insurrectionary Scare is an important site for assessing how enslaved people could reappropriate areas of wilderness into spaces of refuge and reconnaissance. Such spaces would expand the opportunities for creating sites where informal exchange economies could occur, offering autonomy and the increased ability to sustain pathways and spaces for liberation from bondage. In highlighting the material elements of enslaved people's mobility—the unique features of the eastern North Carolina landscape that influenced its transport geography—and placing them in conversation with geographical theoretical considerations, it will become clear that enslaved people's potential for refuge and reconnaissance, which could become a nexus for insurrectionary plots, was tied to the very antebellum industries that sought their labor.

Enslaved Mobility: Theoretical Considerations for Human Geography

The theoretical and practical dimensions of mobility for enslaved people requires considering their subjectivity as enslaved people in motion at various transit points. Beyond the structures that forced their labor within these work regimes, it is also important to disentangle the spatial concepts that ordered the world of slave societies and the individual destinies enslaved people shaped on their own.

The domestic slave trade allowed Manuel to elude sale to Louisiana and remain in North Carolina despite having been accused of being part of the 1821 Insurrectionary Scare.[5] The uneven topography of North Carolina's coastline necessitated the building of transport channels—ventures that aided local, regional, and, international networks of credit.[6] Enslaved people themselves were sold on credit, and when debts were called in, other enslaved people were sold for ready cash.[7] Indeed, an intrinsically American-style capitalism proliferated in a system of exchange and, as Calvin Schermerhorn explains, "chains of debt moved around the Atlantic basin in countermotion to the trajectories of captives, goods, and commodities."[8] This elasticity of credit within the domestic slave market provided moneymaking opportunities for enslavers, but in times of economic uncertainty—which came in cycles throughout the Age of Revolutions—there were also no guarantees. The instability of the domestic slave market meant that, in a moment of economic desperation, Manuel was sold to an Elizabeth City merchant and shipbuilder, rather than being sold down south in New Orleans. This set in motion the turn of events that resulted in Manuel securing the freedom of his enslaved wife, Eve, and, later, her friend Sall.

How Slavery Complicates Mobility as a Geographic Concept

Geographers in the 1950s trained in quantitative methods identified how movement was integral to understanding the theoretical foundations of geography as a field. Geographers looked to economists for ideas. J. H. Von Thunen developed the theory of the Isolated State for understanding agricultural land use. Beginning with a set of assumptions, Von Thunen was able to illustrate "the importance of both land value and transportation costs and their impact on the geography of the areas surrounding cities."[9] Walter Christaller's central place theory was a crowning achievement for quantitative geographers and theoretical geography as a whole, as it provided "a general [explanation] that would allow geographers to predict and explain the sizes, numbers, and distributions of towns." Looking to economists, Christaller argued that geography as a field could produce its own laws of settlement. What resulted was an "ideal spatial order that would serve as a normative pattern for settlement distribution."[10]

Leading geographers in the 1970s and 1980s critiqued the quantitative tilt of their field. Humanistic geographers wished to "rediscover the importance of place and the nature of specificities of places overall." It was a "positive affirmation of worlds of meaning and experience in the human relationship to the earth." Humanistic geography was itself a "critique of positivism." Coined by Yi-Fu Tuan, humanistic geography asked "How do people make the earth into a home?"[11] And for those considered stateless, or captive, how do these people find a home when they are seeking survival and freedom?

Studies of mobility in the field of human geography remain a central focus of practitioners today and profoundly shape reconsiderations of fugitive geographies. Critical geographers have outlined how mobility approaches compel an analysis of movement flows and their genealogies. The geographer Tim Creswell acknowledges the ways that routes are meaningful and laden with power. How can historians and geographers amplify how enslaved people harnessed the power of networks and navigational literacy to move through punitive landscapes?

Susan Hanson has written that "transportation in all its forms is woven into the fabric of everyday life, permeating places and lives with meanings transcending more than just the means of arriving at a destination."[12] Although contemporary analyses of transport often center the logistical infrastructural impact of modern-day transit options, the discussion holds relevance here since walking—the common mode of transport of enslaved people engaged in flight from bondage—is included as a central analytical factor. Scholars of mobility studies have provided a generative analysis stemming from studies in geographies of transport. Geographies of transport analyze the spatial aspects and affordances of movement as they produce social and environmental impacts. Cresswell's discussion of the role of moorings to transport functions of mobility reveals how the plight of enslaved people (who historically were seen as property that satisfied a commercial end for slaveholders) complicates the ideas of mobility. He describes how mobilities of transport require consideration of their required moorings—such as airports for aircraft, parking lots for automobiles, ports for ships. There were no vacant, inanimate spaces designed to hold enslaved people in flight. The symbolic tethering of enslaved people's feet was the domicile of the enslaver and the antebellum legal and political apparatus across the southern slave societies. The architecture of slavery, that power relation, permitted enslaved people to move but not for the actualization of their personal human potential—only for the commercial end required in the labor forced on them. Thus, enslaved movement away from the plantation in pursuit of autonomous power and possibilities for self-definition constituted a radical act of self-liberation.

Wayfinding

Pass laws in the antebellum South closed the roads as pathways to freedom for the runaway. The passes were written by slaveholders, usually for the purposes of moving goods or providing a service for the enslaver that required movement to a specified location. Enslaved people engaged in flight had to obtain a forged pass, which required subverting another power relation (the law forbade enslaved people from learning to read or write). Alternatively, runaways could stay off the road, venturing into the wilderness. This concealment in wilderness spaces while under the threat of discovery by slave patrols represents the site of what Gillian Rose refers to as "paradoxical space."[13] She defines paradoxical space as an occupation of space that combines dimensions of both agency and subordination. Though Rose's analysis is grounded in struggles specific to gender, paradoxical space, as Minelle Mahtani points out, can also be viewed through "a racialized prism."[14] To view the sites of concealment for enslaved people in the swamp wilderness of eastern North Carolina as paradoxical requires understanding that, dangerous as they were, hideouts for enslaved people were sites where they were no longer bound to the hardships of the plantation. At the same time, spaces of concealment were potentially disastrous spaces, as they meant further punishment if the enslaved persons were discovered and, in many cases, also meant being sold away to a locale in another southern state. These spaces of concealment were also places where the responsibility of concealing a pregnancy or birth meant added terror for a woman who hid in the swamp with her children or while she was pregnant. These women had to use additional means to conceal not only themselves but those too vulnerable to walk alone.

Such was the ordering of the antebellum spatial apparatus that undergirded the social space of slave societies. As Stephanie Camp puts it, "the heart of the process of enslavement was a geographical impulse to locate enslaved people in plantation space."[15] Studies of human mobility expand our understanding of enslaved mobility by exploring the "power discourses and practices of mobility in movement and stasis."[16]

Kathryn Benjamin Golden argues convincingly that the swamp wilderness provided cover for insurgent Maroons who ignited insurrectionary attempts. Golden's work also discusses how the "social gravity of the swamp itself stands as a crucial source for reading against the limitations of available written archives."[17] In utilizing the historical landscape of coastal North Carolina and its geographies of transport as a unit of analysis, this chapter is in the tradition of highlighting a historical account with a paucity of written sources. But the material elements of enslaved people's mobility are too important a feature of their experience to ignore.

Sites of the 1821 Conspiracy

White Oak Pocosin, the rumored site of the Maroon insurgents' refuge and reconnaissance, is located within northern Onslow County and southern Jones County. The White Oak River rises in the Pocosin running through a succession of swamps and forested areas. The deep recesses of swampland were inhospitable due to excessive humidity, difficult terrain, mosquitoes, biting flies, and wild animals. These attributes, along with widespread beliefs during the nineteenth century that swamps emitted "noxious vapors," made most people unwilling to enter them.

All of these impregnable sites on land and on the waterways flowing throughout the swamplands developed into areas of hideouts, temporary forays, and conduits to freedom for enslaved people possessing the navigational literacy to move through them. Apart from having knowledge of the wilderness landscape, enslaved people attempting to run away found ways to elude slave patrollers as well as wild animals by moving cautiously through the swamp environment.

The remote nature of the coastal North Carolina swamps posed problems for overseers managing groups of enslaved people. They needed to monitor enslaved movement throughout the landscape as well as devise a system of punishments and rewards to dissuade enslaved people from trying to run away. As early as 1741, slaveholders in North Carolina could rely on laws to uphold their interests

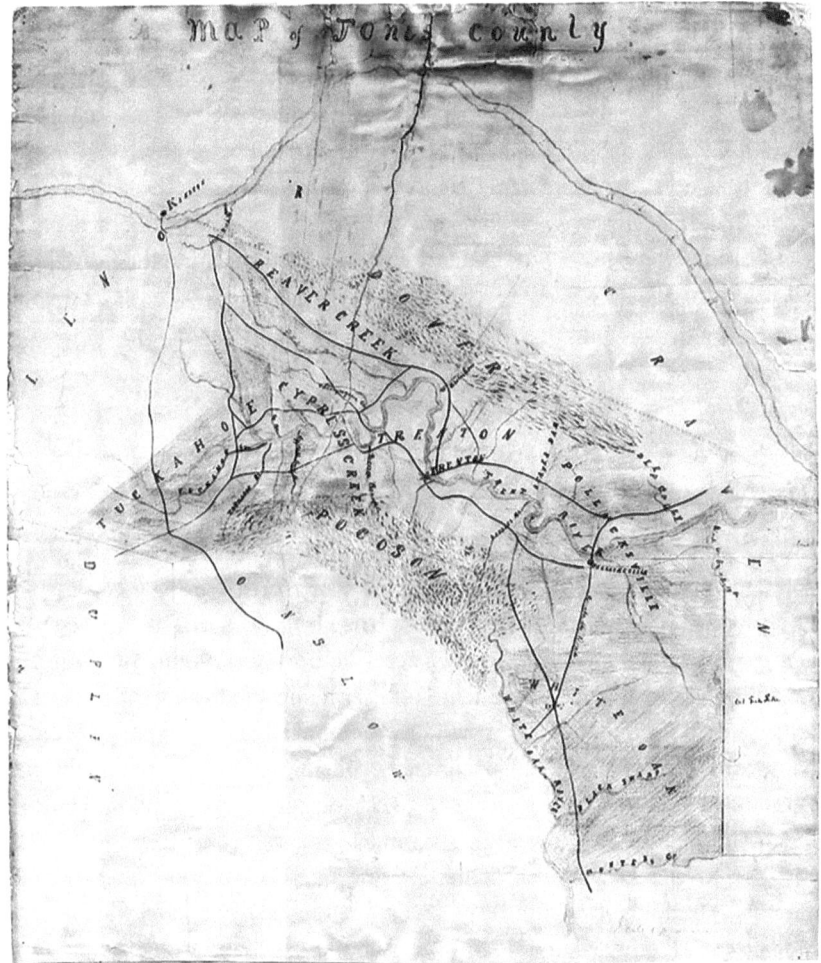

FIGURE 7.1. Map of Jones County, North Carolina. Joseph Kinsey, 1843. https://dc.lib.unc.edu/cdm/singleitem/collection/ncmaps/id/206/rec/3rev. 2/2011.

in keeping enslaved people's movement under control. The law begins by referencing the incidence of enslaved people escaping to wilderness areas while killing "cattle and hogs and committing other injuries to the inhabitants of this State." For any such rebels, the law authorized extraordinary action for ordinary people: "It shall be lawful for any person or persons, whatsoever to kill and destroy such slave or slaves, by such ways and means as he shall think fit, without accusation or impeachment of any crime for the same."[18]

The 1741 law is proof that enslaved people had been hiding out in swamps for some time. The law also gave absolute power to sheriffs of these locales to take any precaution they deemed necessary to contain enslaved people who appeared to be "lurking" in places where they should not be. This establishes a preliminary step within the historical record on how law enforcement of the period controlled the movements of enslaved people as well as those thought to be enslaved. In addition, enslaved people were often prohibited from hiring their time, burning firewood, entertaining free persons of color within their dwellings, and exhorting. These were the forms of containment forced on enslaved people during the antebellum period that served to protect white society from the specter of an enslaved insurgency.

The Need for Navigation in Eastern North Carolina

In the town centers within the coastal region of North Carolina, commercial activity was brought about on flatboat and barge traffic canals that enslaved people were forced to dig, which provided access to ports at Wilmington, Beaufort, New Bern, Elizabeth City, and Edenton. Enslaved people from various parts of eastern North Carolina and Virginia were hired out to canal companies to build them—the Dismal Swamp Canal Company, Albemarle Chesapeake Canal Company, and specific to the area near the 1821 Conspiracy, the Clubfoot Harlowe Canal. The Clubfoot Harlowe Canal, six miles long, was constructed to connect the Neuse River at New Bern with the Newport River and Beaufort Inlet. The first attempts at building it, beginning in 1796, were funded by private investors with the state contributing beginning in 1822. With the state's infusion of funding for the project, the company was able to advertise the need for enslaved laborers, guaranteeing that the work was agreeable:

LABORERS WANTED.

The Directors of the Clubfoot and Harlowe's Creek Canal Company want to hire immediately from 40 to 50 stout, able bodied negro men to work on the canal. Those who have hands to hire will please apply to Samuel Simpson, who can inform them of the wages given. All the hands now at the canal are healthy and well satisfied with their employment.
JAMES MANNEY.
President Canal Co.
Beaufort, April 8, 1822.

FIGURE 7.2. A new map of North Carolina with canals. Henry Tanner, 1833. https://dc.lib.unc.edu/cdm/singleitem/collection/ncmaps/id/476/rec/4. "A new map of Nth. Carolina: with its canals, roads & distances from place to place, along the stage & steam boat routes."

The last sentence, advertising that the "hands" are "healthy and well-satisfied" is a curious statement. Enslavers rarely showed concern about the job satisfaction of their human property, and there is little reason to believe that this was the canal company's advertising strategy. More likely, they were assuaging slaveholder fears that their chattel property might lose value (read, get sick or die) while employed building the canal. But there is another way to read this advertisement. Enslaved watermen and their expert knowledge were important to the North Carolina coastal economy. They were essential workers. The small measure of power they gained by such employment might have necessitated concern for their contentment.

Enslavers had good reason to be concerned. Enslaved watermen charged with navigating coastal transport waters shared information with one another and expanded their navigational literacy while under the yoke of slavery.[19] The widened networks of people they encountered, enslaved or free, were the access point to the Maritime Underground Railroad. David Cecelski has written extensively on enslaved mariners linking cultural traditions that emerged from their navigational practices and travels. His work has illuminated the network of relations among enslaved mariners and how their navigational literacy was expanded through work regimes. On canal building, Cecelski explains in vivid detail just how much geographic knowledge laborers gathered. They had to know where the dangerous shoals and inlets were that would create pathways that connected North Carolina to Chesapeake Bay. Such knowledge was not limited to the larger transportation networks. "When slave laborers dug smaller canals for swamp drainage, water power, or rice cultivation, those narrow channels doubled as local waterways: for rafting goods to market, for floating timber and shingles out of swamp forests, for gaining access to fishing grounds, for visiting neighbors and going to town."[20]

There was a deep irony to the use of enslaved labor to build up North Carolina's coastal transportation network. It required enslaved people to excavate canal beds, cut courses through thick brush containing tangles of vines, trees, and scrub. A nineteenth-century observer remarked on the Great Dismal Swamp that it was a "quagmire of peat filled with dead roots, buried logs, and living trees and roots on and at the surface."[21] It was backbreaking work that would have been difficult to find free white labor to undertake. But entrusting it to enslaved labor carried a risk. The work required axes and spades, tools that could be repurposed as weapons within internal exchange economies that could involve canal laborers, Maroons, and white company agents.[22] Enslaved people were often the first to cut paths through the not-yet-traveled wilderness. This was important geographic knowledge, knowledge that they shared with one another. Bits and pieces of this knowledge could be connected together, helping fugitives and Maroons navigate courses to the swamp wilderness from the farm or plantation where they originated. Systems such as this sustained fugitive slaves seeking freedom in the swamps.

The irony was real. Enslavers attempted domination of enslaved mobility. They constructed surveillance systems, passed laws empowering sheriffs and deputizing literally anyone into policing the enslaved, and attempted to prevent the enslaved from gaining access to the geographic knowledge that one might acquire from consulting maps, almanacs, and books. At the same time, enslavers needed enslaved persons to turn the landscape from a wilderness into a functioning state. Enslavers entrusted them with important jobs that communicated geographic knowledge. And however much they wanted to control mobility, they revealed at important points just how precarious their grip on the enslaved was. The landscape itself had become a "terrain of resistance beyond the racial, spatial, political, and economic rationale of a world built and maintained by chattel slavery."[23]

One group of enslaved runaways discussed and strategized ways to escape the bondage of the plantation. Their story highlights how labor demands and enslavers' attention to natural increase within the community of the enslaved created possibilities for attaining (and maintaining) both kinship channels and avenues for an expanded navigational literacy. The labor and logistical demands on enslaved people by enslavers inadvertently allowed a range of opportunities for enslaved people to plan strategies for self-liberation as they transformed spaces in their work regimes into pathways out of bondage for themselves.

Manuel's Private Revolution

Manuel entered the historical record as an enslaved man who asserted his right to move through antebellum spaces. Enslavers may have asserted absolute control over the bodies of their enslaved laborers, but Manuel circumvented antebellum slave codes by hiring out his time and by running when necessary. Despite being described as having such identifiable physical characteristics as "knock-knees" with light skin and a speech impediment, Manuel was able to use his knowledge of the landscape to maintain kinship ties, earn money, and assist other enslaved people escaping slavery.

The first instance of Manuel escaping was in Chowan County in 1810. William Saunders, Manuel's enslaver, posted a runaway slave advertisement that indicated several attributes of Manuel's appearance. Benjamin Harvey posted another in 1820 which added that Manuel grew up in Perquimans County and still had family members enslaved there. Manuel next appeared in Perquimans County court documents in June 1829. His enslaver, Hugh K. Wyatt, was accused of allowing Manuel to "go at large" and hire out his own time. For the record, Hugh K. Wyatt was listed as part of Captain Ambrose K. Wyatt's Company, which had taken part in the suppression of "certain runaway negroes that were making threats of Persons lives and committing Depredations of various kinds in the County of Perquimans in the Year 1821."[24] If anyone should have been aware of the danger of allowing enslaved people such freedom to wander, it should have been Wyatt.

The chain of enslavers connected to Manuel indicated at least three different residences where he had been held in bondage, in two contiguous counties in eastern North Carolina. He was being sold fairly frequently. Manuel also asserted his mobility, clearly audaciously given that his enslaver had to answer for him in court. This made Manuel a formidable asset for enslaved people seeking freedom. Curiously, Benjamin Harvey was a relative of an Emanuel Harvey of Perquimans County. Because Manuel was described as having been half-white (mulatto), it is possible that his father was a member of the Harvey family. Manuel may have utilized his proximity to whiteness in matters difficult to trace through archival documentation.[25]

Though Manuel was slated to be taken to New Orleans during one of the "negro drives," he was able to elude sale to New Orleans after the 1821 Insurrectionary Scare. He was then able to help his wife, Eve, and, later, her friend Sall escape to freedom.

Manuel's navigational literacy was expanded as a result of his body being purchased and transported between three locales—Chowan, Perquimans, and Pasquotank County where Elizabeth City is located. Though the historical record does not indicate that Manuel ever resided in Onslow County, the site of the 1821 Insurrectionary Scare, some of the white residents fleeing Onslow County reached Perquimans County, alerting authorities that local enslaved people were probably taking part. Manuel, perhaps because of his autonomy, seemed a likely participant in an enslaved uprising. Perhaps Hugh K. Wyatt enslaved Manuel thinking it would be a worthy investment? Manuel's flight from the enslaver Benjamin Harvey in 1820 provides a starting point where Manuel is free or, in the enslaver's view, "at large." What can be documented is that, by 1829, Hugh K. Wyatt was listed as Manuel's owner and, by 1833, Manuel Cluff, an Elizabeth City merchant and shipbuilder, had purchased Manuel. Cluff, much like Wyatt, could not seem to keep control over Manuel's movements.

The insistence of slaveholders on natural increase among their slave chattel allowed Manuel and George the ability to visit Eve and Sall for purposes of creating "wealth" for John Wood. The words Wood expressed in the slave ad for Eve and Sall indicated that he had not fully considered the ramifications of the conjugal visits permitted. While referring to Manuel as a "noted villain," Wood remained amenable to Manuel and George's presence from time to time. These enslaved men, looked to as "studs" for impregnating enslaved women, offered a path out of bondage for them. These passages were fraught for many reasons, chief among them the resonances of suffering and loss that remained with every enslaved person engaged in self-liberation.

Moving through geographies of domination was a perilous journey for enslaved runaways. The environs they crossed carried the invisible markers of enslaved people's suffering—they were "traumascapes." The writer Maria Tumarkin defines traumascapes as a "distinctive category of places transformed physically and psychically by suffering, where the scar tissues remain embedded in the landscape."[26] The flight paths leading to the Great Dismal Swamp included sites of enslaved people's shared suffering, of loved ones sold away or who passed on before their time. This shared trauma undoubtedly created powerful bonds among those determined to survive the ravages of the traumascape and the slavery regime itself. It was through these connections and resonances that enslaved people could recognize how the power of knowing the landscape could be appropriated for their own uses. As a geographic area of historical confinement for enslaved people, the swamplands in the Albemarle region hold significant meaning as a site of contestation between enslavement and liberation seeking.

The Runaway Slave Advertisement

$200 Reward WILL for the apprehension and delivery to me in this place of my two negro women EVE and SALL. Eve is the wife of Manuel, belonging to Mathew Cluff, and Sall is the wife of old George, also belonging to Mathew Cluff. Eve was seduced away by Manuel in September, 1829, and has since that time been kept out with the assistance of white persons at or near Elizabeth, and from there to the head of Pasquotank River. She is a low, thick set woman, about 26 or 27 years of age, bushy head of hair, rather thick lips, smooth, dark skin, though not very black and lisps a little when spoken to. I understand she has changed her name to Mary and has a free pass. Sall is a tall, stout woman, smooth skin, about the same complexion and age of Eve. She was seduced away by old George on the night of the 2nd of this month, on his return from this place to Elizabeth City, where he had been to visit her as usual. His object, no doubt, is to place her with Eve, to be under the protection of that noted villian Manuel and his brothers, who were transported from this County with a view of sending them to New Orleans. I will pay the above reward for both together, with the child or children, of Eve if she has any with her, or $100 for either on delivery. JOHN WOOD. Her ord, July 27, 1833.

John Wood did not hide his disgust for Manuel and George. The ad mentions that Manuel had been "transported from the county with a view to sending them to New Orleans." As the Perquimans County court clerk, John Wood had become acquainted with Manuel during a criminal court case which took place in June 1829.[27] In *State vs. Negro Manuel a Slave*, Manuel's enslaver Hugh K. Wyatt was accused of allowing Manuel to "go at large" which went against "the peace and dignity of the State" of North Carolina. The runaway slave ad indicates that Eve escaped in September of the same year, three months after Manuel's enslaver Hugh Wyatt was scheduled to appear in court.

Manuel did not end up in Louisiana as Wood speculated. He was sold to Matthew Cluff, an Elizabeth City merchant.[28]

Wood's assessment of Eve's appearance and manner is telling. He describes Eve as having a "bushy head of hair, rather thick lips, smooth, dark skin" and that she "lisps a little when spoken to." Such a description indicates that Wood was able to observe Eve routinely. Perhaps Eve was a domestic servant within the house? Wood does not comment on Sall's speech pattern as he does Eve, but both women seem likely to have been tasked with positions that allowed for regular contact with Wood. Domestic positions within the plantation household meant that enslaved women performing those duties had a more limited range of movement than those forced to work in agricultural environments located in outlying areas of the plantation. For this reason, they would require the assistance of their husbands to secure an escape.

The geographer Katherine McKittrick has analyzed disciplinary encounters between Black studies and geography and has observed the prevalent notion that Black people are "ungeographic." McKittrick goes on to explain how "discourses erase and de-spatialize [enslaved people's] sense of place" and thus her work aims to conceptualize Black geographies from that tenuous position of alterity.[29]

To render Black people as "ungeographic" historically is to not question how an enslaved person survived a flight to freedom that landed them seventy miles from the original site of bondage.[30] As the historian Alisha Hines has rightly observed, "black people are allowed to be shapeshifters within the historical record."[31] Adding texture and depth to enslaved movement helps provide evidentiary rigor to the committed yearning for freedom held by enslaved people seeking liberation.

Tools and Symbols Needed in Flight

William H. Robinson, who had run away several times near Wilmington, North Carolina, explains the importance of certain tools for fighting off bloodhounds: "There was always an understanding between the slaves, that if one ran away they would put something to eat at a certain place; also a mowing scythe, with the crooked handle replaced with a straight stick with which to fight the bloodhounds."[32] The image of the scythe for the enslaved runaway signified the potentiality of an event that would test the determination of their plans. One look at the scythe and they knew there was no turning back.

Robinson also spoke of the elderly enslaved woman who gave him onions for use during his flight to freedom:

> She gave me four or five onions, and told me upon the peril of my life, not to eat a single one of these onions, because they would make me

sleepy and I would be liable to be caught. But she said negro hunters came along there every two or three hours in the day; and I learned for the first time how to decoy the blood hounds, for she told me whenever I heard the baying of hounds on my trail, to rub the onions on the bottoms of my feet and run, and after running a certain distance to stop and apply the onions again, then when I came to a large bushy tree, to rub the trunk as high up as I could reach, then climb the tree.[33]

Here, Robinson highlights the way enslaved people worked together to help each other move through dangerous geographies. It must be noted how enslaved people, upon reaching a certain age, were considered of zero value in the slave market. Because slaveholders saw no value in the old slave, they were considered worthless and inconsequential. But little did the enslavers know that the aged, enslaved person possessed a deeper knowledge of place and of practices that helped freedom seekers in their flight out of bondage.

We may never know how long Eve, Sall, Manuel, and George were able to survive as enslaved fugitives. The 1840 census lists John Wood as owning no female slaves in the age range originally indicated on the runaway slave ad for Eve and Sall.[34] Seven years later, the North Carolina assembly passed its most draconian legislation in hopes of reducing the number of enslaved people running away to the Great Dismal Swamp.[35] What cannot be contested is that Eve and Sall, with the help of their enslaved husbands, possessed the courage to move through.

NOTES

1. Petition of Captain John Rhem to the North Carolina Legislature, November 25, 1822, Race and Slavery Petitions Project, PAR 11282206, https://dlas.uncg.edu/petitions/petition/853/.

2. Herbert Aptheker, *American Negro Slave Revolts* (New York: Columbia University Press, 1943); Guion Griffis Johnson, *Ante-Bellum North Carolina: A Social History* (Chapel Hill: University of North Carolina Press, 1937); John Hope Franklin and Loren Schweninger, *Runaway Slaves: Rebels on the Plantation* (New York: Oxford University Press USA); Peter Hinks, *To Awaken My Afflicted Brethren: David Walker and the Problem of Antebellum Slave Resistance* (University Park: Pennsylvania State University Press, 1997).

3. John James Kaiser, "'Masters Determined to Be Masters': The 1821 Insurrectionary Scare in Eastern North Carolina" (MA thesis, North Carolina State University, 2006).

4. Sylviane A. Diouf, *Slavery's Exiles: The Story of the American Maroons* (New York: New York University Press, 2014), 272.

5. The separation of enslaved families brought about by the domestic slave trade and the backbreaking work extracted from enslaved people in southern regions are documented powerfully by Ed Baptist in *The Half Has Never Been Told* (New York: Basic Books, 2014). In reading the numerous accounts Baptist cites from consulting enslaved testimony, we learn of the expansive networks wrought by the slave trader, who lawfully controlled the movement of enslaved people as they were transported in shackles from place to place after a sale.

CHAPTER 7

6. Sven Beckert and Seth Rockman, eds., *Slavery's Capitalism: A New History of American Economic Development* (Philadelphia: University of Pennsylvania Press, 2018).

7. Beckert and Rockman, *Slavery's Capitalism*.

8. Calvin Schermerhorn, *The Business of Slavery and the Rise of American Capitalism, 1815–1860* (New Haven, CT: Yale University Press, 2015), 2.

9. Tim Cresswell, *Geographic Thought: A Critical Introduction* (Chichester, West Sussex, UK: Wiley-Blackwell, 2013), loc. 2774, Kindle.

10. Cresswell, *Geographic Thought*, loc. 2774.

11. Cresswell, *Geographic Thought*, loc. 3794, loc. 3177, loc. 3306.

12. Susan Hanson, "Gender and Mobility: New Approaches for Informing Sustainability," *Gender, Place and Culture* 17, no. 1 (2010), 5–23.

13. Gillian Rose, *Feminism and Geography: The Limits of Geographical Knowledge* (Minneapolis: University of Minnesota Press, 1993), 159.

14. Minelle Mahtani, "Racial Remappings: The Potential of Paradoxical Space," *Gender, Place and Culture: A Journal of Feminist Geography* 8, no. 3 (2001): 299–305.

15. Stephanie Camp, "The Pleasures of Resistance: Enslaved Women and Body Politics in the Plantation South, 1830–1861," *Journal of Southern History* 68, no. 3 (2002): 534.

16. Kevin Hannam, Mimi Sheller, and John Urry, "Mobilities, Immobilities and Moorings." *Mobilities* 1, no. 1 (2006): 1–22, 4.

17. Kathryn Benjamin Golden, "'Armed in the Great Swamp': Fear, Maroon Insurrection, and the Insurgent Ecology of the Great Dismal Swamp," *Journal of African American History* 106, no. 1 (2021): 1–26.

18. Revised Code No. 105, "Slaves and Free Persons of Color: An Act Concerning Slaves and Free Persons of Color," 1816 c 910 s 3 Sheriff's duty when they abscond, Documenting the American South, accessed June 25, 2016, http://docsouth.unc.edu/nc/slavesfree/slavesfree.html.

19. North Carolina State Board of Agriculture, *North Carolina and Its Resources* 1896, Documenting the American South, https://docsouth.unc.edu/nc/state/state.html, 131.

20. North Carolina State Board of Agriculture, *North Carolina and Its Resources*, 93.

21. North Carolina State Board of Agriculture, *North Carolina and Its Resources*, 109.

22. Marcus P. Nevius, *City of Refuge: Slavery and Petit Marronage in the Great Dismal Swamp, 1763–1856* (Athens: University of Georgia Press, 2020).

23. Willie J. Wright, "The Morphology of Marronage," *Annals of the American Association of Geographers* (2019), https://doi.org/10.1080/24694452.2019.1664890.

24. State Archives of North Carolina, Perquimans County Records, Records of Slaves and Free Persons of Color, C.R.077.928.2, accessed June 9, 2021.

25. *Incidents in the Life of a Slave Girl* points to these transgressive intimacies. Jacobs/Brent shares a story about an enslaved woman who knew the father of her child was her enslaver but was sold off when she uttered this information to her husband, who was flogged and also sold off. The enslaver, when confronted by the enslaved woman as she was to be taken off in a coffle, shouted, "Damn you!" Such was the foundation of illicit intimacies with enslavers whose promises were only as stable as their fleeting dalliances. Harriet A. Jacobs, *Harriet Jacobs and "Incidents in the Life of a Slave Girl": New Critical Essays* (Cambridge: Cambridge University Press, 1996).

26. Maria Tumarkin, *Traumascapes: The Power and Fate of Places Transformed by Tragedy* (Carlton, Victoria: Melbourne University, 2005),13–15.

27. *State vs. Negro Manuel, a slave, Hiring His Own Time*, Perquimans County, 1829, Record of Slaves and Free Negroes, State of North Carolina Archives Division.

28. For more on archival silences and the violence embedded within records related to the history of slavery, refer to Jessica Marie Johnson's "Markup Bodies: Black [Life] Studies and Slavery [Death] Studies at the Digital Crossroads," *Social Text* 36, no. 4 (137) (December 1, 2018): 57–79.

29. Katherine McKittrick, *Demonic Grounds: Black Women and the Cartographies of Struggle* (Minneapolis: University of Minnesota Press, 2006).

30. In this memoir the author recalls an instance where an enslaved person ran away from a Bertie County plantation in North Carolina and ended up being captured in Suffolk, Virginia, with no discussion of the range of antagonisms involved in the enslaved person's moving across the antebellum landscape. Robert T. Arnold, *Dismal Swamp and Lake Drummond* (S.l.: Tradition Classics, 2012

31. Alisha Hines, "To Make Her Own Bargains with Boats: Black Women, Steamboats, and Rival Geographies of the Western River World" (presentation, Association for the Study of African American Life and History conference, Atlanta, GA, 2015).

32. William H. Robinson, *From Log Cabin to the Pulpit, or, Fifteen Years in Slavery* (Eau Claire, WI: James H. Tifft, 2010), 32.

33. Robinson, *From Log Cabin*, 61.

34. Sixth Census of the United States, 1840, NARA microfilm publication M704, 580 rolls), Record Group 29, Records of the Bureau of the Census, National Archives, Washington, DC.

35. North Carolina General Assembly, "A Bill to Provide for the Apprehension of Runaway Slaves, in the Great Dismal Swamp, and for Other Purposes" (WR Gales, 1846), 2; Sylviane A. Diouf, *Slavery's Exiles: The Story of the American Maroons* (New York: New York University Press, 2016). For more on the history of Great Dismal Swamp Maroons, see also Nevius's *City of Refuge*; Kathryn Benjamin Golden, "Through the Muck and the Mire: Marronage, Representation, and Memory in the Great Dismal Swamp" (PhD diss., University of California, Berkeley 2018); Dan Sayers, *A Desolate Place for a Defiant People: The Archaeology of Maroons, Indigenous Americans, and Enslaved Laborers in the Great Dismal Swamp* (Gainesville, FL: University Press of Florida, 2014); Eric Anthony Sheppard and Moses Grandy, *Ancestor's Call* (Elkridge, MD: Tech-Rep Associates, 2003); Elaine Nichols, "No Easy Run to Freedom: Maroons in the Great Dismal Swamp of North Carolina, 1677–1850" (PhD diss., University of Southern California, 2018); and Moses Grandy, *Narrative of the Life of Moses Grandy, Late a Slave in the United States of America*, DocSouth books edition, 1843 (Chapel Hill: University of North Carolina, 2011), e-book.

8
MAPPING MYAAMIA LANDOWNERSHIP, 1795–1846 AND TODAY

Cameron Shriver

The possession and dispossession of the continent is a process fundamental to American and Native American histories. On one hand, a narrative of *dispossession* is important for understanding Native history and the Myaamia diaspora since the American revolutionary era. The modern Miami Nation developed in tandem with territorial disputes, while modern justice, reconciliation, and legal claims all depend on a clear-eyed vision of past dispossession. The acquisition of Indigenous territory was essential to the development of a modern United States, both in the postrevolutionary settlement and today. On the other hand, *possession* is critical to the Myaamia axiom: "We are a people *with* a past, not a people *of* the past." Land loss tends to support disappearance narratives in the American mainstream imagination, but Myaamia people, like so many Indigenous communities and their allies, strive to convince people that they still exist.[1] Landownership touches cultural, legal, and economic histories; it forces us to consider loss and survival, change and stasis, at the intersection of US and Myaamia societies.

Recognizing the significance of such a task, scholars at the Myaamia Center—a research collaboration between the Miami Tribe of Oklahoma and Miami University—created the *Aacimwahkionkonci* "Stories from the Land" Project.[2] The Miami Tribe of Oklahoma, the descendant of Myaamia town polities that preceded the United States, is a nation of about seven thousand citizens headquartered in Ottawa County, Oklahoma, with significant populations in eastern Kansas and northern Indiana; it is spread across all fifty states and internationally. Myaamia (the downstream people) and Miami are used interchangeably.

The Aacimwahkionkonci Project is a long-term digital history venture crucial to understanding—and in turn telling—the stories of a complex legal patchwork that has come to define *Myaamionki*, "the land of the Miamis." The Aacimwahkionkonci team, of which I am the lead researcher, is currently accumulating, digitizing, organizing, and mapping Myaamia land holdings up to the present.[3] The goals are twofold. First, Aacimwahkionkonci is an online digital archive focused on land transfer documents. The legal privatization of Indian Country in the two centuries since the American Revolution required a parallel paper trail connecting documents, land parcels, and a bureaucracy to oversee it. The archive, like the continent, is massive. It contains not only treaties but thousands of manuscript pages detailing Myaamia land transfers, including property sales, guardian disputes, allotments, and repossessions. Digital tools, such as archival and mapping platforms, are necessary to reconstruct the atomization of the Myaamia Nation's land. Aacimwahkionkonci uses CONTENTdm, hosted at Miami University's King Library, to display digital surrogates and metadata of original sources. This archive is connected to a bespoke database, within the infrastructure of Miami University Information Technology Services, implemented with a Laravel web framework and a relational database. This database organizes and sorts the archive. It includes biographical information about individuals, legal descriptions of their lands with maps, and events associated with those lands, such as treaties or real estate sales. Currently, an ESRI ArcGIS Hub displays most of the mapping. Fortunately, Myaamia Center programmers have experience building digital tools to meet specific research needs, notably for linguistic research from manuscript sources.[4] The digital archive and database promises browsing, searching, and eventual quantitative investigating of this archive. Aacimwahkionkonci has research needs built into its infrastructure.

Second, and no less importantly, Aacimwahkionkonci is a public humanities project that digitally reconnects people with important places. Through its database and mapping tools, users can explore people, places, and events. A Myaamia person could, for instance, search for their ancestor or family and find places they once owned. By bringing together these archival resources from various locations by means of digital technology and a mapping platform, we are creating a unique opportunity to learn about the Miami Tribe's history of location and relocation. In short, Aacimwahkionkonci aims to fill a specialized research need and a public humanities need. The genealogy of families and of specific places has emerged as a dominant, albeit complex, theme in how Aacimwahkionkonci developers and stakeholders have considered this project about Myaamia land. I will return to that notion at the end of this chapter.

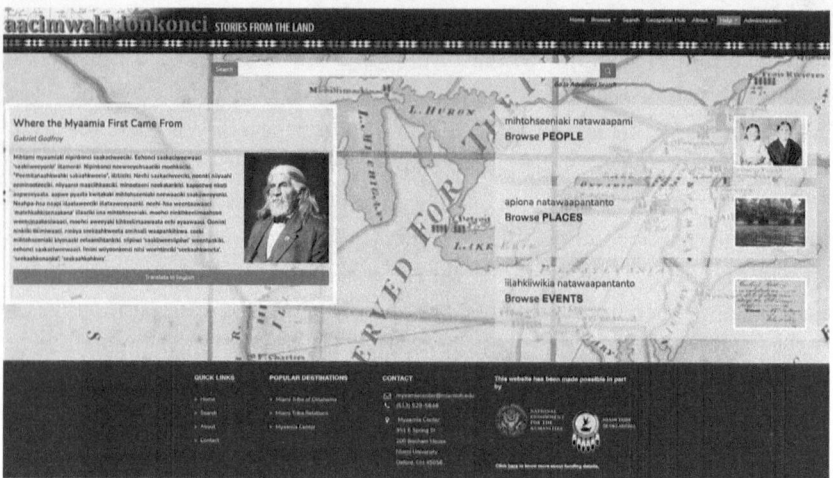

FIGURE 8.1. The Aacimwahkionkonci public home page, https://mc.miamioh.edu/aacimwahkionkonci/

An "American Revolution" Project?

The American Revolution generated an empire. Control over land was a critical cause and consequence of the American Revolution. "The Citizens of America, placed in the most enviable condition," George Washington pronounced in 1783, basking in newfound independence, considered themselves "the sole Lords and Proprietors of a vast tract of Continent."[5] It would take five generations to accomplish that image of what Washington termed "our Empire," so in the long view, property regimes changed more slowly than revolutionary speeches might suggest.[6]

Nonetheless, if the Revolution generated a new empire, then those imperialists quickly latched on to treaties as a powerful weapon. In choosing a starting point for histories of landownership, we could choose the Myaamia story of emergence from *Saakiweeyonki*, or Pope Alexander VI's 1493 Papal Bull, or the 1763 Royal Proclamation. (The Myaamia origin story is featured on the site's home page.) For the purposes of this chapter, it makes more sense to highlight treaties as crucial documents of American landownership. As Allan Greer has demonstrated through comparative analysis, the "land-purchase approach to colonization" was not the only style available for European empires, but it dominated English-speaking colonial projects. The treaty, a genre of agreement implicating international diplomacy, emerged as the fulcrum of the land-purchase approach as it evolved in North America. The long revolutionary era marked a transition. Colonial legal pluralism (to a point) saw all manner of agreements and accords established between and among Native and non-Native communities for diverse objectives not limited to territory transfer. From the period of the Seven Years' War to the War of 1812, colonists-turned-rebels-turned-Americans learned to use the treaty process to "transform Native American homelands into American real estate." US acquisition of Native territory took place, by and large, via treaties. By the nineteenth century, the United States had effectively massaged Indian treaties into instruments of property transfer within its putative borders. In their American Indian context, treaties became tools of colonialism. This colonization route received American legal approval in the 1823 *Johnson v. M'Intosh* Supreme Court decision.[7]

It is worth pausing a moment to consider the potential implications of treaties in the revolutionary moment. Although treaties between the United States and Native American nations developed into coercive contracts aimed at property and jurisdictional transfer, that future was fuzzy for the revolutionary generation. At that time, treaties sought potentially divergent goals. In establishing sovereignty, diplomats of the new United States desperately hoped to enter the world of European nations. From the start, the United States struggled to mold itself into a treaty-worthy sovereign. The era of the American Revolution saw a proliferation of treaties, most of which were not negotiated through the Confederation Congress but rather signed by Native nations, Spain, France, private land companies, and states. Problematically, European public law, enunciated by jurists such as Emer de Vattel, required a minimum level of unitary sovereignty. It was a threshold that Confederation-era America, with its decentralized government of relatively independent states, could not meet until it centralized through a new Constitution in 1787.[8] Facing Europe, American nationalists needed to get their house in order.

Those same Americans also faced Indian Country. From the moment of independence forward, American diplomats sought treaties with Native nations. Indigenous polities were treaty-worthy, of course. Additionally, their diplomatic and military significance created a constitutional crisis for early US founders discovering the hazards of the Articles of Confederation. Having won an unstable independence from Great Britain, American diplomats asserted that Native nations, as allies of Britain, had been conquered during the Revolutionary War. Although the Myaamia vacillated between supporting Britain and asserting neutrality in that conflict, US politicians certainly applied this "defeated with Britain" rationale to the Miamis and their territory. Then, they transitioned from a theory of conquest to a more moderate approach in the late 1780s—using contracts to acquire territory, a pragmatic choice given that most Native communities were not, actually, conquered. "The independent nations and tribes of Indians," Secretary of War Henry Knox told President George Washington in 1789, "ought to be considered as foreign nations, not as the subjects of any particular state." An important policymaker, Knox ruminated that the federal government would de-escalate border tensions if they enacted a law "that the Indian tribes possess the right of the soil of all lands within their limits respectively and that they are not to be divested thereof but in consequence of fair and bona fide purchases" from the United States, including the crucial rule "that their [that is, Indian nations'] authority and consent should be considered as essentially necessary." Like the various European approaches to colonization, the very idea that sovereignty was a territorial endeavor itself is a historical artifact and a precondition to understanding American imperialism. And, similar to their British predecessors, US officials debated whether Native nations were subjects or independent, foreign or domestic, subjugated or autonomous. Of course, Native nations made (and continue to make) clear and sustained arguments in favor of their own independence and sovereignty, both within and outside of the US constitutional order.[9] Sustained by treaties, that debate continues with increasing Indigenous input today, and is almost always wrapped in complicated jurisdictional questions. When it comes to treaties, the past is not the past.

This short description of the multiple meanings of treaties in the revolutionary era forces us to think about sovereignty and property as an unsettled business in an ongoing empire. Treaties established the sovereignty of the United States after its War of Independence; to sign international contracts was to stake a claim of nationhood. Americans and Native Americans also negotiated a colonial property regime in North America using treaties. West of the Appalachian Mountains, nearly all title to territory bounded by the United States descends from an Indian treaty. The 371 Senate-ratified Indian treaties are the founding documents of American real estate.

Indigenous landownership occupies a critical but underinvestigated piece of this American imperial history. In the national master narrative, Native Americans owned land only to lose it to non-Indians. Typically, the American Revolution is not a centerpiece in this story. Instead, US continental expansion is generally enveloped by a nineteenth-century, Western field of vision. An active national state was required to achieve these territorial ambitions, not just for the purposes of military conquest.[10] The idea that land could be a commodity bought or sold in a market was invented piecemeal. Would-be possessors required complicated legal systems to transform a bit of soil into real estate, making the American Empire a massive colonial archive.[11] If the Revolution created a new empire, and if those new imperialists relied on treaties to dispossess Native Americans, then we must also accept that this imperial approach created an important genre of American real estate—"Indian" landowners. Aacimwahkionkonci aims to discover, digitize, and map that messy process for one Indigenous nation and its citizens: the Myaamia Nation.

Treaties are also essential data for understanding Myaamia sovereignty and territory. The United States sent commissioners to negotiate contracts with polities on the southeastern frontiers of Myaamionki "Miami territory" in the 1780s and 1790s. Miamis did not sign these, and most Myaamia families actively protested the influx of American settlers north of the Ohio River. Secretary of War Henry Knox was flummoxed by the failure of American policies and philosophies—the Miamis and Shawnees did not voluntarily relinquish their land in the 1780s or 1790s. Transitioning to military methods, two successive invading armies met defeat from Miami, Shawnee, Delaware, and other defenders. "It became necessary to make an experiment of the effect of coercion," Knox wrote in late 1791.[12] Finally achieving military victory in 1794, then, the United States and its diplomats returned to their notion of colonialism by contract. In the process, US speakers sought to convince Native diplomats of their unitary, federal sovereignty. Although formerly "an association of several separate states, like their several separate tribes," by 1792, the nation was now "a general gover[n]ment embracing all parts of the Union, as it respects foreign Nations and Indian tribes."[13] In 1795 Miami leaders signed the first of fifteen treaties with the United States. It was both a treaty of peace and a contract of land dispossession. As Andrew Cayton contends, the lead American negotiator "got more than land, however; he also got—or, more important, believed he got—legitimacy."[14] American elites at the 1795 treaty conference needed their Native counterparts to recognize US sovereignty. In effect, they purchased their treaty-worthiness through a costly war on the heels of their own fight for independence.

Studies of Indigenous landownership typically focus, understandably, on treaties. Of those works, most highlight the South and West rather than the nations of the Old Northwest or other regions.[15] By accumulating, digitizing (including metadata), mapping, and then mining the documents generated by the bureaucratic empire, I expect to continue to answer questions about Miami landownership, generate new questions, and provide opportunities to reconnect Myaamia citizens with place-based histories and stories. Taking treaties as a starting point, Aacimwahkionkonci moves us beyond them and into the frenzied and complex real estate market—an empire of landowners, Native and not—left in their wake.

The digitization is ongoing, so conclusions based on the evidence remain tenuous, but broad strokes are becoming discernible. We now know that Myaamia families and individuals were active players in the land market, selling land to each other and directly to non-Myaamia buyers and acquiring significant land grants through treaties. The Myaamia are an intriguing, albeit complicated, case study of tribal landownership. Not only were tribal domains ceded via treaties; the Myaamia were one of a handful of nations to receive allotments before the 1850s. (Allotment means to allocate communal land to individuals. The result of the process of allotment is also called "an allotment.") Miamis were among the first Native Americans (in 1818) to negotiate reserves held by tribal individuals.[16] In fact, the Myaamia experienced allotment—the division of their lands into privately held parcels—twice before the "allotment era" of the Dawes Allotment Act in 1887, and most of their allotments, both in number and acreage, were apportioned before the Dawes Act or similar legislation specific to Indian Territory.[17] Miami reservations also underwent allotment successively in Kansas and Oklahoma. Following their first deportation in 1846, the Miami reservation in Kansas was allotted in 1854, 1856, and 1869.[18] Like their tribal relatives the Shawnees, Wyandots, Peorias, Seneca-Cayugas, Potawatomis, and others, the Miami lands in Kansas eventually were deemed too valuable for Indian ownership, and the Miamis were expelled again, their last removal to date, to northeast Indian Territory (Oklahoma) where the Miami national domain was allotted to individuals in 1892. Each of the three Myaamia homelands therefore witnessed the privatization of tribal ownership until, by the 1930s, there was no "Miami" land owned by the tribe.[19]

Dispossession was a form of wealth transfer out of the Miami Nation. Geographic Information Systems (GIS) and digital archiving have allowed scholars to develop techniques to reevaluate old interpretations and analyze evidence in new ways. In Native American history, for example, Claudio Saunt and Robert Lee, respectively, have digitized maps developed for Indian land claims—called "Royce" maps—and analyzed those tribal domains and cessions.[20] There are real

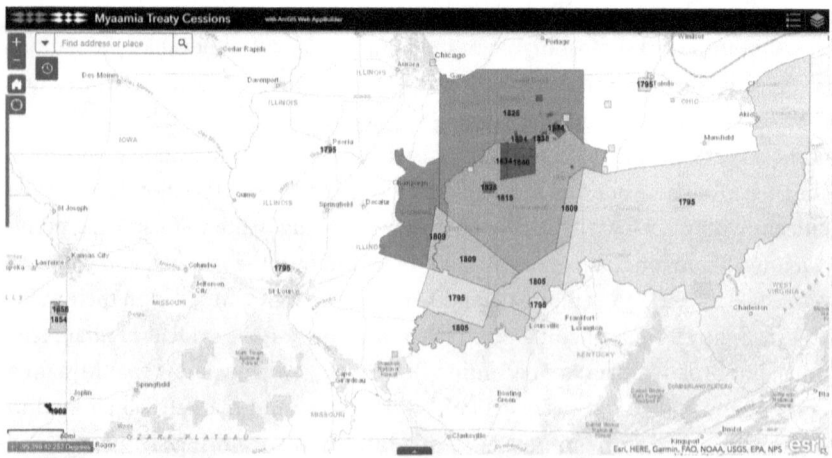

FIGURE 8.2. All Myaamia Cessions Map, 1795–1892. Map by Owen Larson. Basemap sources: ESRI, HERE, Garmin, Intermap, increment P Corp., GEBCO, USGS, FAO, NPS, NRCAN, GeoBase, IGN, Kadaster NL, Ordnance Survey, ESRI Japan, METI, ESRI China (Hong Kong), © OpenStreetMap contributors, and the GIS User Community.

benefits to analyzing a few large territories, and a map of the Myaamia cessions appears as figure 8.2.

Yet Myaamia individuals continued owning land. As the Miami Nation lost a land base, Myaamia people continued ownership and control in multiple homelands. The Aacimwahkionki Project deploys GIS and big data to piece together the more voluminous small reservations that dotted, and in some cases continue to characterize, Indian Country including Myaamionki.

Miami Landownership in the Early Republic

Although the hemispheric view suggests territorial losses, more local, smaller reserve lands demand attention because Myaamia folks continued owning them. Let us now turn our attention to land that Miami people retained in the

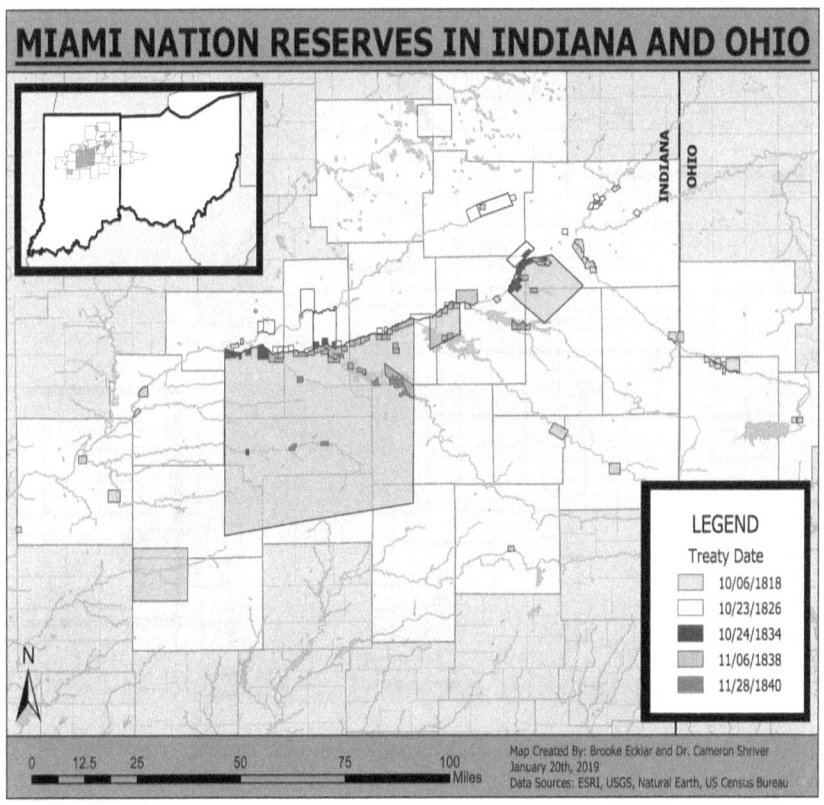

FIGURE 8.3. Miami Nation Reserves in Indiana and Ohio

postrevolutionary generations. Here, the sheer volume of the archive and associated land parcels requires digital tools. As figure 8.2 illustrates, between 1795 and 1840, most of the heritage homeland of the Miami people was transferred to US control for eventual sale to US citizens. But as figure 8.3 shows, treaties negotiated in 1818, 1826, 1834, 1838, and 1840 also resulted in the securing of legal title for the Miami Nation and almost seventy Myaamia individuals across over one hundred distinct reserves. For this time and place, Miami people retained recognized title both to *communal* and *individual* lands described in these five treaties.

In these treaties, Miamis transferred title from their aboriginal or recognized title to the ownership of Myaamia individuals, or the Miami community as a whole. Thus, these lands include both communal and individual lands described

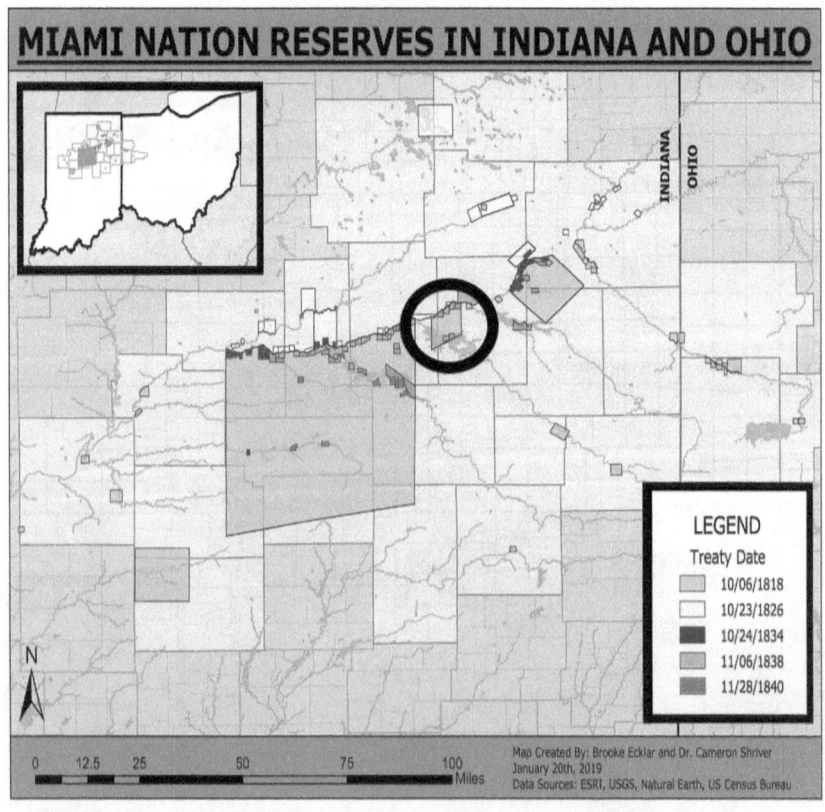

FIGURE 8.4. Six Mile Reserve. Described in the 1818 treaty: "From the cession aforesaid the following reservations, for the use of the Miami nation of Indians, shall be made. . . . One other reservation, of six miles square, on the Wabash river, below the forks thereof."

in treaties signed in 1818, 1826, 1834, 1838, and 1840. What I am calling "communal" lands were acknowledged by the United States in 1818 through the phrase: "reservations, for the use of the Miami nation of Indians," or "for the use of the said tribe," followed by boundary descriptions. These lands were held aboriginally (aboriginal title) but also were recognized as Miami through US documentation such as treaties (recognized title).[21] An example of a communal reservation was known colloquially as the Six Mile Reserve, acknowledged in the 1818 treaty. Based on research to date, all Myaamia communal reserves were ceded to the United States domain through treaties.[22]

Those that I have titled "individual" reserves were granted directly to individuals (sometimes multiple, named individuals as tenants in common). For

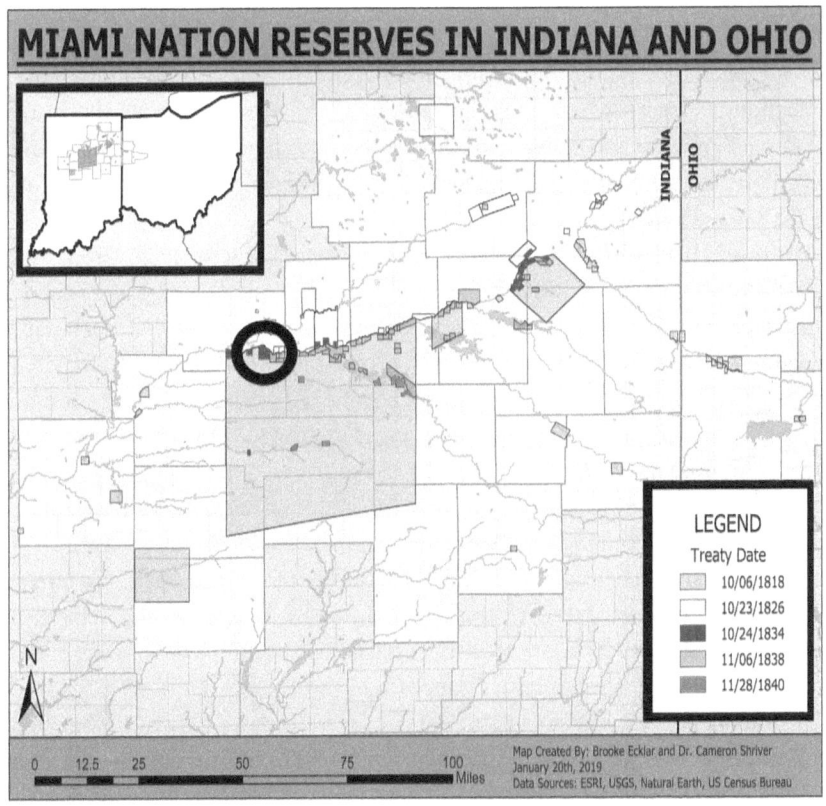

FIGURE 8.5. Mahkateemaankwa's Reserve, modern central Indiana. Described in the 1834 Treaty: "To Mac-keh-teh-maug-guaw, or Black Loon, one section of land to be located on the Wabash river, at the upper line of that part of the big reserve ceded by the first article of this treaty."

example: "To Mac-keh-teh-maug-guaw, or Black Loon, one section of land to be located on the Wabash river, at the upper line of that part of the big reserve ceded by the first article of this treaty." There were 129 individual reserves granted to Miami people in the treaties in Indiana and Ohio from 1818 to 1840, amounting to about 230,422 acres, or 360 square miles (plus three reserves never surveyed because the acreage was sold prior to survey).

Property after Treaties

Having mapped all allotments or individual reserves across the three homelands (Oklahoma n = 100, Kansas n = 300, Indiana/Ohio n = 109), the next, albeit more expansive, goal is to document all transactions of Myaamia reserves in all three geographic contexts. Simply put, we know the "where," but we do not yet know the "what" of possession or dispossession when it comes to transactions *after treaties*. Understandably, this is a multiyear endeavor, and to date, well over one thousand recorded sales have been located by our team, primarily in county recorders' offices and the National Archives in Washington, DC. Conclusions, therefore, will continue to change with new data digitally stored in the Aacim-wahkionkonci archive.

Unlike what the literature would suggest based on Choctaw, Creek, and Chickasaw reserve lands in the same era, Miami reserves did not become alienated rapidly, at least not most of them.[23] With the usual caveat that research is ongoing, current mapped data shows that into the 1880s (the limit of current research on land sales), Myaamia people had retained possession of parts of more than fifty original individual reserves granted in the 1818–1840 treaties in Indiana. (This does not count the allotments in Kansas, or those about to be created in Oklahoma in 1892.) About half of individual reserves were sold in the first decade after the granting treaty, but the other half were held, bequeathed, or otherwise transferred among Miami people for several generations, if not longer. Restricting our view to treaty cessions leaves Myaamia landownership in the past. Documenting a longer history challenges misguided notions that Native families, such as the Myaamia in Indiana, disappeared.

Myaamia participated, as a matter of course, in the tangled legal landscape of American property. The status of "Indian land"—now commonly called "Indian Country"—was important and debated by US policymakers beginning in the 1790s. Enshrined as federal law, treaties delineated (and continue to define) various types of landownership for Native Americans. Most Miami reserves acquired in 1818 and 1826 treaties "shall never be transferred by the said persons or their heirs, without the approbation of the President of the United States."[24]

Senators had decided after a similar treaty made in Ohio with Wyandots, Delawares, Ottawas, Potawatomis, Senecas, and Shawnees in 1817 that fee simple (fully owned and alienable at the owner's will) land grants to Indians was a step too far. Upon deliberation, the Senate, who had the power to ratify treaties, suggested that the 1817 treaty be modified. At the same time and place that Miamis were negotiating a separate treaty in 1818, the signatories of the 1817 treaty saw their property ownership revised. Their reserved acres would be held not as fee simple property but "held by them as Indian reservations have heretofore been held."[25] Property would flow through a federal bureaucracy. In turn, this bureaucracy created the bulk of the archive currently being processed by the Aacimwahkionkonci team. By the 1830s, the Myaamia treaties entailed Miami land in fee simple without restriction. Fee simple land did not require federal oversight as "Indian" land, therefore those archives remain at the county level rather than in Washington, DC.[26]

For two centuries, there have been a series of restrictions placed on Myaamia property rights, a theme consistent with federal oversight of Native American property in all forms. These encumbrances include requiring presidential approval for sale, requiring a twenty-five-year waiting period before sale, passing a "competency" interview before sale, or requiring the approval of the secretary of the interior to sell. Sometimes, the same parcel of land passed through several types of restrictions over time.

Why does this matter? Land status continues to be central in federal Indian law, shaping the local contours of American Indian sovereignty in the United States.[27] One of the difficult tasks the Aacimwahkionkonci programming team has accomplished is allowing title restrictions to be tracked by researchers. This feature, which might at first seem unimportant, represents one of the many ways we have attempted to plan for many users with multiple needs. Aside from understanding the past conditions of Miami landownership, American Indian tribes currently are restricted in land they can assert sovereignty over. If a tribal nation wants to add to its national domain, for example, it must go through a fee-to-trust application with the federal government.[28] Trust land—land entered into a trusteeship with the US federal government on behalf of a federally recognized American Indian nation—is where a tribal nation can exercise its national responsibilities, such as healthcare for citizens or police powers. The fee-to-trust process follows complicated legal distinctions set by federal legislation and case law. In short, understanding the legal history of specific parcels of land—for example, "was this parcel once part of a specific Miami citizen's restricted allotment?"—aids decision making among Miami Tribe leaders. The Aacimwahkionkonci Project is not built for lawyers or land consolidation planners, but we have attempted to not foreclose future uses of the archive.

Administration

Land transactions, and in particular Indian land transactions, were an essential component of the burgeoning administrative state. Myaamia land records sit beside and are interspersed with the land records of Citizen Potawatomi, Creek, Choctaw, Shawnee, Seminole, Peoria, and a whole range of Indigenous landowners in the national archives. The fact that Indian Affairs was a federal bureaucracy (the War Department, and then the Interior Department after 1849) meant that the usual apparatus of local land offices and county recorders was instead, or additionally, centralized to deal with transactions across huge geographic and temporal scales. Aacimwahkionkonci's database architecture allows for storage and discovery of a complicated series of datasets generated by this imperial nation-state.

While soldiers and missionaries animate colonization's stories, the humbler office clerk deserves attention as a category in itself. Myaamia landowners certainly tracked their own lands—Miami families have contributed their nineteenth-century land grants and property deeds to our archives. But the vast majority of bureaucrats were US citizens. At the top were federal officials. Indian Affairs was perhaps the most active federal bureaucracy throughout the early republic.[29] Its employees were paid through the War Department (until 1849), and aside from members of the military, the primary bureaucrats were Indian agents and Indian factory (trading house) workers who submitted periodic reports, tracked merchandise sales, and attempted to "civilize" local communities. The General Land Office also kept files on each restricted reserve, meaning that each of the Miami reserves which required presidential approval for sale had a central file in Washington, DC. In 1837 the General Land Office transferred those files to the Indian Office.[30] More locally, Miami treaties led to the creation of counties, so property transfers were recorded by county recorders and their clerks, confirmed by notaries, and adjudicated in circuit courts, frequently involving estate administrators, guardians, and public auction sales by the local sheriff. Estate decisions are often recorded in a county clerk's office, separate from the recorder's office. County recorders, if they knew the legal status of a Miami Indian reserve, sent land sale documents to the commissioner of Indian affairs, who forwarded them to a presidential secretary, where a president eventually signed the transfer, and his secretary mailed that decision back to the county recorder.

The resulting paperwork is significant. Let us take one example: "To the two eldest children of Peter Langlois, two sections of land, at a place formerly called Village du Puant, at the mouth of the river called Pauceaupichoux." This was a two-mile parcel, jointly owned by Elizabeth and Peter Longlois Jr., located adjacent to modern Lafayette, Indiana, which grew around and into the reserve. (The

Aacimwahkionkonci website will include a place names database that will, for example, allow a user to click on the "river called Pauceaupichoux" and learn that it is the modern Wildcat River, named partially after the Myaamia name referencing a story about a bobcat's stomach.) Because all land sales required presidential approval, all land sales eventually required correspondence with the secretary of war, a commissioner of Indian affairs, or an Interior Department bureaucrat. One real estate agent forwarded his card among the letters—William Levering, a property broker, notary, and administrator of estates, who was also appointed by a circuit court to divide the property of their heirs. The stack of papers is five inches tall, and when each transaction is mapped, the reserve looks like this (ca. 1870s):

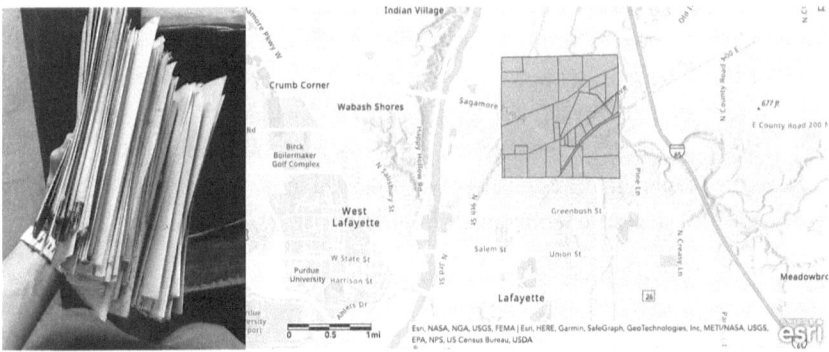

FIGURE 8.6. Two views of the Longlois Reserve. Left: Longlois Reserve packet of documents, National Archives, Washington, DC. Photo by the author. Right: Elizabeth and Peter Longlois Reserve, as parceled between 1818 and 1875.
Basemap sources: ESRI, HERE, Garmin, Intermap, increment P Corp., GEBCO, USGS, FAO, NPS, NRCAN, GeoBase, IGN, Kadaster NL, Ordnance Survey, ESRI Japan, METI, ESRI China (Hong Kong), © OpenStreetMap contributors, and the GIS User Community.

Following the letter of past treaties might have been difficult, but whether buyers or sellers consciously flouted restrictions or were simply ignorant is hard to know. One sale in 1853 was approved by Teddy Roosevelt in 1902, ending a half century of nonlegal settler occupation of the parcel.[31] The lack of official approval, as required by treaty, apparently did not affect "ownership" on the ground. Another sale, this one entered in the register of the local county recorder in 1827, was finally approved over a century later by Franklin Roosevelt in 1943.[32]

The Longlois Reserve in Tippecanoe County, Indiana, is an extreme illustration of a general issue with Indian "dispossession," which is that most of the Myaamia reserves were not sold by the initial owner. Most were transferred by the death of the initial owner and partially sold off over time, a drawn-out process in which Myaamia families continued their relationship, in property rights and otherwise, with their ancestral land. In some cases, Miami wills bequeathed specific parcels. Mihšihkinaahkwa bequeathed "all my estate both real and personal" to his only child.[33] James Godfroy left land to his mother; money to buy land to be given to a friend or relative named Pinšiwa; $200 for the "daughter of Revoy"; pacing horses and a mare sleigh and harness to his brother; and all the rest "divided in common amongst my father's family."[34] Josette Beaubien, when she passed away, transferred various parcels that her children were occupying, including a tract to Mary Ann and her husband; a tract to Margaret and her husband; and "to Joseph Robedoux, Twosign [Toussaint] Robedoux, and James Robedoux the several tracts of land in their possession each to keep and have the tract, they now possess."[35] In the absence of wills, Myaamia landownership followed state law; namely, property descended jointly to a surviving spouse or children of the deceased. In many cases, probate courts assigned administrators—who were sometimes Myaamia—to partition land held by multiple heirs for sale at auction.[36] As these patterns have continued into the twentieth century and even to the present, many extended Myaamia families have ended up living on or very near their ancestral family reserve lands.

Genealogy

It is frequently asserted that American Indian nations are preconstitutional, and that is correct. It is also generally understood that US colonialism—a process fundamentally enjoined to the American founding period—animates the legal, political, and territorial realities of those American Indian nations launched from the Constitution and treaties signed between nations. Genealogy is, I think, the critical nexus between how historical researchers and the public can use the Aacimwahkionkonci Project.

One example suggests the analytical and storytelling power of place-based genealogies. Jean Baptiste Richardville's portfolio illustrates a man with considerable wealth in silver, social and cultural capital, and land. He acquired the most land of any Myaamia person in the treaty years between 1818 and 1840—33.75 sections initially. A communal reservation of ten sections was later granted exclusively to Jean Baptiste Richardville via treaty agreement. It was at Wiipicahkionki, colloquially called the Ten Sections Reserve or the Forks of the Wabash, near modern Huntington, Indiana.

The Ten Sections Reserve at Wiipicahkionki illustrates the challenge that individuals can own multiple parcels of land, multiple individuals can own the same parcel of land, and ownership and boundaries change. We employ a method we are developing called "parcel genealogy," or the evolution and associations between parcels and their owners over time.[37] The Ten Sections Reserve was a Miami Indian Reservation created through the 1826 Treaty. Like all parcels, it is assigned a unique identification number, R052. Because it descends from Myaamia aboriginal title, its parent parcel is parcel 1, the Myaamia homeland. Jean Baptiste Richardville held "Indian title" or a paper grant to this reserve as a trustee for his nation. In 1834 Richardville, with apparent approval from his community, swapped "Indian title" for fee simple property ownership of the reserve. When he died in 1841, Jean Baptiste Richardville bequeathed his considerable property to his family. This meant parceling the Ten Sections property. For example, one half-section transferred to Ah-tah-pah-tah-ne-ah. She and her husband relinquished their 320-acre parcel to John Roche, an Irish-born merchant who often appeared as "guardian" to Myaamia minors and an executor of Myaamia estates, and frequently purchased or otherwise schemed for their land.[38] The property abutting the river, where the chief's impressive house had been built in 1834, was willed to his daughter Catherine Richardville Lafontaine. This property, descending from R052, is logged as R0520007. Also called Catees and Pakankihkwa, Catherine married Francis Lafontaine, a Myaamia chief, and they lived in the house on the property. In turn, in her 1848 will Catherine bequeathed the land to her son and daughter, Joseph and Archangel Lafontaine, and the Lafontaines (married with other Myaamia and non-Myaamia spouses) continued to own, worship, and live in the community. The database architecture allows each sale to be recorded as a distinct event, and each parcel subdivision to receive a new unique identifier, such as R0520070001, and so on through great-great-great-grandchildren of the original Ten Sections Reserve (R052).

Although no communal American Indian reservations were acceptable to state politicians or federal officials, the transition from "Indian" land to private property did not bar an extended Myaamia family from continuing to live at Wiipicahkionki until the present. The colonial archive is filled with instruments of land transfer; it is a part of the story of the place but an incomplete one. Those records tell us who bought and sold the land but are silent about why. On the other hand, the Lafontaine family maintains stories connected to the people and to the old house itself. Illicit deals and disputes linger in family memory, which help connect the stories of the land, and its people, to the real estate archive. Despite the current titleholder of the land being a nonprofit organization, Lafontaine family members (like the larger Miami Nation) continue to keep a relationship to the homestead where the river forks. Like all places in the United States, one could theoretically research the present chain of title back to aboriginal ownership. Every acre has its own genealogy. Like the genealogies of the Myaamia

FIGURE 8.7. Lafontaines at the house at Wiipicahkionki "Huntington, Indiana" in 1923. Archangel Lafontaine Engelmann "Tahkamwa" sits in the center, holding her great-grandson, Glenn Godfroy. Photo used courtesy of Sue Strass.

Nation itself, the genealogy of Myaamia land is an important framework of the past. An impersonal outline concerned with boundaries and prices and dates, it still requires us to make meaning of those people and places.

The Revolution set in motion a particular relationship between land, property, and the United States that continues to animate American Indian sovereignty. This is not unique to the Miami Tribe of Oklahoma. Perhaps, given further development and discussions with other Native nations, the tool could house the archives of their real estate as well. New media, including interactive mapping and archival metadata collection, helps us both historicize and tell new stories about something intrinsically old: Indigenous land.

NOTES

1. Rebecca Kugel, "Planning to Stay: Strategies to Remain in the Great Lakes, Post–War of 1812," *Middle West Review* 2, no. 2 (2016): 1–26, 2; Illuminative, https://illuminatives.org/reclaiming-native-truth/.

2. *Aacimwahkionkonci* "Stories from the Land" Project, https://mc.miamioh.edu/aacimwahkionkonci/. Aacimwahkionkonci was coined in 2019 in a brainstorming session by Daryl Baldwin. Aacim "to speak" + ahki "land, soil" + onk "locative suffix" + onci "to come from a place suffix." We use "Stories from the Land" and "Land of Stories" as typical denotations of the word. The author would like to acknowledge the contributions of the National Endowment for the Humanities, Humanities Collections and Reference Resources Grant, 2019 (PW-264006-19) for funding the development of the Aacimwahkionkonci software and database.

3. Current team members include Daryl Baldwin, project manager; Cam Shriver, lead historian; Doug Troy, technology development; Gabe Skidmore, computer engineering graduate student; Robbyn Abbit, GIS coordinator; Jessica Stoyko, GIS graduate student; Owen Larson, GIS graduate student; Doug Peconge and Julie Olds, cultural resources, Miami Tribe; Aaron Shrimplin, library adviser; Alia Wegner, library adviser; George Ironstrack, research team; Jonathan Fox, communications and publications; Carole Katz, graphic design; and Stella Beerman, transcriptions and metadata.

4. Daryl Baldwin, David J. Costa, and Douglas Troy, "Myaamiaataweenki eekincikoonihkiinki eeyoonki aapisaataweenki: A Miami Language Digital Tool for Language Reclamation," *Language Documentation and Conservation* 10 (2016): 394–410; Miami-Illinois Indigenous Languages Digital Archive, https://mc.miamioh.edu/ilda-myaamia/dictionary. This explanation of the programming aspects of the project was overseen by Doug Troy, the team's programming lead.

5. "A Circular Letter from His Excellency General Washington," June 8, 1783, Founders Online, https://founders.archives.gov/documents/Washington/99-01-02-11404.

6. For American territorial acquisition as a cause of the Revolution, see, for example, Woody Holton, *Forced Founders: Indians, Debtors, Slaves, and the Making of the American Revolution in Virginia* (Chapel Hill: University of North Carolina Press, 1999). For land transfer via treaties as a consequence of the Revolution, see, for example, Patrick Griffin, *American Leviathan: Empire, Nation, and Revolutionary Frontier* (New York: Hill and Wang, 2007); Eric Hinderaker, *Elusive Empires: Constructing Colonialism in the Ohio Valley, 1673–1800* (New York: Cambridge University Press, 1997); and Michael J. Witgen, *Seeing Red: Indigenous Land, American Expansion and the Political Economy of Plunder in North America* (Chapel Hill: University of North Carolina Press, 2022).

7. Colin G. Calloway, "Treaties and Treaty Making," *The Oxford Handbook of American Indian History*, ed. by Frederick E. Hoxie (New York: Oxford University Press, 2016), 539–52, 543; Gregory Evans Dowd, *War under Heaven: Pontiac, the Indian Nations, and the British Empire* (Baltimore: Johns Hopkins University Press, 2002), 177–79; Allan Greer, "Dispossession in a Commercial Idiom: From Indian Deeds to Land Cession Treaties," in *Contested Spaces of Early America*, ed. Juliana Barr and Edward Countryman (Philadelphia: University of Pennsylvania Press, 2014), 69–92; Stuart Banner, *How the Indians Lost Their Land: Law and Power on the Frontier* (Cambridge, MA: Harvard University Press, 2005); Colin G. Calloway, *Pen and Ink Witchcraft: Treaties and Treaty Making in American Indian History* (New York: Oxford University Press, 2013), 2–3. For limited legal pluralism in European colonialism, see Daniel K. Richter, "Intelligibility or Incommensurability?," in *Justice in a New World: Negotiating Legal Intelligibility in British, Iberian, and Indigenous America*, ed. Brian P. Owensby and Richard J. Ross (New York: New York University Press, 2018): 291–305; Jeffrey Glover, *Paper Sovereigns: Anglo-Native Treaties and the Law of Nations, 1604–1664* (Philadelphia: University of Pennsylvania Press, 2014); Daragh Grant, "The Treaty of Harford (1638): Reconsidering Jurisdiction in Southern New England," *William and Mary Quarterly* 72, no. 3 (2015): 461–98.

8. Calloway, *Pen and Ink Witchcraft*, 96–98; Leonard J. Sadosky, *Revolutionary Negotiations: Indians, Empires, and Diplomats in the Founding of America* (Charlottesville: University of Virginia Press, 2009); Eliga Gould, *Among the Powers of the Earth: The American Revolution and the Making of a New World Empire* (Cambridge, MA: Harvard University Press, 2012), esp. 12–13, 113–35; David M. Golove and Daniel J. Hulsebosch, "A Civilized Nation: The Early American Constitution, the Law of Nations, and the Pursuit of International Recognition," *New York University Law Review* 85, no. 4 (2010): 932–1066.

9. Henry Knox to George Washington, July 7, 1789, Founders Online, https://founders.archives.gov/documents/Washington/05-03-02-0067. On Native Americans as constitutional challenges, see Gregory Ablavsky, "The Savage Constitution," *Duke Law Journal* 63, no. 5 (2014): 999–1089. On early American Indian treaty making, see Dorothy V.

Jones, *License for Empire: Colonialism by Treaty in Early America* (Chicago: University of Chicago Press, 1982); Donna L. Akers, "Decolonizing the Master Narrative: Treaties and Other American Myths," *Wizazo Sa Review* 29, no. 1 (2014)): 58–76; Sadosky, *Revolutionary Negotiations*, 76, 121–38; Glover, *Paper Sovereigns*; Bethel Saler, *The Settlers' Empire: Colonialism and State Formation in America's Old Northwest* (Philadelphia: University of Pennsylvania Press, 2015), 29; and Calloway, *Pen and Ink Witchcraft*, 98–100. On territorial sovereignty, see Jeremy Adelman, "An Age of Imperial Revolutions," *American Historical Review* 113, no. 2 (2008): 324–25.

10. Stephen J. Rockwell, *Indian Affairs and the Administrative State in the Nineteenth Century* (New York: Cambridge University Press, 2010).

11. Allan Greer, *Property and Dispossession: Natives, Empires and Land in Early Modern North America* (New York: Cambridge University Press, 2018).

12. Henry Knox, "Statement Relative to the Frontiers Northwest of the Ohio," December 26, 1791, American State Papers, Indian Affairs, 1:197, Library of Congress.

13. Rowena Buell, ed., *The Memoirs of Rufus Putnam, and Certain Official Papers and Correspondence* (New York: Houghton, Mifflin, 1903), 257–61.

14. Andrew R. L. Cayton, "'Noble Actors' upon 'the Theatre of Honour': Power and Civility in the Treaty of Greenville," in *Contact Points: American Frontiers from the Mohawk Valley to the Mississippi, 1750–1830*, ed. Andrew R. L. Cayton and Fredricka J. Teute (Chapel Hill: University of North Carolina Press, 1998), 238.

15. The notable exceptions are John P. Bowes, *Land Too Good for Indians: Northern Indian Removal* (Norman: University of Oklahoma Press, 2016); Stephen Warren, *The Shawnees and Their Neighbors, 1795–1870* (Champaign, IL: University of Illinois Press, 2008); and Mary Stockwell, *The Other Trail of Tears: The Removal of the Ohio Indians* (Yardley, PA: Westholme, 2016). A vital subfield and methodology combining history and anthropology—called ethnohistory—emerged from investigations into these transfers in the 1950s. The Indian Claims Commission, established in 1946 and lasting into the late 1970s, dealt specifically with arbitrating Native land rights and caused pragmatic scholars, lawyers, and politicians to focus on documenting property and delineating between concepts of value, ownership, and groupness over time. For example, see Michael E. Harkin, "Ethnohistory's Ethnohistory: Creating a Discipline from the Ground Up," *Social Science History* 34, no. 2 (2010): 113–28.

16. Paul Gates asserts that a Choctaw Treaty of 1805 was the first, and there is one 8-mile reserve "for the use" of two Choctaw women, 7 Stat. 98 (1805). A Chickasaw Treaty in 1805 granted one reserve of one section to a chief for his use, 7 Stat. 89 (1805).

17. Oklahoma/Indian Territory experienced allotment via legislation distinct from the Dawes Act, although the rationale was consistent. The Miami Nation, along with the Osages, Peorias, Sacs and Foxes, Chickasaws, Cherokees, Muscogee Creeks, Seminoles, and Choctaws were exempted from the Dawes Act. Myaamia allotment in Indian Territory followed the Act of March 2, 1889, 25 Stat. 1013, "An act to provide for the allotment of land in severalty to United Peorias and Miamis in Indian Territory."

18. Mary Young contends that the 1854 treaties—which outlined allotment for the Miamis—was a resumption of policy after a lull in allotments. Mary Elizabeth Young, *Redskins, Ruffleshirts and Rednecks: Indian Allotments in Alabama and Mississippi, 1830–1860* (Norman: University of Oklahoma Press, 1961), 193.

19. This statement should not be read to mean that the Miami Nation controls no territory, or has given up claims to all land, or that citizens of the Miami Tribe of Oklahoma did not or do not own land in a US legal sense.

20. For example, Robert Lee, "Accounting for Conquest: The Price of the Louisiana Purchase of Indian Country," *Journal of American History* 103, no. 4 (2017): 921–42; Claudio Saunt and Sergio Bernardes, "The Invasion of America," https://usg.maps.arcgis.com/apps/webappviewer/index.html?id=eb6ca76e008543a89349ff2517db47e6.

21. Aboriginal and Recognized titles are explained in David E. Wilkins, *Hollow Justice: A History of Indigenous Claims in the United States* (New Haven, CT: Yale University Press, 2013).

22. For this chapter, I am leaving aside the significant and interesting case of the Meshingomesia Reserve, a more complicated history because it was allotted by legislation in the 1870s, and is thus an outlier among Miami reserves. Scott Shoemaker, "Trickster Skins: Narratives of Landscape, Representation, and the Miami Nation" (PhD diss., University of Minnesota, 2011). This does not mean that all land owned or controlled by the Miamis was ceded, only that reservations agreed to with the United States were ceded.

23. Cf. Paul W. Gates, "Indian Allotments Preceding the Dawes Act," in *The Frontier Challenge: Responses to the Trans-Mississippi West*, ed. John G. Clark (Lawrence: University Press of Kansas, 1971), 141–70; Young, *Redskins, Ruffleshirts, and Rednecks*.

24. 1818 Treaty, 7 Stat., 189.

25. Treaty with the Wyandot, etc., 7 Stat., 178, 1818. Agents also negotiated a treaty with Cherokees, which included the article that granted reservations of one section to each head of family that remained east of the Mississippi River, which would revert to fee simple when it passed to the head of family's children or heirs. 7 Stat., 156, 1818.

26. Unlike some of the southern treaties, Miamis did not need to reside on their land for five years to receive title. Cf. Young, *Redskins, Ruffleshirts, and Rednecks*, 29. Although the US federal government did not maintain land records for fee simple land, the Myaamia landowners continued in their legal status as Indians. See, e.g., *Wau-pe-man-qua v. Aldrich*, 28 F. 489 (1886); *Swimming Turtle, aka Oliver Godfroy v. The Board of County Commissioners of Miami County*, Civil Action S. 74-98, Northern District of Indiana (1977).

27. "Trust land" throughout Indian Country indicates land owned by the federal government as trustee for a Native individual or more often federally recognized nation. This includes Indian reservations and many allotted parcels.

28. US Department of the Interior, Office of Indian Affairs, Fee to Trust Land Acquisitions, https://www.bia.gov/bia/ots/fee-to-trust.

29. Rockwell, *Indian Affairs*; Claudio Saunt, *Unworthy Republic: The Dispossession of Native Americans and the Road to Indian Territory* (New York: W. W. Norton, 2020), 117–18.

30. Register of Cases in Reserve File A, Department of the Interior, Office of Indian Affairs, Land Division, Record Group 75, Records of the Bureau of Indian Affairs, National Archives, Washington, DC.

31. Monique Bondee Mittee to Louis Sims, April 9, 1853, Carroll County Recorder, book R: 58, Carroll County, Indiana.

32. Jean B. Richardville to Joseph Holman, October 17, 1827, Allen County Recorder, book A: 94–95, Allen County, Indiana.

33. Will of Me-she-nac-quah, August 17, 1839, Probate Will Record v. 1–2a, 1831–1875, 30, Allen County Probate Records, Allen County, Indiana.

34. Will of James Godfroy, January 27 1840, Probate Will Record v. 1–2a, 1831–1875, 30–31, Allen County Probate Records, Allen County, Indiana.

35. Will of Josette Beaubien, 1825, Allen County Recorder, book A: 211, Allen County, Indiana.

36. For example, see Jean B. Richardville's St. Mary's reserve, parceled in 1854 by probate court (Probate Order Book F: 127, Allen County, Indiana, available at https://archive.org/stream/abstractoftitleo00inds#page/24/mode/2up) and again in 1870 (Allen County Recorder, book 52: 54–55).

37. All credit goes to Robbyn Abbitt, GIS team lead and member of Miami University's Department of Geography, in translating this concept for our use.

38. Bert Anson, "John Roche—Pioneer Businessman," *Indiana Magazine of History* 55, no. 1 (1959): 47–58.

Part III
DATA AND DATABASES

9

(COUNTER-)REVOLUTIONARY DISCOURSE IN THE AGE OF REVOLUTIONS

Brad Rittenhouse, Christian Boylston, and Afshawn Lotfi

The purpose of this chapter is to investigate the rhetoric of revolution during the Age of Revolutions and the continuities and differences in that discourse in different contexts. To begin the investigation, we acquired a large corpus of texts from HathiTrust covering all works in English published between the years 1750 and 1875.[1] This original corpus represented 626,167 pieces of writing. The date range was selected to give about a twenty-five-year buffer on either side of what might strictly be considered the Age of Revolutions so that nascent and trailing revolutionary speech would be included. After cleaning and assembling metadata on the English-language corpus, we were left with about 165,000 works that we felt would be appropriate to include in the analysis, representing a total of approximately 50,000 authors.[2] "Appropriate" in this case meant that the works were out of copyright, we had exact dates of publication, and the OCR was acceptably clean.[3]

Because we cast such a wide net in corpus acquisition, this still left us with an exceptionally large object of study, encompassing works of fiction and nonfiction tracts, pamphlets, government documents, newspapers, and myriad other artifacts, which could yield insights into global trends in political discourse in the English language. For context, in the field of quantitative literary analysis, many major studies use corpora consisting of several dozen, several hundred, or, very rarely, several thousand works. See, for instance, the sixty-four-work corpus used by Dennis Yi Tenen in "Toward a Computational Archaeology of Fictional Space" (which uses the same corpus as "Extracting Social Networks from Literary Fiction," the 352-work corpus created and used by Mark Algee-Hewitt and Mark McGurl in "Between Canon and Corpus: Six Perspectives on 20th-Century

Novels," or the 1,327-work corpus used in "The Emotional Arcs of Stories are Dominated by Six Basic Shapes.")[4] There are some notable exceptions, particularly from interdisciplinary research teams and interlocutors from computer science and related fields, including Smitha Milli and David Bamman, but overwhelmingly, quantitative text analysis in the humanities has not often met the potential for scale in digital studies.[5]

Of course, that is not to say that bigger is better in this type of study: often large corpora prove unwieldy and invalidating errors can abound. In this case, though, the inclusion of so many works should provide insight into trends, conflicts, and continuities that may have been overlooked by human researchers, or allow us to discover discourse that may have been excluded or censored from the historical record, and thus was not included in a more curated corpus. Studies of large corpora like this, however, will never be comprehensive. A close reading of 165,000 documents could not be done in 3,000 pages, let alone the 30 allotted here. What we have done is use quantitative measures to find pertinent works in the corpus, explore those works more closely, and connect those works to larger trends that we have also measured quantitatively. We develop a larger argument around what we think is a surprising feature of the corpus, but there are many other equally valid stories that could be told about these works.

Process

Once the final corpus was assembled, we performed a series of Natural Language Processing (NLP), machine learning, and neural net processes on it.[6] Because the corpus represented nearly a billion sentences, we needed to establish a methodology simply to find revolutionary rhetoric within it. To better understand how revolutionary utterances within the corpus differ across time, we needed to extract the relevant segments of text for more specific analysis. We decided to make the unit for analysis a sentence containing any relevant keyword(s). The choice to bound the largest unit of analysis at the sentence level stems from the number of token limitations of text representation models we use further downstream.

To bootstrap our list of relevant revolutionary terminology, we started by finding the 20 most similar words to "revolutionary" as determined by a large pretrained FastText word-embedding model.[7] We repeated this process by again collecting the 20 most similar words to the first set of 20 yielding 400 words. Next, we filtered out all duplicate words and then manually removed irrelevant or modern terms such as "horsepower" or "nazi."[8] Using this generalized word vector similarity approach, we generated the following topically broad collection of fifty-one revolutionary terms ("revolution" and its fifty "closest neighbors"):

TABLE 9.1 Revolutionary terms

coups	insurrection	fascism	dictator
revolution	radical	nationalism	mutiny
authoritarianism	movements	uprisings	rebellions
totalitarianism	reformers	rebellion	repressive
anarchists	militant	revolutions	uprising
revolutionaries	dictatorship	revolution	protests
riots	leftism	demonstrations	dictatorships
tyranny	revolt	revolutionary	oppressive
capitalism	revolted	totalitarian	leftist
unrest	movement	repression	proletarian
revolts	coups	resistance	freedom
liberation	guerilla	regime	
upheavals	dictatorial	overthrow	

Subsequently, we extracted all sentences from our corpus that contained any of our revolutionary keywords. To account for the issue of misspellings that is all too common in OCR scans of physical books, we utilized fuzzy string matching to catch sentences containing keywords with slight misspellings or different variations of the keyword such as altered verb tenses or simply plural/singular versions of the keyword. This procedure resulted in the collection of just under eight million sentences from the billion in the corpus.

With the collection of "revolutionary" sentences complete, we sought to discover relationships among the sentences across time. To unearth these relationships, we embedded our extracted sentences utilizing a large, pretrained sentence BERT (Bidirectional Encoder Representations from Transformers) model. This model leverages recent advancements in NLP to represent sentences as points in a high-dimensional space such that semantically similar sentences represent similar vectors in this space. We will henceforth refer to these high-dimensional, numeric representations of sentences as "embeddings." Next, we applied a clustering algorithm, which grouped like points together based on their distance from one another in the semantic space.

After that, we generated a corresponding embedding for each sentence using a sentence-level implementation of Facebook AI's robustly optimized BERT approach (aka RoBERTa). The pretrained RoBERTa was chosen for its ease of use and near-state-of-the-art performance on the General Language Understanding Evaluation (GLUE) benchmark, beating out the standard BERT model. Using the most robust sentence-embedding model readily available, we hoped to be able to circumvent some of the topic extraction limitations of less sophisticated, word-level models like Latent Dirichlet Allocation (LDA) topic modeling.

Our embedding procedure resulted in 8 million sentence embeddings, each of which was of length 768. Having so many embeddings, each with a large number of dimensions (768), posed a problem for clustering algorithms that often take exponentially more time to run with increases in dimensions and embeddings. A common technique to mitigate this problem is using specific algorithms that reduce the dimensions of the embeddings while retaining as much information as possible, typically measured through variance. We struggled to reduce the dimensions of the embeddings while still retaining their expressiveness, so we resorted to other methods to speed up model training.

We decided it would be best to train the models on a stratified sampling of the data. We considered each work to be its own population of sentences. From each of these works we sampled 1 percent of the extracted sentences originating from the work. If a given work contained fewer than one hundred sentences containing one of our revolutionary terms, we randomly selected one of the revolutionary sentences from the work. Obviously, this is not a perfect stratified sampling, but we hoped that sampling this way would help us avoid cultivating a sampling dominated by a few works with many sentences meeting our extraction criteria.

Using this method on the reduced sentence-embedding dataset, we were able to run an iteration of k-means in a reasonable amount of time. K-means requires that we specify the number of clusters (k) that we want to find in our dataset. To determine the best k for our dataset, we calculated the average Silhouette Score for each k-means model with differing values of k up to k = 2000. The Silhouette Score essentially uses distance metrics to create a combined calculation of how similar points in a cluster are to each other and how different they are compared to points in another cluster.

This method yields a Silhouette Coefficient for each cluster and we averaged these scores to find the average Silhouette Coefficient for a model. We then used the model with the k value that maximized the Silhouette Coefficient of the model. Ideally, we would utilize other clustering algorithms, but given the size of our data and time limitations, we will leave more robust clustering methods for future work.

With our Silhouette Coefficient optimized, we chose a suggested k value of 300. We ran our entire sentence-embedding dataset through the model to classify each sentence into a cluster including those it was not trained on. Next, we utilized n-gram counts to aid in quantitatively interpreting the meaning of the clusters. Finally, we utilized this information, along with other metrics to be discussed shortly, to identify topically relevant clusters for further qualitative analysis and interpretation.

Results

Overview

Before giving an overview of the findings, it should be emphasized that working with data is messy, especially when that data is meant to represent something as multivalent and ambiguous as human communication. As such, we primarily use quantitative measures of our texts not as the basis for empirical arguments but rather to spark further human investigation.

We will begin with an overall description of the corpus, inputs, and outputs of the study. The number of texts acquired from HathiTrust per year increased at a near exponential rate, which can most likely be attributed to rapidly increasing publication rates in general over the course of this period (fig. 9.1). At first glance, the total number of sentences and total number of revolutionary sentences per year seem to follow a similar rate of change (figs. 9.2 and 9.3). However, by calculating the proportion of revolutionary sentences to all sentences, we can see

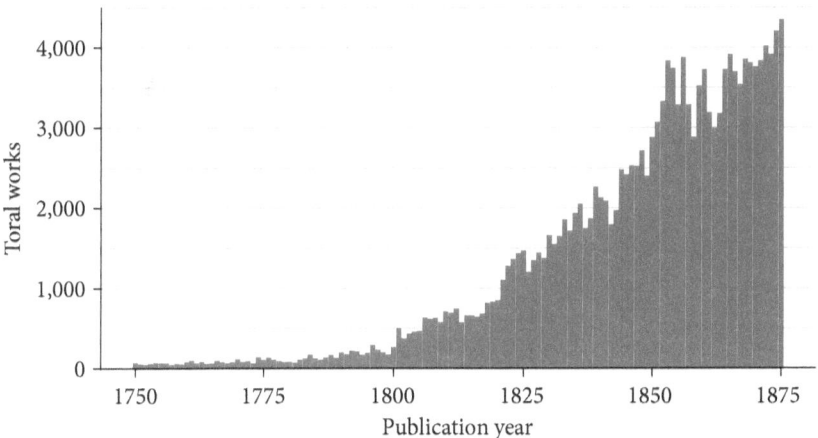

FIGURE 9.1.   Total acquired texts per year

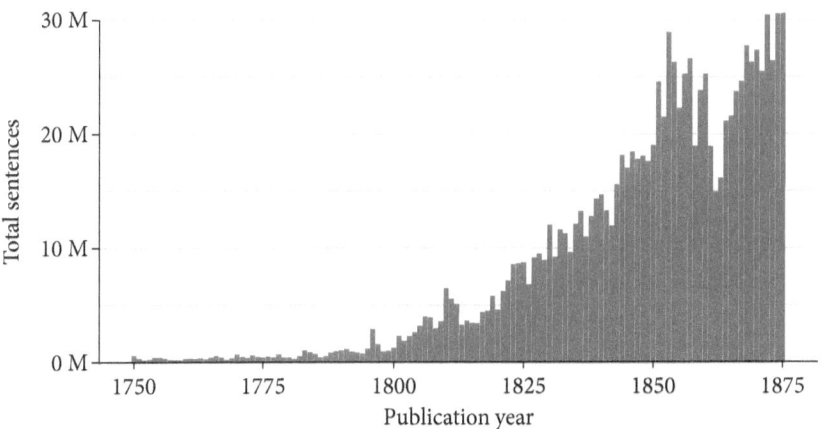

FIGURE 9.2.   Total sentences per year

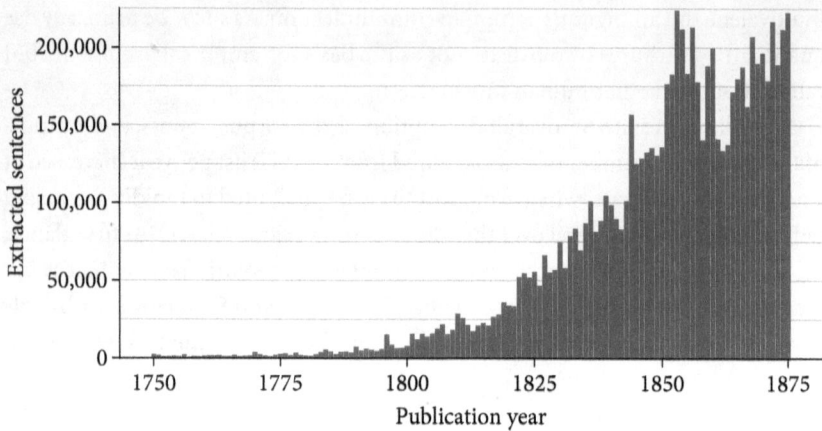

FIGURE 9.3. Total revolutionary sentences per year

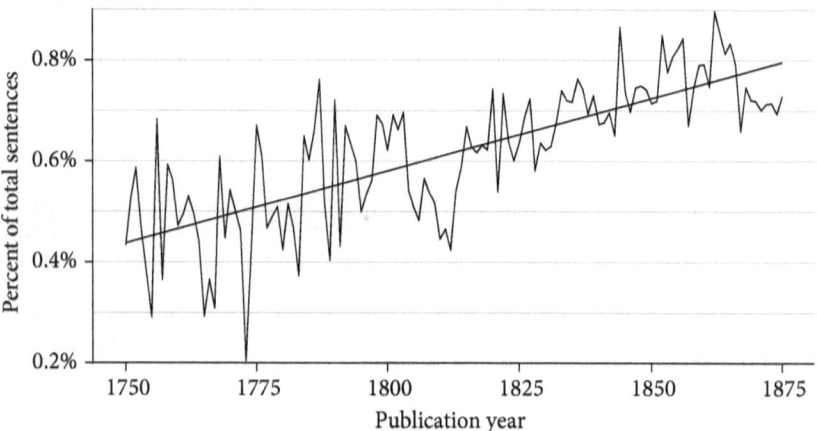

FIGURE 9.4. Percentage of revolutionary sentences over time

that percentage of revolutionary utterances rose more linearly than exponentially (fig. 9.4).

While a rather tiny portion of statements overall, it nonetheless seems that writers focused more and more on the topic of revolution as time progressed. Summary statistics of the regression model strongly confirm this, with a p-value of less than 2.2e-16 suggesting a very strong correlation between the passage of time and an increase in the proportion of statements about revolution.

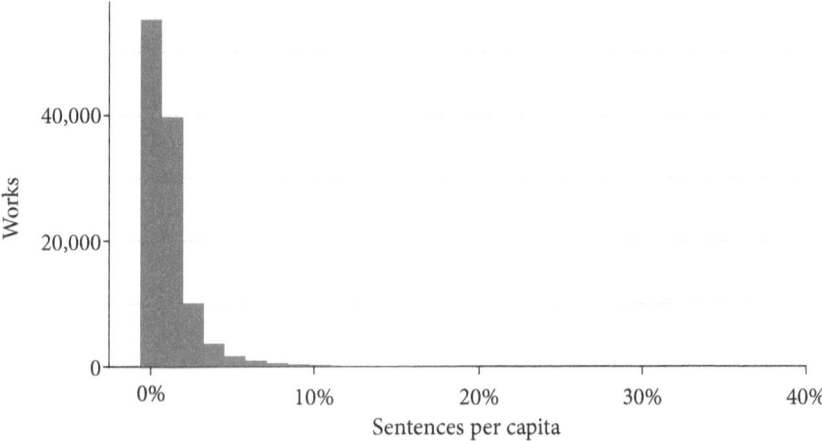

FIGURE 9.5. Percentage of sentences over time

But as we might expect from these low overall percentages, the distribution of revolutionary statements in the corpus trace a very long tail, meaning that the vast majority of texts in the corpus contain few or no references to revolution or similar topics, but there are also texts at the high extreme of the dataset: while it is hard to see, observations extend to the far right side of the graph (fig. 9.5).

In terms of authors and works, the most represented authors as measured by revolutionary statements were as shown in table 9.2. The general prominence of many of these authors suggests that the number of revolutionary utterances is not so much an indication of the radical nature of an individual but rather of how prolific or popular they were, and how often their works were published and republished. Much the same phenomenon can be observed in the works with the most revolutionary sentences (table 9.3).

Histories and reference works proliferate, which makes sense as they dominated printing presses during the period of study. The sole identifiable works of fiction are Walter Scott's *Waverley Novels*, which were both voluminous and wildly popular, factors, as we have discussed, that can bias these metrics. A final feature to note here is the inclusion of "Maryland Reports: containing cases adjudged in the Court of Appeals of that State" as the twenty-fifth most revolutionary work, a fact that will become important as we tweak our analytical measures to control for the simple volume of text produced by authors and works in the corpus. While one might expect government entities and bureaucracies to be able to produce an immense amount of text (and they do), a more consequential phenomenon also seems to be at work.

TABLE 9.2 Revolutionary sentences by author

AUTHOR	SENTENCES
Scott, Walter, Sir	53305
Macaulay, Thomas Babington Macaulay, Baron	43571
Hallam, Henry	35744
Gibbon, Edward	32763
Cooper, James Fenimore	25213
Irving, Washington	23254
Hume, David	23014
Shakespeare, William	21255
New York (State)	18845
Thiers, Adolphe	18832
Smith, William	18723
Chitty, Joseph	18176
Channing, William Ellery	17455
Byron, George Gordon Byron, Baron	17260
Robertson, William	17074
Motley, John Lothrop	16829
Prescott, William Hickling	16660
Goldsmith, Oliver	16359
Lossing, Benson John	15237
Alison, Archibald, Sir	15070
Milton, John	14666
James, G. P. R.	14421
Burke, Edmund	14063
Lytton, Edward Bulwer Lytton, Baron	13895
Maunder, Samuel	13663
Parsons, Theophilus	13503
Goodrich, Samuel G.	13270
Brougham and Vaux, Henry Brougham, Baron	13024
Story, Joseph	12378
Russell, William	12263
Dickens, Charles	12250
Blackstone, William	11948
Parton, James	11940
Story, Joseph	11936
Johnson, Samuel	11932
Coleridge, Samuel Taylor	11739
Abbott, John S. C.	11218
Mackintosh, James, Sir	11163
Frost, John	11052
Rose, Hugh James	11012
Ranke, Leopold von	10858
Taylor, W. C. (William Cooke)	10579
Pollard, Edward Alfred	10386
Milman, Henry Hart	10014
Mommsen, Theodor	9741
Carlyle, Thomas	9712
Plutarch	9666
Phillipps, S. M.	9582
Woodhouselee, Alexander Fraser Tytler, Lord	9575
Guizot, M. (Francois)	9501
Tocqueville, Alexis de	9371
Jenkins, John S.	9337

TABLE 9.3 Revolutionary sentences by work

WORK	SENTENCES
The living age	37900
The history of the decline and fall of the Roman empire	10583
The history of Rome	10445
The history of the French revolution	9633
The monthly magazine, or, British register	9413
The Encyclopaedia britannica: or, Dictionary of arts, sciences, and general literature	9270
New moral world	8616
A new general biographical dictionary	8386
Works	7886
The history of the decline and fall of the Roman Empire	7800
Encyclopaedia Britannica; or, A dictionary of arts, sciences, and miscellaneous literature, enlarged and improved	7377
The law of contracts	7350
The works of the English poets, from Chaucer to Cowper	7101
A history of England principally in the seventeenth century	6743
Johnson's (revised) universal cyclopaedia; a scientific and popular treasury of useful knowledge	6251
The cyclopaedia; or, Universal dictionary of arts, sciences, and literature	6245
Waverley novels	6147
The Edinburgh encyclopaedia	5940
The life of Napoleon Bonaparte, Emperor of the French. With a preliminary view of the French Revolution	5738
Chambers's encyclopaedia; a dictionary of universal knowledge	5677
The works of Jeremy Bentham	5644
Chambers's encyclopaedia: a dictionary of universal knowledge for the people	5529
Commentaries on equity jurisprudence: as administered in England and America	5493
Chamber's encyclopaedia: a dictionary of universal knowledge for the people	5375
Maryland reports: containing cases adjudged in the Court of Appeals of that State	5318

Revolutionary Focus and Its Necessity

In examining the most represented works and authors in the corpus, it became obvious that a more precise metric was needed. In other words, we needed to measure how apt a person or work is to write about revolution, rather than how much they do, since the latter can be a simple product of how much they write (or are published) in general. Table 9.4 shows authors with the highest percentage of total sentences using one of our revolutionary terms.[9]

This has the effect of devaluing the length of works and number of works published, and valuing the proportion of space an author or work dedicates to revolution. As Audre Lorde reminds us, there are "enormous differences in the material demands between poetry [or other short forms] and prose.... A room of one's own may be a necessity for writing prose, but so are reams of paper, a typewriter, and plenty of time."[10] Compared to the first measure of revolutionary

TABLE 9.4 Revised revolutionary sentences by author

AUTHOR	REVOLUTIONARY SENTENCES	TOTAL SENTENCES	PERCENT REVOLUTIONARY
Boyer, Benjamin M.	35	137	25.5
Alger, Edwin Alden	48	202	23.8
Tarbell, John P.	30	130	23.1
Republican party, Louisiana. State campaign committee.	18	87	20.7
Pomeroy, Samuel Clarke	37	180	20.6
Hutcheson, Robert	48	234	20.5
Carpenter, James W.	17	84	20.2
Herrick, Anson	59	293	20.1
Georgia Republican Association, Washington, D.C.	11	56	19.6
Wadsworth, William Henry	123	632	19.5
Stevens, Aaron Fletcher	60	311	19.3
Sprague, Achsa W. (Achsa White)	16	84	19
Davis, Thomas	186	1009	18.4
Greenwood, Alfred B.	34	188	18.1
Bennett, Hendley S.	37	205	18
Eastman, Alfred W.	44	244	18
Ewing, Thomas	66	366	18
Rankine, David	82	460	17.8
New York. General committee of Democratic Whig young men.	23	130	17.7
Fish, Hamilton	39	225	17.3
Sitgreaves, Charles	41	237	17.3
Schieffelin, Samuel Bradhurst	32	186	17.2
Knapp, A. L. (Anthony Lausett)	45	263	17.1
Democratic party. Wisconsin.	54	324	16.7
Pennsylvania Select Committee relative to the admission of Kansas into the Union.	17	102	16.7
Clark, Joseph	249	1505	16.5
Democratic party. Co. New Jersey. Gloucester	16	97	16.5
Rodgers, James H.	64	394	16.2
Wright, D.	38	234	16.2
Dunne, Henry C.	26	161	16.1

focus, this revision in our analytical framework should return more inclusive results that ignore the raw number of sentences one is able to write about revolution. And on the whole, we do find far fewer extremely prominent people of letters, suggesting that this measure of revolutionary focus privileges one's popularity and access to publishing far less: an encouraging sign.

In more closely researching these individuals, though, we found a preponderance of politicians and lawyers, citizens who, broadly speaking, had a vested interest in the status quo. The specific entities are largely different organizations and individuals from the first list, but they still represent the legitimizing classes of hegemonic power. Almost all the men on the list were US legislators;

Alfred B. Greenwood was both a US and Confederate legislator and an attorney; E. A. Alger plied the latter trade.

Ignoring them for a moment, though, the new metric did bring at least one differing voice to the forefront. Achsa Sprague, the twelfth-ranked entity on the list, was a female spiritualist and medium. What prompts an occultist's inclusion on a list of lawyers and politicians? Her sole work in the corpus is a poem called "I Still Live: A Poem for the Times," described in a biography by Leonard Twynham in the December 1941 *Proceedings of the Vermont Historical Society* as "a cry for freedom, a treatment of the contemporary scene, dedicated to hearts 'offering their lives at the shrine of liberty.' It is," he continues, "a moving didactic piece in pentameter couplets. It extols the names of Washington, Adams, Jefferson, and Webster; it refers to civil war, to 'a house divided against itself,' to the Union; it is an intense exhortation in behalf of emancipation in America, a vigorous denunciation of slavery and oppression."[11] The coincidence of the biography with the month of America's entry into the Second World War seems to weave a tidy thread through periods of great upheaval in the country's history. While Twynham likely would have been working on the biography prior to the Japanese invasion of Pearl Harbor, Americans were nonetheless thinking more and more about issues of patriotism, which Twynham identifies as a major theme of Sprague's poetry.[12] And while patriotism often serves as a thin veil for anti-revolutionary activity, "I Still Live" does prioritize self-sovereignty, a sentiment in line with her work as a women's rights advocate and abolitionist. Sprague, for instance, expresses solidarity with the Poles who, in the nineteenth century, fought a series of unsuccessful rebellions to preserve Polish sovereignty from the Prussians, Austrians, and Russians:[13]

> Though Poland fell while struggling with her chains,
> The love of freedom in her sons remains,
> Though other lands—in past and present hours
> Too weak to rise above the tyrant's powers—
> Yet every effort that the patriot gave,
> His country from the tyrant's hand to save,
> Has lived: though for his land has rung the knell,
> Its tolling struck anew great Freedom's bell.

Sprague, however, explicitly connects her and America's struggles for freedom and self-determination with those of other persecuted groups, an internationalist perspective on freedom, revolution, and emancipation that is very unique in this upper quadrant of the corpus.

Contrast this with the rhetoric of someone like Benjamin M. Boyer, the first-ranked entity on the list and a Democratic US House Representative from

Pennsylvania serving from 1865 to 1869. His only work in the corpus, 1869's *The Results of the Presidential Election*, as one may expect from the title, proves all too relevant to the United States today. Boyer's prose is perhaps most characterized by his frequent use of the word "radical" not to promote radical change or substantively dispute progressive policies but rather to paint nonreactionary policies as radical while at the same time normalizing reactionary points of view. As residents of Atlanta, we saw this tactic used extensively in the 2020 elections to discredit the "radical liberals" John Ossoff and Raphael Warnock. Boyer offers a series of rhetoricals in building his argument against racial justice in the South:

> But has the Radical policy of reconstruction itself been so approved and established that it can never be disturbed by future elections? Is there nothing to be apprehended from the continued violation of natural laws and a possible collision of races? Are the reconstruction laws themselves so firmly intrenched [sic] upon constitutional grounds that a general revulsion of feeling among the superior race might not find a ready excuse for sweeping from its foundations the whole work of Radical reconstruction? Radicalism has not itself been overscrupulous in the use of means. Usurpation is a dangerous game for any party to play if it would have its work outlast the passions from which it derived its power to tyrannize and proscribe.[14]

At the risk of flattening the historical record,[15] one can hear echoes of the rhetoric politicians developed to provoke and justify the January 6 insurrection at the US Capitol, stirring up ire against a cabal of usurping politicians denying the supposed will of the people. To be fair, contemporaries did critique this type of inflammatory speech, as in the anonymously penned 1863 document *A Few Words for Honest Pennsylvania Democrats*—an antislavery doctrine appearing on the most revolution-focused works list (table 9.5)—which documents some of the antidemocratic theories and actions being suggested to usurp federal power in the slavery question.

A focus of the document was William B. Reed's "favorite scheme, equally revolutionary in its tendencies, of erecting the banner of revolt against the United States by means of State conventions which would," as he puts it, "restore the Union, and by the same operation establish on a firm basis the rights of the States forever." The pamphlet is cogent in its critique of demagogues who, "with more or less openness, carry out the teachings of Calhoun and Jefferson Davis, and stimulate you to armed resistance to the Government, inflaming your passions by complaints of oppression, the falsity of which is best demonstrated by the freedom with which they are allowed to spout their incendiary harrangues [sic]." The writer also condemns aspiring elites like George Northrop, "another of the

TABLE 9.5 Revised revolutionary sentences by work

WORK	REVOLUTIONARY SENTENCES	TOTAL SENTENCES	PERCENT REVOLUTIONARY
Agreement between the city of Toronto and the Grand Trunk Railway Company of Canada	47	108	43.5
Agreement between the city of Toronto and the Trunk Railway Company of Canada	48	140	34.3
Half-pay to officers of the revolutionary war	3	9	33.3
An oration . . . in commemoration of the anniversary of national independence	22	77	28.6.3
Mexico: its present government, and its political parties	56	200	28
Letter from Horace Binney	7	27	25.9
The results of the presidential election	35	136	25.7
Lecture on the evils . . . that flow from the party divisions	51	200	25.5
Resolutions respecting the present war, and the causes leading thereto	7	28	25
Address of Democratic members of Congress to the Democracy of the United States	32	129	24.8
Speech of E. A. Alger, esq., delivered before the Democrats of Lowell	48	200	24
Oration delivered before the democratic citizens . . . of Middlesex at Groton	30	130	23.1
A popular exposition of the effect of forces applied to draught	82	359	22.8
Rules of the Circuit courts of the state of Michigan	40	177	22.6
A few words for honest Pennsylvania Democrats	45	204	22.1
The status of Rebel states	60	275	21.8
The churches of the Middle Ages	17	78	21.8
An address to the American people	17	78	21.8
The party of freedom and its candidates. The duty of the colored voter.	37	170	21.8
Revolution against free government not a right but a crime	104	480	21.7
Article from the New Orleans Bee of August 20, 1871	48	226	21.2
A defence of Republicanism	55	261	21.1
The opinions of old Jonathan Faneuil on modern politics in the Unites States	50	238	21

men who are endeavoring to rise from obscurity into prominence by luring you to destruction."[16]

The continuities here are frightening, especially when we reflect that January 6 represented perhaps the most serious internal threat to the Union since this pre–Civil War period 170 years earlier. Political upstarts luring the masses into life- and reputation-threatening action (one thinks of Stephen Ayres's testimony from the committee hearings) and legislative theories involving alternative state

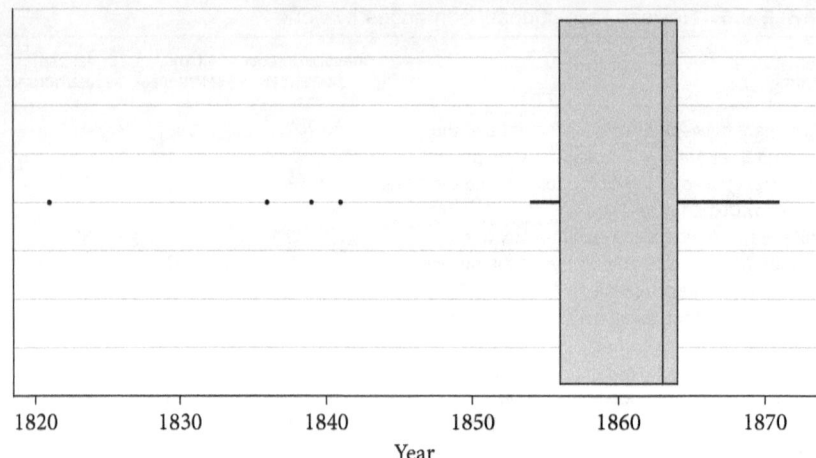

FIGURE 9.6. Box and whisker plot of publication years for the most Revolution-focused works

committees formed to exert one's extralegal desires (John Eastman). When we look at the publication dates of the documents in table 9.5, we see that the works in the corpora were clearly most concerned with this phenomenon at the time it was most relevant: when the American democracy last collapsed (figure 9.6). While we do not have quantitative evidence to link today's United States to this period, the semantic similarities and the strength with which these documents cohere to the Civil War period present frightening implications for the present.

Bleeding Kansas and Institutional Rhetoric on Revolution

The revolution-focused "Pennsylvania Select Committee relative to the admission of Kansas into the Union" points to a prominent vein of writing in the corpus related to "Bleeding Kansas," a series of violent voting disputes in Kansas over whether the state would be free or slave, famously including the caning of Charles Sumner and John Brown's revolt, among other incidents. There are over 350 works in the corpus mentioning "Kansas" in the title, nearly all of them discussing either the Kansas-Nebraska Act or Bleeding Kansas. While it would be hard to say exactly how large this subdiscourse is compared to others within the corpus, the further echoes it shares with the current day certainly warrant further examination.

Another juridical body in the corpus, the "House Committee to Investigate the Troubles in Kansas," represents the broadest accounting of this subject. All entries for this entity come from different publications of a house investigation into the violence in Kansas. While the violence seems to have been provoked by, according to the KSGenWeb Project, "thousands of pro-slavery Missourians [who] crossed the border and posing as Kansans, demanded the right to vote at gun point," and also "browbeat judges, stuffed ballot boxes, and otherwise transformed the election into a grim farce," there was also an abolitionist opposition being armed and supported by abolitionists across the country.[17] As a brief overview, in 1854, around two thousand antislavery settlers from New England began arriving in Kansas "to make that their home and . . . at all elections vote against the institution of slavery."[18] At the time of the vote, around one thousand proslavery Missourians crossed the border and engaged in a combination of illegally voting for slavery in Kansas (they were not settlers like the abolitionists) or (often violently) preventing the settlers and other antislavery Kansans from voting.

The House report includes both minority and majority accounts, so ideas about what exactly constituted revolution and revolt in this murky situation shift quite a bit throughout the report. The document consists of a sixty-seven page majority report and a forty-two page minority report, so that split should be kept in mind as we examine its discourse.[19]

The "House Committee to Investigate the Troubles in Kansas" would seem to be much more concerned with democracy and process than what one may expect from a "revolutionary" document. A quick glance at a word list for the document reveals the predominance of words like "party" (probably more of a legal than political sense in this case, though), "election," "state," "vote," "district," and so on. A bit further down the list, words like "brown" (John Brown), "resistance," and "night" hint at more subversive contexts. Closer examination of the extracted sentences reveals a text primarily focused on reporting on (and likely containing) rather than fomenting revolutionary acts.

Much of the document consists of testimonies by involved individuals who, unsurprisingly, inject their own biases into the proceedings. Matthew R. Walker, for instance, testified that "the people of Missouri acted upon the principle of self-defence, and to counteract the unusual and extraordinary movements which were being made at the north."[20]

The committee as a whole, though, found that "the settlers took but little interest in the election, not one-half of them voting. This may be accounted for from the fact that the settlements were scattered over a great extent, that the term of the delegate to be elected was short, and that the question of free or slave institutions was not generally regarded by them as distinctly at issue." As a result, they concluded, "under these circumstances, a systematic invasion from an adjoining State, by which large numbers of illegal votes were cast in remote and sparse settlements, for the avowed purpose of extending slavery into the

Territory, even though it did not change the result of the election, was a crime of great magnitude."[21]

This crime looked like a large-scale armed insurrection, an "unusual and extraordinary movement" on behalf of the proslavery Missourians, rather than the largely uninterested settlers:

> They came in wagons (of which there were over 100) and on horseback, under the command of Col. Samuel Young, of Boone county, Missouri, and Claiborne F. Jackson, of Missouri. They were armed with guns, rifles, pistols, and bowie-knives; and had tents, music, and flags with them. They brought with them two pieces of artillery, loaded with musket-balls. On their way to Lawrence some of them met Mr. N. B. Blanton, who had been appointed one of the judges of election by Gov. Reeder, and, after learning from him that he considered it his duty to demand an oath from them as to their place of residence, first attempted to bribe him, and then threatened him with hanging, in order to induce him to dispense with that oath. In consequence of these threats he did not appear at the polls the next morning to act as judge.[22]

In light of the January 6 insurrection, during which rhetoric about the rioters being "good people," "unfairly treated" compared to largely peaceful BLM and Antifa demonstrators, it is heartening to note that the revolutionary rhetoric in this document primarily refers to the subversion of voting laws in favor of slavery, and that the report, despite containing both majority and minority sections, almost exclusively treats the proslavery activists as in the wrong.[23] One testifier asserts, for instance, "I am well satisfied that there exists in this territory a secret military organization which has for its object among other things resistance to the laws by force," with another plainly stating "I think that the Missourians who came here came in consequence of counteracting the abolition movement of the north and those who voted with that understanding." Conversely, when John Brown, who would likely be painted with an "Antifa" brush today, is mentioned, it is predominantly to note his group's nonviolent and law-abiding behavior: one person reports "I did not hear any of Brown's party say that day that there were no laws in the territory," while another assures the committee that "Brown's party were not fighting at all."[24] Even as "law and order" has come to function over the twentieth century as code for the suppression of progressive, antiestablishment forces and/or people of color, in this Civil War–era report, at least, partisanship seems less at play and definitions more stable across the political spectrum.

That said, this investigation of revolutionary rhetoric challenges the results we were expecting to find, and instead confirms what we likely should have expected all along. In a corpus of texts that were selected to be published, were important enough to be collected in a university library, and were still on the shelves at least 125 years later to be digitized by Google Books and Hathi Trust, one is more likely to find talk of containing revolutions and other destabilizing forces than of progressive change to the status quo.

Cluster Analysis

So far, we have primarily focused on works that scored highly in a metric measuring how much they talked about revolution. While this is certainly an effective way of finding those authors who most concern themselves with revolution, it does not do much to systemically inform us of what they say about it, nor does it do much to explore the works beyond the extremes of the corpus.

A Note on Methods and Corresponding Results

For this portion of the study, we employed a corpus analysis methodology called term frequency-inverse document frequency (TF-IDF) analysis. The output of this process is a list of texts, words, and corresponding "scores," the latter of which tells one how statistically overrepresented a specific word is in a particular document (or cluster in this case) as compared to the corpus as a whole.

Usually, these scores would correspond to individual words in individual texts, but before deploying TF-IDF, we clustered the texts to improve readability of our results across such a large corpus. Texts were grouped together by a machine-learning algorithm based on similarity, and the "keyness" of terms to a specific cluster was then calculated cumulatively for that cluster rather than for individual texts contained therein.

Revolutionary Rhetoric over Time

One of the major phenomena we wanted to study with this analysis is the evolution of revolutionary rhetoric over time. We took a few approaches to combat the chronological skew of the inputs. First, given that all the clusters skewed toward later mean dates, we decided it would be prudent to individually examine the earlier clusters which otherwise do not weigh heavily on the results: no cluster with a mean publication date before 1845 exceeded sixty-six revolutionary sentences, and no clusters had a mean year earlier than 1800.

As one can see in table 9.6, the earliest clusters in the corpus are both not terribly early and exceedingly small given that the mean revolutionary sentence count of clusters (when clusters = 300) was around 26,000 sentences. Of the early clusters with double-digit sentence counts (a statistically arbitrary but more

TABLE 9.6 Earliest clusters

CLUSTER	MEAN YEAR	SENTENCES
Cluster 191	1800.67	3
Cluster 71	1801.00	1
Cluster 215	1808.44	9
Cluster 149	1822.87	23
Cluster 276	1824.07	28
Cluster 43	1826.00	5
Cluster 258	1828.00	3
Cluster 283	1830.00	1
Cluster 299	1833.75	8
Cluster 13	1834.00	2
Cluster 69	1835.17	18
Cluster 111	1835.25	4
Cluster 15	1835.40	5
Cluster 166	1835.50	26
Cluster 237	1835.90	21
Cluster 28	1836.75	4
Cluster 133	1837.56	9
Cluster 122	1837.67	3
Cluster 243	1838.29	17
Cluster 89	1838.33	3
Cluster 59	1839.40	47
Cluster 20	1840.25	4
Cluster 36	1840.58	66
Cluster 187	1841.16	57
Cluster 266	1841.95	19
Cluster 27	1843.19	27
Cluster 160	1843.53	30
Cluster 269	1843.56	25
Cluster 113	1843.67	33
Cluster 124	1844.90	10
Cluster 32	1845.00	3
Cluster 263	1845.73	90653

manageable number for individual examination), only about half were at least somewhat related to political revolution. Cluster 276, for instance, centered on printings and reprintings of Oliver Goldsmith's *History of Greece*.

Cluster 59 centered on eighteenth-century revolutions and rebellions in the Western world, in particular those in France, England, and states like New York and New Hampshire. The coherence of these geographically and linguistically disparate revolutions into a single cluster suggests the development of international discourses of revolution, and the interchange and adoption of discourse from one national context to another. Matthew Lockwood's *To Begin the World Over Again: How the American Revolution Devastated the Globe* traces the echoes of the US Revolution in the far reaches of the British Raj, Russia, and Spain, to name a few of the many locales he highlights. He warns, though, that "most previous accounts of the American Revolution have by and large restricted their attention to the thirteen colonies.... In so doing they have limited their focus to the stone that caused the splash rather than the waves and ripples that radiated out from its epicenter.... As compelling as these heroic stories are, limiting our focus to the familiar, even comforting, tales of the American Revolution not only skews our understanding of what was in fact a global crisis, but also molds our understanding of America's national history in dangerous ways."[25] The linking of these discourses from different national contexts suggests that, at least in the pamphlets, reports, and speeches that make up a large portion of this corpus, there was a significant interchange of rhetorics about revolution in different national contexts.

For example, nearly all the works in Cluster 59 are historical accounts of people and places involved in revolutions: memoirs, biographies, and histories of George III, Napoleon Bonaparte, England in general, and various of the American colonies. Furthermore, nearly all the works retrospectively discuss punishments or pardons for revolutionary acts. The clustering of these historiographical texts points not so much to a contemporary community of international revolutionaries (somewhat in the Andersonian sense) but rather to postrevolutionary discourse on the exercise of power after rebellion. While some rhetorics certainly found contemporary international audiences—the transference of "life, liberty and the pursuit of happiness" from the French "liberté, égalité, fraternité" comes to mind—the language of Cluster 59 overlaps specifically in its acknowledgment of state power and the role of pardons in the consolidation of said power during periods of instability.

In *Sanderson's Biography of the Signers of the Declaration of Independence*, for instance, John Sanderson circumspectly considers Howe's offer of pardon to the American rebels:

> They had tried a pretended spirit of reconciliation in the year 1776, when congress had deputed Dr. Franklin, Mr. Adams, and Mr. Rutledge, to meet Lord Howe, at the request of the latter. The palpable intention was to lessen the enthusiasm of the people in favour of liberty, and bias their sentiments against revolutionary principles, and not to come to an equitable accommodation: the commission of Lord Howe did not contain any other authority than that expressed in the act of parliament, which was that of granting pardons, with such exceptions as the commissioners might think proper to make, and of declaring America, or any part of it, to be in the king's peace, upon submission.[26]

While Sanderson was looking back on the American Revolution from just before the Civil War, in *The Right Way for Restoring the Late Rebel States to the Federal*

Union, Robert Ruffin Collier, Esq. considers pardons from a legal standpoint, this time with an eye toward restoring the American Union after that war: "It is a salutary power, with promptness, as a single executive may, to bring to quiet the turbulent in the act of their violence. But then nevertheless it is in this instance by the constitutional prescription, only and not beyond a promise of pardon and to that extent and in terms the presidential proclamations to persons in rebellion should be that if they will at once desist though convicted thereafter they shall be pardoned."[27]

Both accounts recognize the role of pardons in quelling rebellion but find themselves on opposite sides of the debate because of the hegemonic power exercised on themselves by the US state. Michael A. McDonnell suggests that early Revolutionary War historians "were all nationalists. They were all committed to the new nation and to promoting the unity they thought vital to it. So they rewrote colonial and Revolutionary history to promote the idea of a gradual, inevitable, and orderly evolution of a new nation based on a set of ideas."[28] Much like in the outliers explored earlier, in the clusters we also see a continual and concentrated effort on the part of historians, politicians, lawyers, and other institutionalists to reify nation-states in the face of resistance.

Raymond Williams's description of hegemony in *Marxism and Literature* reminds us that "it has continually to be renewed, recreated, defended, and modified. It is also continually resisted, limited, altered, challenged by pressures not at all its own ... [but] the decisive hegemonic function is to control or transform or even incorporate them.... Any hegemonic process must be especially alert and responsible to the alternatives and opposition which threaten its dominance. The reality of cultural process must then always include the efforts and contributions of those who are in one way or another outside or at the edge of the terms of the specific hegemony."[29]

While our initial expectations of the corpus may indeed have been faulty, it is perhaps the key takeaway of this study that a subcorpus systemically curated to focus on "revolutionary" texts ends up demonstrating the overwhelming and overriding written efforts to tamp down "alternative and oppositional initiatives and contributions."[30]

Clustering Coherence

Another key conclusion about the study and methodology can be illustrated by Cluster 113, which suggests the suitability and utility of BERT for discovering and identifying continuities in international discourses. While not very substantively focusing on revolution, nearly all extracts in this cluster come from historical summaries and historical passages in guidebooks, particularly for Paris, but also some for Italy. In particular, they describe objects of interest to tourists that had been moved in the past either to escape or as a result of revolutionary violence. For instance, "when in the phrenzy of the French Revolution many churches were reduced to ruins most of the monuments they contained were mutilated and many of them destroyed it was in the Musee des Monumens Francais those that escaped the general wreck were deposited" from *The Modern Voyager and Traveller through Europe, Asia, Africa, and America* or "during the revolution it was removed to the Church des Petits Peres then to the Gaurie de Virginie in the Palais Royal and lastly to the temporary structure in the Rue Feydeau where it is still held" from *The History of Paris from the Earliest Period to the Present Day: Containing a Description of Its Antiquities, Public Buildings, Civil, Religious, Scientific, and Commercial Institutions.*

While not substantive rhetoric around revolution, the cluster is nonetheless remarkable in its ability to identify a mundane generic formation of writing about revolution that humans would likely never conceive of or perceive without computational aid. It calls to mind the principles of Latour's Actor-Network Theory, emphasizing the importance of objects in human history, and the ways they intersect with, travel, and affect human history. The writers of this cluster's passages obviously had some sense of this importance, but very few would conceptualize it as an overarching trend in historical and travel writing.

Cluster 222 also demonstrates this, but in perhaps a more relevant context. Most of the works here are about revolution, but all of them are about trauma. In Johannes Von Müller's *An Universal History, in Twenty-Four Books*, he writes that "the remembrance of the civil wars and of the tyranny of Cromwell was not yet forgotten and was recalled with terror." In *Crests from the Ocean-World*, Alonzo Tripp suggests that "the terrible scenes of the revolution continually haunted his imagination." But also, in Emily Brontë's *Wuthering Heights*, Catherine reflects on Heathcliff's death: "I felt stunned by the awful event and my memory unavoidably returned to former times with a sort of oppressive sadness." A remarkable formation, connecting domestic tragedy with the tragedy of war based on a syntax of trauma. The old saying "Love is war" may hold more truth than we might think.

A quick look at Cluster 124 also points to an international discourse around resistance to hegemony. Many praise efforts against various Russian imperial aggressions, but others celebrate the "skiful [sic][31] and brave and ... gallant resistance" of the Danes to Lord Nelson in the Battle of Copenhagen of 1801, and Richard Cameron's resistance to Charles II's Anglican Church. While harder to find than texts shoring up the state, it is clear that some international discourses did exist around resistance to hegemony, and that people in far corners of the globe conceived of sovereignty in similar ways.

Chronological Bigram Analysis

While cluster analysis can be useful in pointing to macrotrends in a large corpus like this, clustering can also simultaneously obscure less prominent trends. Another methodology we used to get a better glimpse at eighteenth-century rhetoric was to abandon clusters in favor of a content analysis that focused on discourse by year. While this methodology can make it difficult to tell what texts were participating in similar discourses, it does become easier to trace dominant discourses over time. Specifically, rather than using clusters, which group texts thematically, we grouped texts by years, and used n-gram analysis (bigrams in this case) to get a better idea of the content for each year. Bigrams were chosen because they gave a better sense of text content: 1-grams can be a bit too granular, and trigrams are unwieldy because there are not a large quantity of three-plus word phrases.

While we produced a total of the top ten bigrams per year, perhaps the best way to examine dominant and salient discourses over time is to look at the most prominent bigram topics per year. As we can see from figure 9.6, these are remarkably stable, with only fifteen top bigrams across the 125-year span. "United States," perhaps unsurprisingly in an English language corpus, represents the most dominant bigram in "revolutionary" statements, both in magnitude and chronologically, first rising to prominence in 1786 and drowning out nearly all other bigrams from about 1850 on ("said party" and "one party" make four appearances total in the span between 1850 and 1875. "Revolutionary War" makes two appearances in 1824 and 1842, perhaps as that term became more widely used. "One party" competed steadily with "United States" in the late eighteenth and early nineteenth centuries but fell off abruptly in 1857, never to be seen again. Prior to 1800, more radical terms, and terms connected to specific revolutions dominated: "coup de," "French revolution," and different variations of "radical," perhaps as yet another sign that early in the Age of Revolutions, there was more talk of actual revolution until state rhetoric to reify power drowned it out.

Discussion

As scholars such as Louis Althusser, Noam Chomsky, and Benedict Anderson have suggested, in different contexts, the results of this study point to an overriding trend in revolutionary rhetoric, particularly in print, toward co-optation and governmental (if not hegemonic) consolidation. The written word has, especially as time has progressed, not been involved in the project of promoting revolution but in shoring up ideological state apparatuses. The chronological bigram analysis, for one, showed a discourse converging toward a Bakhtinian monologism solely focused on history's largest hegemony, the United States. While it is possible that a sizable portion of this speech is concerned with challenging power, the examination of revolutionary focus showed that discussion on these matters has often been dominated by institutions and institutionalists who, it would seem, never tire of talking about revolution with an eye toward preventing it. Though this study is not comprehensive—there is far too much text to make a conclusive argument about all of it—initial findings suggest that, at scale, progressive revolutionary rhetoric barely registers.

There are, however, also discursive formations among those resisting hegemony, ones that can both confirm and extend our human understanding of the continuities among revolutions on the international scale. As Michael McDonnell reminds us, historical accountings of revolution are often no more than war stories, fabrications meant to serve a purpose, often that of power.

Quantitative analysis does not change that, and as human interlocutors with the data, we must be careful not to reify the prior mistakes we have made as historians and citizens in constructing and propagating rhetorical formations around revolution. However, this type of study can help us to question the stories we have told by exposing us to new sources, revealing new connections, and simply allowing us to look at the data of history differently and at a broader scale.

NOTES

1. HathiTrust Digital Library," https://www.hathitrust.org.

2. This type of counting at scale is a bit fraught, as authors may be listed several times with slightly different metadata attached to their names. We tried to normalize names as much as possible before counting, but there may be authors counted multiple times such as "Paine, Thomas" and "Paine, Thomas [old catalog]," as an example.

3. Optical Character Recognition, the process by which a computer recognizes and translates text on a scanned page into plain text. While this process has improved dramatically in recent years, many common errors still occur during translation such as the letter *m* being transposed as *rn*, or even total failures that render entire works as collections of ASCII characters. Works with high error rates were excluded from the study.

4. Dennis Yi Tenen, "Toward a Computational Archaeology of Fictional Space," *New Literary History* 49, no. 1 (Winter 2018): 119–47; Nicholas Dames, David K. Elson, and Kathleen R. McKeown, "Extracting Social Networks from Literary Fiction," in *Proceedings of the 48th Annual Meeting of the Association for Computational Linguistics* (Uppsala, Sweden: Association for Computational Linguistics, 2010), 138–47; Mark Algee-Hewitt and Mark McGurl, *Between Canon and Corpus: Six Perspectives on 20th-Century Novels*, pamphlet 8 (Stanford, CA: Stanford University Literary Lab, 2015); Dilan Kiley, Christopher M. Danforth, Andrew J. Reagan, Lewis Mitchell, and Peter Sheridan Dodds, "The Emotional Arcs of Stories Are Dominated by Six Basic Shapes," *EPJ Data Science* 5, article 31 (2016): 1–12, 1.

5. Smitha Milli and David Bamman, "Beyond Canonical Texts: A Computational Analysis of Fanfiction," in *Proceedings of the 2016 Conference on Empirical Methods in*

Natural Language Processing (Austin, TX: Association for Computational Linguistics, 2016), 2048–53.

6. Natural Language Processing, a collection of processes by which humans use computational methods to analyze language.

7. Word embeddings, in general, plot words as vectors in n-dimensional space based on their syntactic similarity in that corpus. BERT is thought to approach semantic knowledge in its embeddings given certain advancements in its modeling.

8. Some other more modern words such as "totalitarianism," "authoritarianism," "fascist," "leftist," and "nationalism" did sneak into the list. Closer examination of the works in question suggests several reasons for this: misattributed publication dates, postpublication prefaces, a very small number of later works mistakenly included in the dataset, or the words may simply have been in wider use earlier than previously thought. Given the size of the corpus, we did not identify these issues prior to running the word embedding, but none of these words appear in subsequent analyses. Of the words in question, "nationalism" appears the most, though only in fewer than seventy of the nearly 8 million extracted revolution-concerned sentences (0.000875 percent). They are statistically insignificant given the size of the corpus.

9. For reference, these are the summary statistics of this measure: Min: 0.0000061, 1st Qu: 0.0034771, Median: 0.0075158, Mean: 0.0122937, 3rd Qu: 0.0151728, Max: 0.4351852.

10. Audre Lorde, "Age, Race, Class, Sex: Women Redefining Difference," in *Sister Outsider* (Trumansburg, NY: Crossing Press, 1984), 110–19, 111.

11. Leonard Twynham, "Achsa W. Sprague (1827–1862)," in *Proceedings of the Vermont Historical Society* 9, no. 4 (December 1841): 271–79, 274.

12. Twynham, "Achsa W. Sprague (1827–1862)."

13. Achsa W. Sprague, *I Still Live: A Poem for the Times* (Oswego, NY: Oliphant, 1862), 6.

14. Benjamin M. Boyer, *The Results of the Presidential Election* (Washington, DC: F. and J. Rives and G.A. Bailey, 1869), 4.

15. Here, I am thinking of Carolyn Dinshaw's discussion of alterity and mimesis in *Getting Medieval: Sexualities and Communities, Pre- and Postmodern* (Durham, NC: Duke University Press, 1999).

16. We were unable to find more information on Northrop. Anonymous, *A Few Words for Honest Pennsylvania Democrats* (Philadelphia: King and Baird, 1863).

17. "Troubles in Kansas," KSGenWeb, 1996, accessed August 18, 2020, http://www.ksgenweb.org/archives/troubles.html#cre.

18. United States Congress House Committee to Investigate the Troubles in Kansas, *Report of the Special Committee Appointed to Investigate the Troubles in Kansas, with the Views of the Minority of Said Committee*, Government Report (US House of Representatives, 1856).

19. As context, in the House of Representatives during the 34th Congress, the majority was the "Opposition Party," an alliance between former Whigs and nascent Republicans, with the minority being the Democrats who, at this juncture, were primarily proslavery.

20. United States Congress House Committee to Investigate the Troubles in Kansas, *Report of the Special Committee*.

21. United States Congress House Committee to Investigate the Troubles in Kansas, *Report of the Special Committee*.

22. United States Congress House Committee to Investigate the Troubles in Kansas, *Report of the Special Committee*.

23. This chapter was initially penned soon after January 6, 2020. While the January 6 Committee has gone far to correct some of this rhetoric, it was relatively widespread at the time of the insurrection. One should also note more partisan accounts in the corpus,

including William Addison Phillips's (described in the *Encyclopedia Americana* as "almost radically progressive in his views") *Conquest of Kansas by Missouri and Her Allies* or, conversely, Charles Sumner's speech "The Crime against Kansas." William A. Phillips, *The Conquest of Kansas, by Missouri and Her Allies: A History of the Troubles in Kansas, from the Passage of the Organic Act until the Close of July, 1856* (Boston: Phillips, Sampson, 1856); Charles Sumner, *The Crime Against Kansas: Speech of Hon. Charles Sumner, of Massachusetts, in the Senate of the United States, May 19, 1856* (New York: Greeley and McElrath, 1856).

24. United States Congress House Committee to Investigate the Troubles in Kansas, *Report of the Special Committee*.

25. Matthew Lockwood, *To Begin the World Over Again: How the American Revolution Devastated the Globe* (New Haven, CT: Yale University Press, 2019), 4.

26. John Sanderson, *Sanderson's Biography of the Signers to the Declaration of Independence*, rev. and ed. Robert T. Conrad (Philadelphia: Thomas, Cowperthwait, 1844), 174.

27. Robert Ruffin Collier, *The Right Way for Restoring the Late Rebel States to the Federal Union* (Petersburg, VA: A. F. Crutchfield, 1865), 88.

28. Michael A. McDonnell, "War Stories: Remembering and Forgetting the American Revolution," in *The American Revolution Reborn*, ed. Patrick Spero and Michael Zuckerman (Philadelphia: University of Pennsylvania Press, 2016), 9–28, 11.

29. Raymond Williams, *Marxism and Literature* (Oxford: Oxford University Press, 1977), 113.

30. Williams, *Marxism and Literature*, 114.

31. This is from the *Edinburgh Review*, so possibly Scots and not an error.

10

BY CONVERSATION WITH A LADY

Women's Correspondence Networks in the Founders Online Database

Maeve Kane

In the spring of 1759, the young lawyer John Adams wrote that "By Conversation with a Lady, and Tryals of her Temper, and by Inquiry of her Acquaintance, a Man may know, whether her Temper will suit him or not" before marrying.[1] Later that summer, he would meet the fifteen-year-old Abigail Smith, whom he would eventually marry. Together, their correspondence has helped shape the gendered memory of the founding generation from "remember the ladies" through recent scholarly and popular biographies of Abigail Adams and other founding mothers.[2] In many ways, both Abigail Adams and the men in her sphere were exceptional precisely because of her prominence in the Adams family correspondence, and the way archival collection practices in the Adams Papers have preserved a historical memory of her prominence.

The Founders Online database (https://founders.archives.gov/) has digitized this correspondence as well as the papers of George Washington, James Madison, Thomas Jefferson, Alexander Hamilton, Benjamin Franklin, and John Adams and other members of the Adams family, from the authoritative transcriptions originally published in print by the University of Virginia Press, Princeton University Press, Columbia University Press, Yale University Press, Harvard University Press, and the Massachusetts Historical Society. Examining the metadata of the more than 165,000 records of the Founders Online database written between 1730 and 1830, in this chapter I argue that women writers of the founding generation like Abigail Adams were more horizontally enmeshed in their correspondence networks than contemporary male writers were.[3] Women writers and their correspondence with male leaders have long been recognized as important parts of the revolutionary era, but digital network analysis helps reveal the growth of women's connections with one another in the early republic.

The scholarship of women's literacy, correspondence, and political activism in the early republic raises a question for the Founders Online data: How and when did women form networks that became foundational to the growing political activism of the antebellum era?[4] As Kate Davies has shown, women's correspondence networks began to flourish in the 1770s and 1780s, forming a previously unrecognized literary canon of women's letters between the likes of Mercy Otis Warren and Catherine Macauly (both of whom are poorly represented in Founders Online). Cassandra Good has argued that friendships between women and men were an important part of the revolutionary era and ethos before the rise of republican motherhood, the two-party system, and universal manhood suffrage inscribed cross-gender friendships with domestic rather than political meaning.[5] Susan Branson, Catherine Allgor, Cynthia Kierner, and others have shown that American women participated widely in political activity such as partisan rallies and political patronage during the revolutionary era before the "revolutionary backlash" a generation later, as Rosemarie Zagarri has asserted.[6] Mary Kelley and Jeanne Boydston have shown that revolutionary-era elite women created a space for themselves in civil society that was neither political nor domestic, where women could fashion a "civic self" and participate in political discourse despite legal exclusion from formal politics.[7] This place in civil society is the space in which women's letters in the Founders Online database flowed.

The networks formed by correspondence in the database largely conform to the patterns shown by previous scholars who have shown a growth in women's connections and political activity in the early republic, with a few significant departures. Although the networks of the Founders Online correspondence do show growth in cross-gender connections after the Revolution, women were more likely to have cross-gender correspondents and female correspondents than men were. Women's civil society included both women and men, but the barrier of formal, legally sanctioned political participation still formed a gendered barrier that made a male space of official state business that is apparent in the gendered structures of the correspondence networks.

The Founders Online networks are also most suggestive about women's political discourse and the foundations for later activism for what they do not show. Many of the women's networks discussed below have abruptly severed edges because of the archival practices that preserved these collections, and these jagged network edges are suggestive for the shape of both elite and non-elite women's networks in the revolutionary era. Martha Washington, for example, is notably underrepresented in the networks discussed here despite copious documentation of the "republican court" or salon in her account books.[8] As a primarily oral space, a face-to-face salon culture lacks archival documentation of the kind best suited for digital network analysis. Other, less elite women who entered political discourse in this era, like those examined by Lori Ginzberg, also likely shared primarily face-to-face networks that went undocumented in the archive like the elite letters examined here precisely because those networks were local and in person.[9]

The network created by the Founders Online metadata is a necessarily incomplete, changeable object of study for several reasons. First, the collected papers included in the published volumes and database are centered on the correspondence of Washington, Franklin, Jefferson, Hamilton, Madison, and John Adams. The documentation practices that created the correspondence, the archival practices that selected and preserved the documents, and the political and historical value placed on editing and later digitizing certain genres of correspondence all shaped the Founders Online corpus.[10] Second, the version of the corpus analyzed for this chapter also likely differs from the database the reader may consult in the future. The database includes "early access" draft transcriptions of many letters as well as the authoritative, edited transcriptions that appear in the print editions; these early access transcriptions are replaced with the authoritative transcription as each edited volume becomes available.[11] The analysis for this chapter relies on the relatively stable metadata information (date, author, and recipient/s) about each letter rather than the slightly more changeable letter contents.[12] Despite these limitations, the Founders Online database offers a window on both women's writing and the perception of women's writing during the revolutionary era. By comparing networks centered on individual women and their correspondents, then juxtaposing them against the networks of individual men who also sent or received similar numbers of letters in the database, this chapter examines the density of gendered connections within the overall network.

Men's correspondence networks tended to be less dense and primarily composed of other men, while women's correspondence networks were denser and of mixed gender. Further, men's correspondence networks remained similar in form and density before and after the Revolution. Women's networks before the Revolution were smaller, mostly male, and less dense, with a shift to more dense mixed-gender networks over the course of the Revolution. Women writers corresponded with more women, who were themselves connected to other women, while male writers primarily corresponded with only one or a few women who were not connected to others. This change in women's network structure lasted into the decades following the Revolution. The shift suggests a fundamental change in the role and perception of women writers by their female and male peers during the revolutionary era. Male writers remained enmeshed in primarily male networks, their ability to perceive women writers largely unchanged from before the Revolution, while women's networks grew larger and more intermixed with both women and men. Network analysis helps confirm the argument from previous scholarship that cross-gender friendships were an important facet of life in the early republic but also shows that the bar to women's formal political participation nevertheless created starkly gendered correspondence networks to which only a few exceptional women gained entry.

Data and Methodology

Data for this project was downloaded from the National Archive's Open Government repository (https://www.archives.gov/open/nhprc/dataset-founders-online) and cleaned to standardize spellings. Cleaning was as minimal as possible: for example, the spellings "Jacquelin Ambler," "Jaqueline Ambler," and "Jacquelin (Jaquelin) Ambler" in the transcriptions were merged into simply "Jacquelin Ambler." For the purpose of tracing network connections, letters with multiple authors or recipients were split into multiple records with only one author and one recipient each. For example, the letter "Sarah Read to Benjamin and Deborah Franklin, April 10, 1734,"[13] was split into two records: one with "author: Sarah Read" and "recipient: Benjamin Franklin," and the other with "author: Sarah Read" and "recipient: Deborah Franklin." Links were not made between multiple authors or recipients of a letter (in this case, Benjamin and Deborah Franklin). In most cases, multiple authors or recipients shared so many other links that the omission of these multiple authored letters was negligible to the final network. No effort was made to remove duplicate copies of a letter from the dataset, but number of connections was not a point of analysis for this study and so does not affect the results.

Gender of individuals in the network was assigned by first names or historical fact, or if ambiguous for a pseudonym, group of people, or unidentifiable, was marked N/A. These individuals with unassigned gender were not included in the totals for female and male correspondents in each network, but it is notable that female correspondents had nearly no correspondence with pseudonymous or corporate entities like "A Friend" or "The German Citizens of Philadelphia." The male writers in the database who received these pseudonymous and corporate-authored letters were public figures who held office. The absence of these pseudonymous letters in women's correspondence indicates that even prominent, elite women were not viewed as public figures who could be addressed by strangers or public appeals.

The Founder's Online database includes correspondence between more than 17,000 individuals, most of whom are connected to each other through correspondence with others. In this larger overall network, women correspondents were relatively insignificant. Only four women—Mary Cranch Smith, Louisa Catherine Johnson Adams, Abigail Amelia Smith Adams, and Abigail Smith Adams—number in the top fifty authors or recipients of letters in the database by number of letters sent or received. Sheer number of letters sent or received is not always a good indicator of connectedness, but sometimes it is. Occasionally, an individual who connects otherwise unconnected parts of a network but only corresponds with a few other people can be more influential than others who send letters to many people because they control the flow of information between parts of a network. This measure is called "betweenness" and calculates how often an individual is on the shortest path between themselves and any other member of the network. As measured by their betweenness, the four prominent Adams women are unimportant to the overall cohesion of the larger network because they did not appear often on the shortest path between themselves and others. If these women were removed, the overall structure of the network would remain largely unchanged.

However, in smaller "ego networks," the gendered experience of revolutionary correspondence becomes apparent. An ego network considers only an individual, the individuals directly connected to them, and the connections between all of them. In these networks that focused around individual women and their correspondents, the structure of their correspondent networks is very different from the structure of the larger Founders Online network and the ego networks of their male contemporaries. Considered separately from the larger, male-dominated network, women's and men's individual ego networks show starkly gendered differences in structure and change over time.

The women selected for this project were the fourteen with the most correspondents (rather than the greatest number of letters alone). Abigail Adams, with 214 correspondents in the database, was among the most prolific female and male correspondents, but Martha Washington, with only four correspondents, was among the top female correspondents in the database because there were so few women overall. The men chosen for comparison were identified primarily to give a representative range of ego networks similar in size to the women's networks. These men included the diplomat Arthur Lee, who had the closest number of letters and correspondents to Abigail Adams, whose network dwarfed almost all other writers in the network except John Adams, Washington, Jefferson, Hamilton, and Franklin. These major figures were excluded from consideration because, as collections of their personal papers, their correspondence is overrepresented in the database and their ego networks correspondingly dwarf all other networks. Other men were selected both for the size of their networks and for their similarities or connections to female figures; Philip Schuyler, John Payne Todd, and James Warren were chosen for their connections to their respective daughters, mothers, and wives.

Absences and Collection Practices

Abigail Adams and her daughter-in-law Louisa Catherine Adams are notable outliers in the database whose networks are significantly larger than most of their female and male contemporaries. Women connected to the Adams family are seven of the thirteen largest women's networks. These include Abigail and Louisa, spouses of presidents; Abigail Amelia Adams Smith, daughter of Abigail Adams; Elizabeth Smith Shaw Peabody, Caroline Smith De Windt, and Mary Smith Cranch, sisters of Abigail Adams; and Harriet Welsh, a distant Adams relation. The other major female networks in the database are those of Dolley Madison, Deborah Franklin, and Martha Washington, spouses of men whose personal collections form the database; Angelica Schuyler Church, the sister-in-law of another; and Ellen Coolidge, granddaughter of Thomas Jefferson. There are many other women in the database, but they are unconnected to others. Mercy Otis Warren is the only other female figure in the database who corresponded with more than one other person in the database; the only woman with a large network of her own who was not related by blood or marriage to one of the male figures whose papers form the database; and the only one whose correspondence comes from multiple collections rather than primarily a single collection like the Adams family papers.

The profiles of the women who appear to have their own large networks in the Founders Online database are due to the histories of the collections themselves. The collections were shaped on one level by the broader challenges of women's history, in which fewer women writers in the historical period translates to fewer letters by and to women in the database. Women like Elizabeth Schuyler Hamilton, wife of Alexander Hamilton, are notably absent from the Founders Online database: there are only thirty-one letters to Elizabeth Schuyler Hamilton in the database, all of them from Alexander Hamilton. Consequently, Elizabeth Schuyler Hamilton has been excluded from consideration here because she only had one correspondent in surviving letters. This dearth of letters is not because women did not send or receive letters. Although Elizabeth Schuyler Hamilton likely corresponded extensively with many other figures in the Founders Online collection, someone's experience of correspondence in life is not necessarily reflected in the archive for a variety of reasons. Elizabeth Schuyler Hamilton heavily edited, redacted, and destroyed her family's papers after her husband's death, and she likely destroyed many of her own letters and letters sent to her.[14] Whether from lack of production or deliberate omission, many women's correspondence is likely underrepresented.

Likewise, because the database was created from the papers of John Adams, George Washington, Benjamin Franklin, Thomas Jefferson, Alexander Hamilton, and James Madison, the correspondence of other contemporary figures is not completely represented. Philip Schuyler, the father of Elizabeth and Angelica Schuyler, has 444 letters with thirteen correspondents in the Founders Online database, making him one of the more minor male correspondents in the database. However, if one were to consider the Schuyler family correspondence held by the New York Public Library, Philip, Elizabeth, and Angelica Schuyler's networks would all likely look very different.[15] The Schuyler family collection at the New York Public Library includes more than 100,000 letters across forty years, but without a robust API (application programming interface) that includes machine-readable author/recipient metadata for individual letters, it is impossible as of this writing to compare how network structure for individuals varies across different archival collections.

Similarly, the diplomat Timothy Pickering's network from the Founders Online database includes no women, but the letters in his personal papers held by the Massachusetts Historical Society include copious letters to and from his wife, daughters, sisters, and female friends and relatives during his travels.[16] Pickering's Founders Online network represents his political and diplomatic work, from which his many female correspondents in his own personal papers have been omitted. As metadata for letter collections becomes increasingly available at scale, comparing the structure of archival networks across collections will be a productive direction for future research.[17]

Other men's correspondence networks are entirely male precisely because of their domestic interactions with women. John Parke Custis, son of Martha Washington and stepson of George Washington; John Payne Todd, son of Dolley Madison and stepson of James Madison; and John Barker Church, husband of Angelica Schuyler Church, all have small, entirely male correspondence networks and no correspondence with their more famous female relations. Custis died young, but the lack of women in all these men's networks point to the way the editors of the respective Washington, Madison, and Hamilton papers shaped the historical record and memory of women's involvement. Men whose papers were collected as part of the more familial-oriented John Adams collection, such as John Quincy Adams, Thomas Boyleston Adams, Charles Francis Adams, and George Washington Adams, have many more women in their correspondence networks, though still not as many as their female peers. In the more political affairs–oriented Washington, Madison, and Hamilton collections, the correspondence that Custis, Todd, and Church had with women may simply not have been included.

The six major collections of the papers of Washington, Adams, Jefferson, Franklin, Hamilton, and Madison had very different collection practices before being brought together as the Founders Online database, and these histories shaped the gendered structures of the networks now available. The Adams Family Papers at the Massachusetts Historical Society were collected and held by the Adams family until 1954, and include papers of the extended family through 1889.[18] The Washington, Jefferson, Hamilton, Madison, and Franklin Papers were all started between 1943 and 1953 as scholarly projects. Both the Jefferson and Hamilton paper collections only indexed routine and "domestic" correspondence, rather than including it in the collection, which significantly shaped the collections and the networks evident in them.[19]

The correspondence networks in the Founders Online database might best be considered an imperfect snapshot of how Adams, Washington, Franklin, Jefferson, and Madison (and their editors) understood their contemporaries. As a record of letters that entered the personal collections of these five major figures, the Founders Online networks are something of a proxy for what Adams, Washington, Franklin, Jefferson, and Madison saw and remembered about how the women and men in their spheres interacted. These imperfect snapshots were then further shaped by twentieth-century conceptions of gendered significance.

Gendered Network Structure

Structurally, the individual networks created by this edited correspondence have some similarities as well as one major gendered difference. Correspondence networks are formed when one person sends a letter to another. The directionality of this correspondence can affect some computational measures of network structure, such as measures of influence within the network, but because of the highly artificial construction of these networks, this directed measure of influence was not used in the current study. Women's networks were much smaller on average than men's networks, due to the factors considered above that exclude women's letters from the collection.

Very few individuals in the database had extensive networks in this collection before 1775. Of the women, only Deborah Franklin, Abigail Adams, and her sisters Elizabeth Smith Peabody and Mary Smith Cranch's correspondence shows an extensive network before 1775, and Deborah Franklin's was shared almost entirely with her husband. These early networks were small and sparse, without many connections between correspondents. Women's correspondence networks grew more extensive during the course of the Revolution. By the end of the war in 1783, both women and men's networks within the database grew larger and more extensive. Gendered differences between women's and men's networks began to appear at this point. The networks of Abigail Adams and her sisters had been small and sparse before 1775, and grew both larger and more dense over the course of the war, but not as dense as the networks of their male contemporaries.

Network density refers to the number of connections all individuals in a network have to all other individuals in a network. If all individuals in a network are directly connected to all possible other individuals, then the network is maximally dense (expressed as a decimal between 0 and 1, where 1 is maximally dense with every person in a network connected to every other person). A maximally dense network is easiest to achieve in a small network. If Benjamin Franklin sends a letter to Deborah Franklin, Deborah Franklin sends a letter to John Adams, and John Adams sends a letter to Benjamin Franklin, then that very small network of three people is maximally dense because each individual is directly connected to all other people in the network. Network density in this case reveals the connections between an individual's other correspondents and the degree to which those other correspondents were connected to one another. In a sparse network, if Benjamin Franklin sent a separate letter to each Deborah Franklin and John Adams, all individuals would be indirectly connected, but because they are not all connected directly to each other, the network density would be lower than in the first, maximally dense example.

This network density is a proxy for how well a person is integrated with their wider network. By 1783 the diplomat Arthur Lee had the densest network of any of the ego networks considered here, indicating two things. First, that he had spent much of the Revolution in London, Paris, Spain, and Prussia, communicating with his American contemporaries largely by letters that could be archived and captured by the database. Second, those he corresponded with were in communication with each other, making him densely enmeshed in revolutionary-era correspondence. In contrast, the sixteen-year-old John Quincy Adams had a very sparse network in 1783. During his travels in Europe as part of John Adams's diplomatic missions, the younger Adams had many ties to his aunts and female cousins, who were also densely connected to one another, but both he and his female relations were less well-connected to the young Adams's wider network.

As a young man, John Quincy Adams corresponded with influential figures like his father, who in turn corresponded with Washington, Hamilton, and Jefferson during the war years. John Quincy Adams's ego network was relatively low density because the women he corresponded with were not connected to the men he corresponded with, and he remained tied primarily to his own family members until he entered politics himself.

The connections of women like Abigail Adams or Mercy Otis Warren fall into more of a gray area. The Smith sisters—Abigail Adams, Mary Smith Cranch, and Elizabeth Smith Peabody—developed dense connections within the relatively small networks of their relatives and in-laws. Mercy Otis Warren, despite being one of the most prolific writers of the revolutionary period, had only a small, sparse network up until 1783 in the Founders Online database. Before 1783 women's networks in the database and individual women in other ego networks, such as the network of John Quincy Adams, had high clustering coefficients. Clustering coefficient refers to the tendency of individuals within larger networks to form small subnetworks that are denser than the rest of the network. For women's sparse networks with high clustering coefficients, this means that a network like Abigail Adams's was composed mainly of small clusters loosely connected to one another. Abigail Adams also had a high clustering coefficient in other networks in which she appeared, like Arthur Lee's, but some women like Martha Washington had very low clustering coefficients in otherwise very dense networks like Henry Knox's because she was otherwise unconnected to the larger network.

Prior to 1800 Abigail Adams and her sisters were the exception to the rule of women's small, sparse networks and low clustering coefficiency in men's networks, likely due to the collection practices of the Adams and other paper collections. After 1800 men's and women's networks begin to converge structurally, with the notable exception that many more women enter women's networks, but women remain isolated within men's networks and relatively unconnected. Women's networks become both larger and denser, and this was sustained through the later years of the Founders Online database to 1830. By 1800 all women's networks were composed of 20 percent or more women correspondents, with most women's networks made up of 30 percent or more women. Only two men's networks—those of John Quincy Adams and Charles Francis Adams, both sons of Abigail Adams and correspondents with their maternal aunts—had 20 percent women correspondents.

The profound absence of women in most networks of the revolutionary era is hinted at in Dolley Madison and Louisa Catherine Johnson Adams's networks. In the wider Founders Online database network, most individuals had a connection to at least two other people. The major figures whose personal papers compose the database—Washington, Adams, Jefferson, Madison, Franklin, and Hamilton—had many correspondents whose only connection was with them. That is, the six major figures either sent or received letters from many people who had no other documented connection with anyone else in the network. Part of this is because, as mentioned earlier, men like John Adams received letters from "A Friend of Justice," for example, just after the passage of the Alien and Sedition Acts, while women like Abigail Adams did not receive letters from pseudonymous correspondents. These pseudonymous letters or letters from individuals who were otherwise not involved in politics made up a large portion of Washington, Adams, Jefferson, Madison, Franklin, and Hamilton's papers because these men were public figures. But prominent men also received letters from non-anonymous individuals without other connections to the wider network, and most women also received a few of these as well.

Dolley Madison and Louisa Catherine Johnson Adams were the exceptions. Both they, and to a lesser extent Abigail Adams, corresponded with individuals who otherwise had no other connections within the network. Unlike major male figures—who received letters from otherwise unconnected individuals who were mostly male—female figures like Madison, Louisa Adams, and Abigail Adams mostly received disconnected letters from other women. Some of these are unsurprising—such as Louisa Johnson Adams's correspondence with her mother, Catherine Nuth Johnson, and her sister Nancy Johnson Hellen after her marriage—whose letters to others were not included in the Founders Online collection. However, others suggest a diplomatic correspondence conducted by first ladies that is invisible in the wider network. Louisa Johnson Adams corresponded with the Russian grand duchess Maria Pavlovna, the Russian ambassador Princess Dorothea Lieven, the exiled French baroness Anne Marguerite Hyde de Neuville, and Elizabeth Grenville Lady Carysfort in England during Adams's time as First Lady, suggesting the contours of a diplomatic correspondence otherwise made invisible in the gendered politics and collection practices of the Founders Online collections.

Dolley Madison's correspondence with women like Hannah Nicholson Gallatin, wife of the diplomat Albert Gallatin; Elizabeth Parke Custis Law, granddaughter of Martha Washington and wife of a prominent East India Company administrator; and Theodosia Burr, wife of Aaron Burr, also suggest a domestic diplomacy among leading women of the early republic that is not otherwise captured in the Founders Online database.[20] Dolley Madison's correspondence in the Dolley Madison Digital Edition (DMDE) shows much more extensive correspondence with other women.[21] The network from this collection has been excluded from consideration here because much of the DMDE correspondence dates from a later period than the Founders Online correspondence. However, the much greater volume and frequency of Madison's correspondence with women in the DMDE as well as the abruptly severed edges of her network in the Founders Online database help illustrate the artificiality of women's isolation in the Founders Online networks. Women's connections with one another and their male contemporaries grew ever denser in the early republic, and the edges of Adams's and Madison's archival networks are one hint at the historical density and reach of women's correspondence networks.

Although the structure of women's and men's networks varied somewhat, the most notable difference between them is who their contemporaries and later editors perceived women's and men's other correspondents to be. Before the Revolution, women's correspondence networks were smaller and sparser than men's, with fewer women. After the Revolution, women's networks became on average denser, with more connections between all correspondents and more women. Men's networks remained structurally similar before and after the Revolution with very few women.

After 1800 women's and men's networks converged structurally, although women's networks became denser. The one point where women and men's networks did not converge was the inclusion of women correspondents. Women were perceived to be connected to both women and men, but men were only perceived to be connected to men. Gendered collection practices—such as those that distinguish the Adams Papers from other collections in the Founders Online database—make the most significant difference to the presence of women in both women and men's correspondence networks. However, even within networks composed primarily of Adams Papers letters like those of Abigail Adams and John Quincy Adams, women's networks show a shift to greater density after the Revolution and a larger percentage of women correspondents. In men's networks that did include higher numbers of women correspondents, the women in them tended to be either isolated or clustered in family groupings, rarely perceived to be connected to or influential in men's wider networks. Women's networks with high numbers of isolated correspondents—such as those of First Ladies Louisa Johnson Adams and Dolley Madison—suggest the profound absence of significant portions of women's correspondence or face-to-face networks.

On one level, these gendered conclusions are obvious: women's archival records were long neglected because of women's perceived irrelevance to the historical study of diplomacy and politics, and these absences shape the historical narratives that can be told from archival collections.[22] However, in an era of increasing scholarly reliance on digitized collections due to both ease of access and limited travel because of budget constraints, lack of institutional access to subscription databases, or COVID-19 concerns, digitized collections and their limitations shape the contours of scholarship.[23] Putting aside the issue of digital versus traditional archival access, acknowledging and investigating the deliberate production of archival and historical absence can help us analyze its source and the politics of its creation, and provide new avenues for investigation.[24]

Absence and its meaning have historical significance.[25] Were a project like this one even able to incorporate the Schuyler family papers or the Dolley Madison papers for a fuller picture of those women's networks, multimodal, traditional scholarship working through multiple repositories is necessary to fully contextualize women like Mercy Otis Warren or Martha Washington, who appear as minor figures in the correspondence networks considered here, but who were deeply enmeshed in the politics of their time. This is less urgent for elite women like Warren and Washington, on whom such scholarship has been done, but perhaps more necessary for non-elite women seemingly on the fringes of the political networks considered here, as well as for reconsideration of men's networks. Women's historians already routinely search beyond a single male figure's papers for individual women's correspondence, but the sharply gendered divide in collection practices evident in the Founders Online database networks suggest further investigation is also needed in considerations of prominent men, whose connections with women and the historical memory of those connections has been artificially shaped.

Addendum

Interactive network visualizations for this project are available at maevekane.net/founders-online. These visualizations abstract individuals and their correspondence into nodes or dots (an individual person) and a single line between them representing one or many letters. The physical layout of these networks is not related to time, geography, or physical distance between correspondents; the distance between nodes in the network simulates gravity with push and pull between nodes determined by the frequency of their correspondence with each other and other nodes in a network. These nodes are sized by the number of individuals they personally corresponded with, and as the year slider moves forward and backward in time, the nodes are resized accordingly. Some nodes appear in multiple networks and are sized differently depending on how many individuals they corresponded with who also appear in those other networks. Hovering over a node highlights where it appears in all other networks.

TABLE 10.1 Comparison of women's and men's ego networks

	GENDER	NUMBER OF CORRES-PONDENTS	NUMBER OF WOMEN	PERCENT WOMEN CORRES-PONDENTS	INDIVIDUALS CONNECTED TO ONLY ONE OTHER INDIVIDUAL	NUMBER OF LINKS IN NETWORK	CLUS-TERING COEFFI-CIENT	NETWORK DENSITY
Adams, Abigail Smith	F	215	85	40%	40	20860	0.551	0.021
Adams, Louisa Catherine Johnson	F	72	18	25%	17	8358	0.517	0.045
Madison, Dolley	F	35	13	37%	13	3947	0.448	0.084
Smith, Abigail Amelia Adams	F	19	9	47%	0	5844	0.788	0.292
Peabody, Elizabeth	F	12	5	42%	0	4764	0.677	0.481
Franklin, Deborah	F	10	3	30%	2	523	0.66	0.209
Cranch, Mary	F	10	8	80%	1	1864	0.649	0.382
Warren, Mary Otis	F	7	2	29%	0	5546	0.763	0.679
De Windt, Caroline	F	7	2	29%	0	4764	0.677	0.481
Welsh, Harriet	F	5	3	60%	0	3288	0.839	0.833
Church, Angelica Schyler	F	5	1	20%	0	3151	0.781	0.667
Coolidge, Ellen	F	5	1	20%	1	3431	0.619	0.467
Washington, Martha	F	4	1	25%	0	2793	0.75	0.75
Lee, Arthur	M	201	4	2%	0	18451	0.867	0.027
Adams, John Quincy	M	93	20	22%	7	18118	0.688	0.064
Monroe, James	M	59	2	3%	10	12294	0.741	0.081
Knox, Henry	M	33	1	3%	0	15586	0.783	0.235
Pickering, Timothy	M	32	0	0%	2	12874	0.717	0.241
Smith, William Stephens	M	19	3	16%	0	9259	0.758	0.4
Warren, James	M	16	1	6%	2	5067	0.629	0.32
Gallatin, Albert	M	16	1	6%	3	6265	0.584	0.29
Morris, Gouveneur	M	13	0	0%	0	4570	0.603	0.346
Schuyler, Philip	M	12	0	0%	2	6733	0.449	0.295
Gates, Horatio	M	12	0	0%	0	7566	0.804	0.474
Rush, Benjamin	M	11	2	18%	0	9727	0.817	0.576
Randolph, Thomas Mann	M	10	1	10%	0	3034	0.634	0.309
Church, John Barker	M	9	0	0%	0	6237	0.662	0.433
Todd, John Payne	M	5	0	0%	1	2426	0.361	0.367
Custs, John Parke	M	3	0	0%	0	123	0.75	0.667

NOTES

1. John Adams diary 3, spring and summer 1759, 38, Adams Family Papers: An Electronic Archive, Massachusetts Historical Society, http://www.masshist.org/digitaladams/archive/doc?id=D3.

2. Woody Holton, *Abigail Adams* (New York: Simon and Schuster, 2009); Edith B. Gelles, *Portia: The World of Abigail Adams* (Bloomington: Indiana University Press, 1992); Richard N. Cote, *Strength and Honor: The Life of Dolley Madison* (Mt. Pleasant, SC: Corinthian Books, 2004); Martha Saxton, *The Widow Washington: The Life of Mary Washington* (New York: Farrar, Straus and Giroux, 2020); Jeanne E. Abrams, *First Ladies of the Republic: Martha Washington, Abigail Adams, Dolley Madison, and the Creation of an Iconic American Role* (New York: New York University Press, 2018); Patricia Brady, *Martha Washington: An American Life* (New York: Penguin Books, 2006); Betty Boyd Caroli, *First Ladies: From Martha Washington to Michelle Obama* (New York: Oxford University Press, 2010); Diane Jacobs, *Dear Abigail: The Intimate Lives and Revolutionary Ideas of Abigail Adams and Her Two Remarkable Sisters* (New York: Ballantine Books, 2014); Helen Bryan, *Martha Washington: First Lady of Liberty* (New York: Wiley, 2002); Flora Fraser, *The Washingtons: George and Martha, "Join'd by Friendship, Crown'd by Love"* (New York: Random House, 2015); Tilar J. Mazzeo, *Eliza Hamilton: The Extraordinary Life and Times of the Wife of Alexander Hamilton* (New York: Gallery Books, 2019); Erica Armstrong Dunbar, *Never Caught: The Washingtons' Relentless Pursuit of Their Runaway Slave, Ona Judge* (New York: 37 Ink, 2017); Edith B. Gelles, *Abigail Adams: A Writing Life* (New York: Routledge, 2002); Rosemarie Zagarri, *A Woman's Dilemma: Mercy Otis Warren and the American Revolution* (Malden, MA: Wiley/Blackwell, 2015).

3. The Founders Online database includes more than 183,000 records from between 1720 and 1840, but many of these records were not letters. Family baptismal records, receipts, accounts that had only an author and no recipient, and records without both an identifiable recipient and author were excluded from consideration here.

4. This question is raised by Rosemarie Zagarri in her review of Mary Kelley. Rosemarie Zagarri, "Politics and Civil Society: A Discussion of Mary Kelley's *Learning to Stand and Speak*," *Journal of the Early Republic* 28, no. 1 (2008): 61–73.

5. Kate Davies, *Catharine Macaulay and Mercy Otis Warren: The Revolutionary Atlantic and The Politics of Gender* (New York: Oxford University Press, 2005); Cassandra A. Good, *Founding Friendships: Friendships between Men and Women in the Early American Republic* (New York: Oxford University Press, 2015), 10–11.

6. Susan Branson, *These Fiery Frenchified Dames: Women and Political Culture in Early National Philadelphia* (Philadelphia: University of Pennsylvania Press, 2001); Catherine Allgor, *A Perfect Union: Dolley Madison and the Creation of the American Nation* (New York: Henry Holt, 2007); Cynthia A. Kierner, *Southern Women in Revolution, 1776–1800: Personal and Political Narratives* (Columbia: University of South Carolina Press, 1998); Susan Juster, *Disorderly Women: Sexual Politics and Evangelicalism in Revolutionary New England* (Ithaca, NY: Cornell University Press, 1996); Catherine A. Brekus, *Strangers and Pilgrims: Female Preaching in America, 1740–1845* (Chapel Hill: published by the Omohundro Institute of Early American History and Culture and the University of North Carolina Press, 1998), 18, 23–67; Sandra Gustafson, *Eloquence Is Power: Oratory and Performance in Early America* (Chapel Hill: published by the Omohundro Institute of Early American History and Culture and the University of North Carolina Press, 2000); Rosemarie Zagarri, *Revolutionary Backlash: Women and Politics in the Early American Republic* (Philadelphia: University of Pennsylvania Press, 2008).

7. Mary Kelley, *Learning to Stand and Speak: Women, Education, and Public Life in America's Republic* (Chapel Hill: published by the Omohundro Institute of Early American History and Culture and the University of North Carolina Press, 2008); Jeanne Boydston,

"Civilizing Selves: Public Structures and Private Lives in Mary Kelley's *Learning to Stand and Speak*," *Journal of the Early Republic* 28, no. 1 (2008): 47–60.

8. David Shields and Fredrika J. Teute, "The Republican Court and the Historiography of a Women's Domain in the Public Sphere," *Journal of the Early Republic* 35, no. 2 (2015): 169–83; Amy Hudson Henderson, "Material Matters: Reading the Chairs of the Republican Court," *Journal of the Early Republic* 35, no. 2 (2015): 287–94; Abrams, *First Ladies of the Republic*.

9. Lori D. Ginzberg, *Untidy Origins: A Story of Woman's Rights in Antebellum New York* (Chapel Hill: University of North Carolina Press, 2005).

10. Itza A. Carbajal and Michelle Caswell, "Critical Digital Archives: A Review from Archival Studies," *American Historical Review* 126, no. 3 (September 2021): 1107–8; Michel-Rolph Trouillot, *Silencing the Past: Power and the Production of History* (Boston: Beacon Press, 1995), 27; Elizabeth Edwards and Janice Hart, "Introduction: Photographs as Objects," in *Photographs Objects Histories: On the Materiality of Images*, ed. Elizabeth Edwards and Janice Hart (New York: Routledge, 2004), 1–15.

11. See "About Early Access Documents," Founders Online, https://founders.archives.gov/about/EarlyAccess. As of August 2022, 38,637 documents of 184,323 total were in "early access," about 20 percent of the total collection.

12. A version of the metadata analyzed for this chapter is available for download. Maeve Kane, "Founders Online Correspondence Metadata," *Magazine of Early American Datasets* (Philadelphia, PA: distributed by McNeil Center for Early American Studies, 2022), https://repository.upenn.edu/mead/55/.

13. Sarah Read to Benjamin and Deborah Franklin, April 10, 1734, Founders Online, http://founders.archives.gov/documents/Franklin/01-01-02-0110.

14. For the younger Alexander Hamilton discussing editing of his father's papers with his mother Elizabeth, see Alexander Hamilton to Elizabeth Schuyler, September 6, 1780, mss24612, box 1, reel 1, *Alexander Hamilton Papers: General Correspondence, 1734–1804*, Library of Congress, Washington, DC. I am deeply indebted here to my student Danielle Funicello for her research on the absence of Elizabeth and Angelica Schuyler's letters in her ongoing dissertation project at the University of Albany, "The Indelible Angelica Church: Recovering a Woman in the Age of Atlantic Revolutions."

15. Philip Schuyler's papers are digitized and available online through the New York Public Library, which as of July 2020 makes some metadata available via API, but not at the level of author/recipient needed to construct a correspondence network. Philip Schuyler Papers, New York Public Library Digital Collections, Manuscripts and Archives Division, New York Public Library, https://digitalcollections.nypl.org/collections/philip-schuyler-papers#/?tab=about.

16. This collection has not been digitized and is thus not currently available for large-scale social network analysis as of this writing. Timothy Pickering Papers, Massachusetts Historical Society, https://www.masshist.org/collection-guides/view/fa0256.

17. Projects like the Social Networks and Archival Context cooperative, which catalogs and makes available metadata across multiple institutions, may make it possible to bring together individuals' correspondence held in multiple repositories for a more comprehensive view of correspondence networks. However, the limitations of collection practices and gendered memory and collecting still shape network structure in these more complete networks. Social Networks and Archival Context, https://snaccooperative.org/.

18. Microfilm edition of the Adams family papers, Massachusetts Historical Society, http://www.masshist.org/collection-guides/view/fa0279.

19. "About the Project," Papers of Benjamin Franklin, Yale University, https://franklinpapers.yale.edu/about-project; "About the Papers of Alexander Hamilton," Founders Online, National Archives, https://founders.archives.gov/about/Hamilton; "The Papers of

Thomas Jefferson," Princeton University, https://jeffersonpapers.princeton.edu/; "Papers of James Madison," University of Virginia, https://pjm.as.virginia.edu/; "Project History," Washington Papers, University of Virginia, https://washingtonpapers.org/about/project-history-awards/.

20. Allgor, *Perfect Union*; Catherine Allgor, *Parlor Politics: In Which the Ladies of Washington Help Build a City and a Government* (Charlottesville: University of Virginia Press, 2002).

21. *Dolley Madison Digital Edition*, ed. Holly C. Shulman, University of Virginia Press, https://rotunda.upress.virginia.edu/dmde/default.xqy.

22. Michel-Rolph Trouillot, *Silencing the Past: Power and the Production of History* (Boston: Beacon Press, 1995), 15.

23. Daniel J. Cohen, Michael Frisch, Patrick Gallagher, Steven Mintz, Kirsten Sword, Amy Murrell Taylor, William G. Thomas III, and William J. Turkel, "Interchange: The Promise of Digital History," *Journal of American History* 95, no. 2 (September 2008): 452–91; Molly O'Hagan Hardy, "Archives-Based Digital Projects in Early America," *William and Mary Quarterly* 76, no. 3 (July 31, 2019): 451–76; Lara Putnam, "The Transnational and the Text-Searchable: Digitized Sources and the Shadows They Cast," *American Historical Review* 121, no. 2 (April 1, 2016): 377–402. The problem of discoverability as discussed by Putnam is perhaps especially acute in a post–COVID 19 research landscape.

24. Robert Proctor, "Agnotology: A Missing Term to Describe the Cultural Production of Ignorance (and Its Study)," in *Agnotology: The Making and Unmaking of Ignorance*, ed. Robert Proctor and Londa L. Schiebinger (Stanford, CA: Stanford University Press, 2008), 1–35; Marisa J. Fuentes, *Dispossessed Lives: Enslaved Women, Violence, and the Archive* (Philadelphia: University of Pennsylvania Press, 2016).

25. Ann Laura Stoler, *Along the Archival Grain: Epistemic Anxieties and Colonial Common Sense* (Princeton, NJ: Princeton University Press, 2010); Fuentes, *Dispossessed Lives*; Proctor and Schiebinger, *Agnotology*.

11

IDENTIFYING "A SLAVE"

The Iona University Text Analysis Project Explores a Mystifying Letter to Thomas Jefferson

Gary Berton, Michael Crowder, Lubomir Ivanov, Smiljana Petrovic

Among the large volume of official and private correspondence President Thomas Jefferson handled in the last months of his presidency, a letter dated November 30, 1808, reached the president's desk, signed pseudonymously "A Slave."[1] The eighteen-page letter overflowed with antislavery sentiment, taking Jefferson to task for the well-known contradiction between his philosophical critique of chattel bondage and his status as one of the young nation's most prominent and influential enslavers. Unamused, Jefferson scrawled a notation in the margin that dismissed the letter as a "rhapsody of inconsistencies."

While conducting research into letters sent by the American public to Jefferson during his presidential administrations between 1801 and 1809, Thomas N. Baker uncovered the letter by A Slave archived in the mid-twentieth century, somewhat misleadingly, in files unrelated to Jefferson's letters received from the American public. Baker subsequently published a lengthy research note in the *William and Mary Quarterly*, detailing tantalizing clues about the possible racial, political, and socioeconomic identity and background of the unknown author. Solving the identity of A Slave, Baker wrote, would be a difficult task given the letter's internal evidence and the absence of a conclusive handwriting comparison with a known author. Inspired by the mystifying problem of identifying A Slave, the history philanthropist Sid Lapidus offered encouragement to scholars willing to take on the unique challenge. The Lapidus Query presents a fascinating case study and opportunity to marry historians' archival training with text attribution software, and in this chapter we offer an interdisciplinary effort that hypothesizes a well-known, and potentially unconventional candidate for the letter's authorship: Thomas Paine.[2]

Utilizing the Text Analysis Project's computational tools, analysis of references deployed by A Slave, and the connections of the letter's references to both transatlantic revolutionary ideological currents and the internal politics of Democratic-Republican factions in Pennsylvania in the early republic, the authors conclude that Paine is a compelling authorial candidate of A Slave's missive to President Jefferson.

The authors' collaboration in this chapter is an effort to apply text attribution software developed by the Iona University Computer Science professors Smiljana Petrovic and Lubomir Ivanov, to a question of early nineteenth-century text attribution. We refer to this software package and accompanying methodology as the Text Analysis Project (TAP).[3] TAP, we suggest, has the potential to open up new vistas to humanities scholars reliant on painstaking archival research and teasing out meanings from (as in the case of A Slave's letter) frustratingly vague textual clues. It is not the authors' contention that TAP software must replace the research methodologies historians have previously utilized to explore questions of authorial attribution. Rather, it is our hope that TAP can provide a valuable tool to complement these approaches. The chapter begins with a brief explanation of Text Analysis computer science methodology as well as a description of the testing parameters of the TAP software. The next section presents an application of TAP research methodology, developed by Gary Berton, to the question of A Slave's identity. In that section, TAP methodology emphasizes three core considerations in applying TAP software to nineteenth-century author attribution questions: content, context, and chronology. These are utilized to support the hypothesis that Thomas Paine dictated the anonymous letter to his friend and companion Marguerite Brazier Bonneville (Madame Bonneville), then mailed it to President Jefferson.

The authors' hypothesis that Thomas Paine may have anonymously authored the letter by A Slave offers a potential historiographical insight into Paine's seeming ambivalence toward expressions of antislavery and abolitionist ideology. Given Paine's extensive record of publicly articulated stances against social and legal distinctions derived from monarchical power, aristocratic privilege, and structures of religious authority, in addition to the Enlightenment natural rights roots of his political philosophy, the absence of an extended piece of antislavery writing has long perplexed historians.[4] While not a piece of writing intended for publication and public consumption, if the authors' hypothesis is correct, A Slave's letter would constitute Paine's most extensive and direct expression of antislavery beliefs.

Text Analysis Methodology

Authorship attribution is the task of identifying the writer of a text whose authorship is unknown or disputed from among a set of candidate authors. Automatic authorship attribution is, essentially, a classification problem: given a list of possible authors and samples of their work, determine with a high degree of accuracy who among the candidate authors is most likely to be the author of the text in question.

Stylistic Features

Some authorship attribution applications analyze document content as an important indicator for classification.[5] However, in the case of historical document attribution, all candidate authors write about similar topics and use similar, topic-specific words and phrases. The machine-learning component of our work relies on stylistic features rather than the document content to carry out authorship attribution. Independently, the content is analyzed by our humanities colleagues, who look for historical facts and correlate them with the automated attribution findings.

Stylistic features are elements of a person's writing style and tend to be used unconsciously and relatively consistently. The frequency of use of stylistic features in known texts can be used to train machine-learning classifiers to recognize an individual's writing style. We extract stylistic and linguistic information from the collection of known works of each potential author and use it to generate models based on a variety of machine-learning classifiers as well as classifier ensembles. Once trained, the models can be applied to the document of unknown/disputed authorship to determine which candidate author's writing style most closely resembles the style of the text under consideration. The following surveys provide an overview of the field.[6]

A variety of stylistic features have been studied, and some have proven to consistently provide strong attribution results (function words, character- and word n-grams, part-of-speech (PoS) tags, PoS n-grams). Other features tend to perform well only for specific types of corpora and authors—sentence length, suffices, prepositions, vowel-initiated words, and rare words, to name a few examples. Lately, the most frequent words (MFW) feature has gained much prominence, particularly as part of the Stylo R package.[7] MFWs can be very useful when the amount of available text is large, but less so for smaller corpora. A new direction in investigating stylistic features for attribution is the use of prosodic features as stylistic markers for authorship. The role of lexical stress, alliteration, consonance, and assonance as stylistic features for authorship attribution has been thoroughly investigated.[8] Prosodic features appear to be moderately successful as stylistic markers, yielding strong attribution results when the number of candidate authors is relatively small.

In this work, we consider the seventeen stylistic features outlined in table 11.1.

TABLE 11.1 Features used in our analysis and their descriptions

STYLE MARKER	ABBREVIATION	DESCRIPTION
MW Function Words	MWFW	Function words as defined by Mosteller-Wallace in their Federalist Papers study (Mosteller and Wallace 1964)
Word n-grams	WG2	Sequences of n successive words from a text (in our case, n = 2)
Character n-grams	CG2, CG3	Sequences of n characters from a text (in our case, n is 2 or 3)
Part of Speech	POS	Nouns, verbs, prepositions, adjectives, etc. We use the Maxent Tagger developed by the Stanford NLP Group (Toutanova et al. 2003)
POS n-grams	POSG2, POSG3	Sequences of n parts-of-speech tags (n = 2 or 3) (e.g. "adjective noun", "noun verb", etc.)
First Word in Sentence	FWIS	The first word in each sentence
Prepositions	PREP	The prepositions occurring with the highest frequency in the text
Vowel initial Words	VIW	Words beginning with vowels
Suffices	SUF	The last three letters of every word
Coarse POS Tagger	CPOST	A simplification of the normal part-of-speech tagger, neutralizing minor variations such as plural inflection (singular/plural words are grouped)
Lexical Frequencies	LFREQ	Log-scaled frequencies of words from the general purpose HAL corpus as recorded in the English Lexicon Project (ELP) database (Balota et al. 2007)
Naming Reaction Times	NRT	Naming times from the ELP database; Each word is converted to the time it takes to name that word in the database (Balota et al. 2007)
Sorted Character n-grams	SCG2, SCG3	Alphabetically sorted characters in each n-gram (in our case, n = 2 or 3)
Word Stems	WS	Stems of the words obtained from Porter's stemming algorithm (Porter 1980)

Let us illustrate the use of three of these features—function words, word-n-grams, and character-n-grams:

- *Function words* are the most common connective words (articles, prepositions, pronouns, such as "to," "upon," "and," and so on) in the English language. Since they are topic-independent, they are usually excluded from the feature set of a topic-based text classification. However, since function words are often used in an unconscious manner, they reflect the author's style and are among the best features for authorship attribution. In this work, we used function words as defined by Mosteller-Wallace in their Federalist papers study.[9]
- Word n-grams consider sequences of n (n = 2, 3, etc.) words from a given text. For example, the word-2-grams of the text "Author Attribution of

Paine and His Contemporaries" are "Author Attribution," "Attribution of," "of Paine," and so on.
- Character-n-grams consider sequence of N characters from a given sequence of characters. For example, character-2-grams associated with the text "Author Attribution" are "au," "ut," "th," and so on.

Learning Methods

The second component of the attribution methodology is the choice of machine-learning methods. These algorithms consider the frequency of use of select stylistic features in each sample document in the corpus, and develop a model that can then be employed to tackle the issue of unknown/disputed authorship.

Numerous machine-learning methods exist and have been explored in the context of attribution:

> The *Linear Support Vector Machines* (LSVM) method seeks a hyperplane in the n-dimensional input space, such that the hyperplane best separates points corresponding to different candidate authors. The best separator is the hyperplane that maximizes the distance to the closest training data points of different authors. To attribute a disputed document, we evaluate on which side of the hyperplane the point corresponding to that document lies. We used two implementations, traditional Linear SVM and the more efficient Sequential Minimal Optimization (SMO).
> *Centroid Nearest-Neighbor* approaches represent each author by its centroid vector—a vector whose coordinates are averages of coordinates of all training instances. An unknown document is associated with the author with the nearest centroid. Distance can be measured using different metrics. In our work, we used cosine distance.
> The *Multilayer Perceptron* (MLP) is an algorithm that implements a backpropagation neural network. Inspired by brains, which are biological networks of neurons, artificial neural networks consist of interconnected layers of artificial neurons, referred to as perceptrons. Each perceptron layer receives input from the previous layer, calculating and passing its output to the next layer. During training, the output of the last layer is compared with the correct ("desired") output and any observed difference (error) is propagated backward through the layers, adjusting the perceptions' interconnection weights. Over repetitive passes through the training data, the error is gradually minimized as the system "learns."

Methodology Used in This Study

Our methodology is based on combining a stylistic feature with a supervised machine-learning method to create a so-called base classifier. After training each base classifier, a weighted average of classifier predictions is used to make the final authorship determination.[10] For each stylistic feature, we extract the fifty most frequent values of that feature from each document. We then take the union of these values from all documents and draw from that union the fifty most frequently used values in all the documents. These form a fifty-dimensional feature vector. The normalized vectors of frequencies of those stylistic feature values represent our training examples. For instance, if the selected vector of the most frequent function words is ("in," "our," "at," . . .), the vector of normalized frequencies of those words in Paine's "Forester Letters" (.0013, .00011, .00316, . . .) is labeled as "Paine" and considered one training instance. The labeled vectors of all documents of known authorship represent training data for one experiment.

To attribute a document of unknown or disputed authorship, a normalized frequency vector of that document is analyzed with respect to the training vectors. Based on these vectors, the machine-learning algorithms determine the most likely author of the unattributed text. Each learning method has a different approach to selecting the most likely author. Thus, different learning classifiers may produce different attributions.

Evaluating the Accuracy of Attribution

To evaluate a base classifier (stylistic-feature/learning-method pair), we adopt "leave-one-out" testing: *n-1* of the available *n* documents are used for training, and testing is carried out on the single remaining document. This procedure is repeated *n* times, in such a way that every document is used for testing exactly once. As a result, for each document, each base classifier selects an author based on its learning from the remaining (*n-1*) documents. We record the accuracy (percentage of correctly classified documents) of each base classifier.

Choosing and Combining Classifiers

To further improve performance, we use a weighted sum of supports of different base classifiers for different authors. We implemented a voting procedure where each base classifier votes for the author it selected. The approach is supported by the work of Marquis de Condorcet, who established in his Jury Theorem that, for independent and competent voters, the accuracy of the majority vote improves as the number of voters grows.[11] Each method independently makes a choice ("supports" one author). Each base classifier assigns to its top-choice author the support proportional to its accuracy in leave-one-out testing (see equation 1).

$$support_{classifier}(Author) = \begin{cases} accuracy_{classifier} & \text{if classifier selects Author} \\ 0 & \text{otherwise} \end{cases} \quad (1)$$

Only the base classifiers whose accuracies are larger than the average accuracy of all base classifiers are kept for further consideration (we refer to them as *accurate classifiers* in equation 2). The *weighted sum* method associates with each author the normalized sum of weights of all base classifiers that selected that author as their top choice (see equation 2).

$$support(Author) = \frac{\sum_{accurate\ classifiers} support_{classifier}(Author)}{\sum_{accurate\ classifiers, Authors} support_{classifier}(Author)} \quad (2)$$

The author with the largest weighted sum is declared as the recommendation of the weighted sum method. In our experiments, the weighted sum usually outperformed any individual base classifier.

Our ongoing research further considers different procedures that aggregate the predictions of multiple classifiers. For example, the Jury Theorem guarantees improved accuracy when the number of independent, equally accurate voters increases. Our experiments demonstrate that base classifiers are not independent. To reduce the advisers' correlations and improve the jury's accuracy, we eliminate voters that agree on a wrong choice with some regularity.[12]

Another direction we are pursuing is considering the predictions of the base classifiers as a learning data on a "meta-learning" level. Instead of aggregating predictions made by the base classifiers using a voting procedure, we treat these predictions as input data and infer the correct authors using machine-learning algorithms.[13]

Analysis of the A Slave Letter: Thomas Paine as a Possible Candidate

In the corpora of late eighteenth- and nineteenth-century American and British abolitionist and antislavery writings, the language, argumentation, and cultural and intellectual wellsprings employed by A Slave in the letter to Thomas Jefferson stand out for their militant abolitionism and vituperation of Jefferson, prophet of liberty and enslaver of fellow humans. The most obvious question concerning the identity of A Slave, one that the historian Thomas N. Baker notes Jefferson seems not to have considered, was whether A Slave was in fact a slave, or a Black freed person, or perhaps even a white abolitionist. Unusually for contemporary Black abolitionist writings, A Slave made little use of scripture or antislavery convictions rooted in Protestant Christian theology, primarily hewing to secular moral and political arguments against the horrors of American chattel slavery and the transatlantic slave trade. White abolitionists, moreover, rarely used such militant language in their published works.

Rather than opening with a reference to biblical scripture, common to both Black and white antislavery literature of the period, A Slave began with a reference to the Democratic-Republican journalist and printer William Duane's *Politics for American Farmers*, which Duane intended to inspire an American revolutionary, republican, anti-British political movement at a time of escalating transatlantic tension in the context of the Napoleonic Wars.[14] In addition to citing Duane, A Slave demonstrated deep knowledge of national but especially Pennsylvania politics, in addition to citing and drawing from works by the white British antislavery advocate Thomas Wilkinson and the British gentleman-turned-radical Charles Pigott. Most trenchantly, however, A Slave liberally quoted Jefferson against himself, including passages from *Notes on the State of Virginia* decrying slavery but affirming Black racial inferiority as well as the liberatory language of the Declaration of Independence.[15] Whoever A Slave was, and whatever their racial background, they were well-read, familiar with high-brow antislavery literature, politically conversant, and astute enough to provide Jefferson with a mirror in which to reflect his hypocrisy. Further, judging from A Slave's reading list, they clearly supported the American radical democratic political movements of which Paine was a direct inspiration.[16]

Unfortunately for modern scholars attempting authorial identification, one of the most concrete clues A Slave dropped that might help determine their identity are simultaneously the vaguest and most vexing. A Slave, unusually and radically for their time, demanded that the American state and national government abolish slavery and pay reparations to the freed people pegged to the very specific start date of November 30, 1781—exactly twenty-seven years before the date of the letter. Baker provides several possible explanations for this date, none of which are admittedly satisfactory.[17] A Slave's demand is striking as a very early example of slavery reparations, purportedly from the pen of an enslaved person, a concept that has recently gained traction as a proposal to rectify the iniquities of American historical slavery and subsequent modern racial oppression.[18]

Experimental Design

The results discussed below were obtained using the open source software JGAAP (the Java Graphical Authorship Attribution Program) and programs written by

TABLE 11.2 Authors considered in our experiments of "A Slave"

AUTHOR	LIFESPAN
Benjamin Rush	1746–1813
Henri Gregoire	1750–1831
Thomas Branagan	1774–1843
Thomas Clarkson	1760–1846
Granville Sharp	1735–1813
David Rice	1733–1816
Russell Parrot	1791–1824
Peter Williams, Jr.	1786–1840
John Jay	1745–1829
Gouverneur Morris	1752–1816
William Duane	1760–1835
Joel Barlow	1754–1812
Lemuel Haynes	1753–1833
Hugh Henry Brackenridge	1748–1816
Adam Carman	?
John Dickinson	1732–1808
John Parrish	1729–1807
Richard Allen	1760–1831
Daniel Coker	1780–1846
James Forten	1766–1842
Absalom Jones	1746–1818
Thomas Wilkinson	1751–1836
Thomas Paine	1737–1809

Smiljana Petrovic and Sean Campbell from Iona University and the Institute of Thomas Paine Studies, respectively.[19]

Each experiment utilizes sixty-eight base classifiers, built by combining each of seventeen lexical features with each of four learning methods (see table 11.1 and the Learning Methods section). The authors included in our testing are outlined in table 11.2.

There are several indications that Thomas Paine covertly wrote the letter signed A Slave. Practical application of the TAP software involves a testing methodology incorporating the calculation of three components of analysis, each of which can negate the other two: content, context, and calculation.

The calculation begins by creating writing samples of the possible authors, and then engages the context and content of the letter. In the example of A Slave, the author clearly wrote from a militant abolitionist perspective, and displayed deep familiarity with Pennsylvania politics, primarily Philadelphia. Further, the author is well-educated, well-read, and evinces experience in United States national politics. Finally, the author harbored strong feelings about both Thomas Jefferson and William Duane, and exhibits a familiarity that suggests possible interactions with them. Therefore, abolitionist authors still alive and active in 1809, in addition to other near contemporaries who expressed strong feelings against the institution of slavery and were known to write on the subject, were included in the author base for testing.

The methodology we employ results from numerous tests of documents of known authorship (see the list of potential authors tested in table 11.2). We took works of undisputed authorship—such as portions of Thomas Paine's *Common Sense* or James Madison's "Helvidius" letters—and tested them alongside dozens of other authors to reveal patterns. When testing A Slave, we used the corrected spellings of common words in the letter. The possible reasons why A Slave misspelled words are discussed in the content and context analysis section.

In our experience, the supports of over 40 percent attained repeatedly by the same author, tested against a wide range of authors, warrants serious content and context analysis. An author consistently achieving majority support, while all other authors have low supports (below 20 percent) is very likely to be the real author.

Figures 11.1, 11.2, 11.3, and 11.4 are representative of tests performed with different combinations of authors, as can be seen from a more comprehensive sampling of tests in table 11.2. Numerous tests are necessary for analysis as each test includes only a subset of possible authors. No test showed any other author than Paine as the dominant author, a very significant result. The tests against sixteen of the active abolitionist writers as of 1808, as shown in table 11.2, reveal

TABLE 11.3 Experiments with the list of candidate authors, attribution, and support to the selected author

NUMBER OF CANDIDATE AUTHORS	AUTHORS	CLASSIFICATION OF THE LETTER SIGNED A SLAVE	HIGHEST SUPPORT FOR AUTHORSHIP
7	Benezet, Branagan, Hopkins, Paine, Parrot, Rush, Williams	Paine	58%
8	Benezet, Clarkson, Gregoire Hopkins, Paine, Parish, Rice	Paine	51%
7	Clarkson, Gregoire, Haynes, Hopkins, Paine, Palmer, Parish	Paine	54%
8	Branagan, Carman, Forten, Haynes, Paine, Parrot, Wilkinson, Williams	Paine	67%
7	Allen, Clarkson, Coker, Jones, Paine, Priestley, Rice	Paine	50%
6	Allen, Carman, Clarkson, Gregoire, Paine, Sharp	Paine	62%
5	Forten, Jones, Lundy, Paine, Rice	Paine	67%
5	Allen, Gregoire, Lundy, Paine, Palmer, Parrish	Paine	83%
6	Coker, Jones, Paine Parrot, Rice, Williams	Paine	75%
16 sans Paine	Allen, Branagan, Carman, Clarkson, Coker, Forten, Gregoire, Haynes, Jones, Lundy, Palmer, Parish, Parrot, Rice, Wilkinson, Williams	Clarkson	28%
17	The 16 above plus Paine	Paine	76%

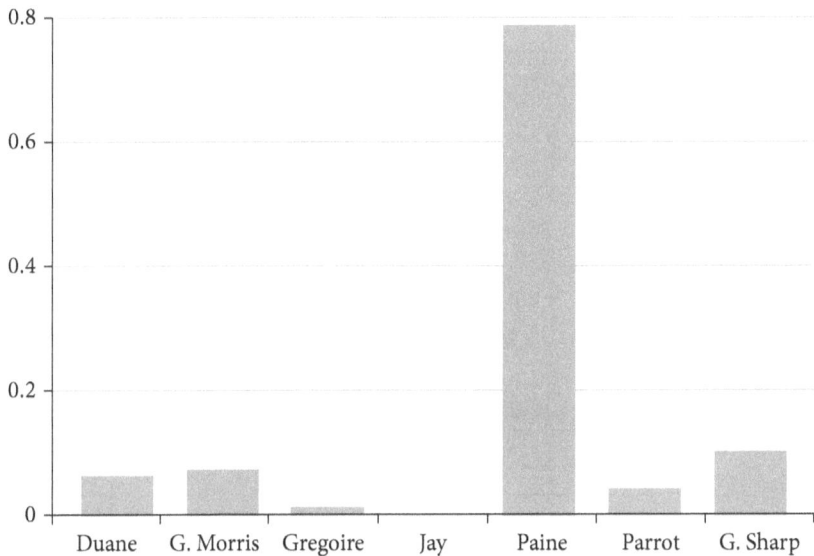

FIGURE 11.1. Comparison of supports for "A Slave" 1808 edited for spelling

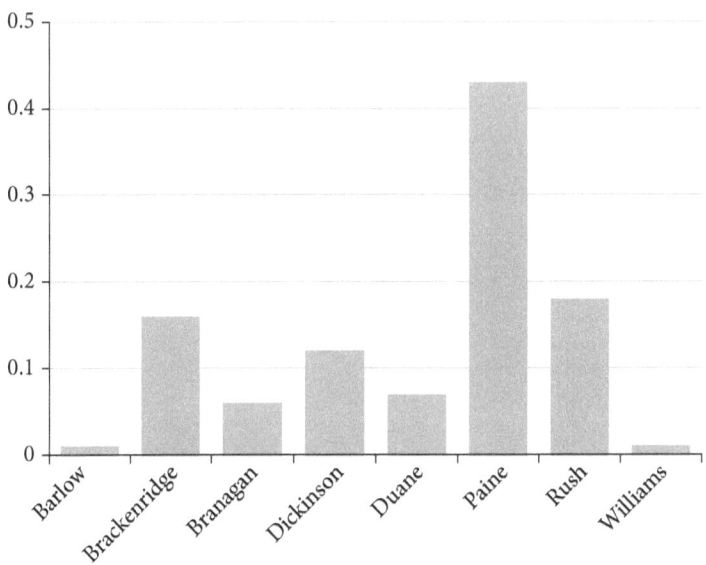

FIGURE 11.2. Comparison of supports for "A Slave" 1808 edited for spelling

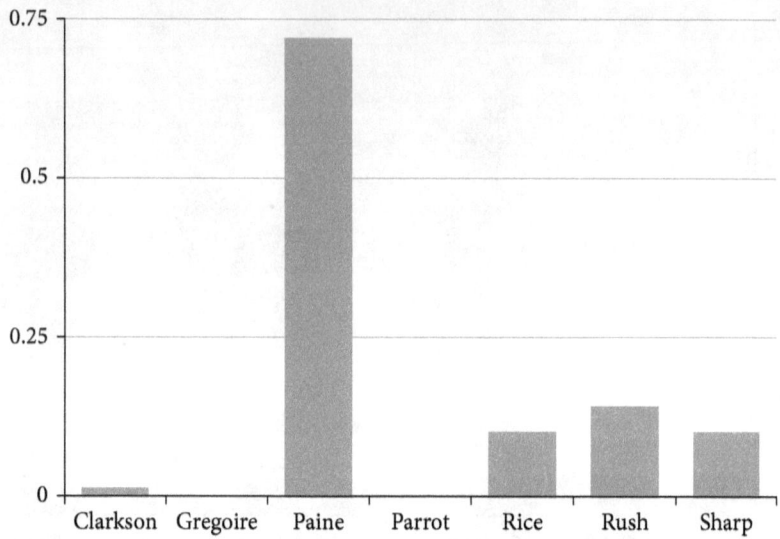

FIGURE 11.3. Comparison of supports for "A Slave" 1808 edited for spelling

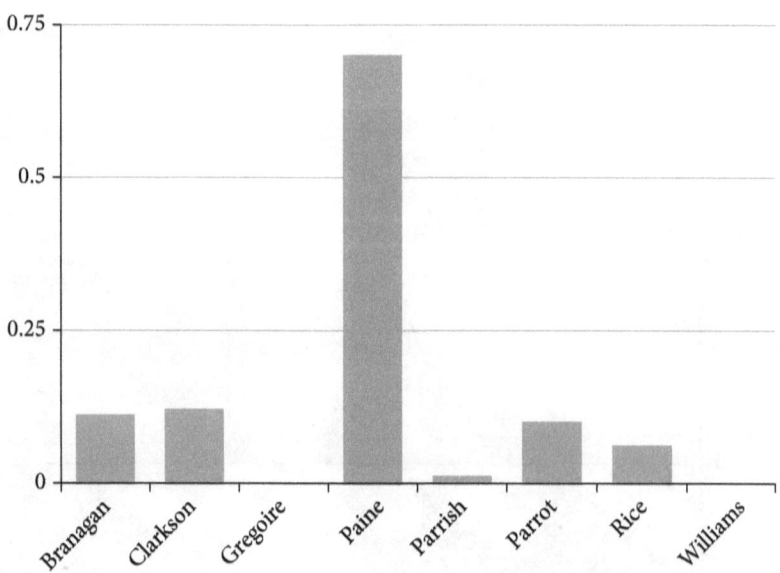

FIGURE 11.4. Comparison of supports for "A Slave" 1808 edited for spelling

a 76 percent support in favor of Paine. Note that with seven authors, the average support is $100/7 = 14.3$ percent; with eight authors $100/8 = 12.5$ percent, and with seventeen authors only 5.9 percent.

The question arises of whether the actual author is present in any particular test and author grouping. Our position is that one author, repeatedly achieving the results described above, rules out the absence of the actual author. In our experiments, when an established author is missing from the testing of a known and attributed document, a pattern of graphs will typically present itself in one of two ways: either different authors will win when the test groups of authors are varied, or multiple authors will simultaneously achieve higher supports with no one author dominating. Neither of these results appeared in the testing of A Slave: Thomas Paine consistently dominated all other candidate authors. Other possible authors appear as leaders in tests that omit Paine, though no author outside Paine repeatedly tested at or above the 40 percent threshold (data omitted). This strongly suggests that Paine is a leading contender for the authorship of A Slave's letter.

It should be noted that for many eighteenth- and early nineteenth-century political writings, there is always a possibility of collaborative writing. There are indications of collaboration when two presentations in the testing occur: if two or more authors repeatedly show as the leading contenders whenever those authors are present; and second, when one of them is absent the other contenders improve in percentage. This did not happen in any of the tests on this letter. There is only one author.

Content and Context Analysis

The calculation indicating a sufficient basis for considering Paine as the author, we then turned to further textual analysis of the content and the context in which it was written. As Baker stated in his careful analysis, the letter "sound[s] much like Thomas Paine."[20]

The content of the letter is a denunciation of slavery based on natural rights and political economy. It brings to bear the deist argument that natural law, the laws inherent in nature and its God, prohibits such systems as slavery, and the deist nature of the "original" draft of the Declaration of Independence opposed slavery on the same basis.[21] Further, in the early nineteenth century, it was typical of American abolitionists—from white Quakers to Black Methodist Episcopalians—to root their antislavery beliefs firmly in Christian tenets. This letter does not, and in this sense it is consistent with Paine's philosophy. In *Old Truths and Established Facts* (1792), for example, Paine laid out this deist argument against slavery, as he did in a newspaper article in March 1776.[22]

There are also more specific words and phrases in the letter that mirror Paine's writing: the letter uses "riot on" several times, a phrase that appeared in letter #3 of "To the Citizens of the United States" (1803) in the same context. The use of "priesthood" and "priest-craft" in the A Slave Letter is another indication of the deist character of the Letter, phrases Paine used repeatedly throughout his entire writing career.

The politics of A Slave's letter also align closely with Paine's positions: his disputes with William Duane triggered the writing of this letter, and he takes Duane to task on many issues. Further, Paine's negative attitude toward Adams appears in the letter, in which Adams's "Nobility" is repeatedly mocked as a false repository of virtue, alongside Aaron Burr and religious clergy. The letter favors Charles Pigott's analysis of corrupt religion and corrupted government being the source of all tyranny, a philosophy Paine shared as well. A Slave quotes Pigott twice in the letter, both referring back to England as the prime example of corruption of all kinds, a consistent trait of Paine's works that used England as the model of monarchical corruption. Moreover, the letter's defense of revolutionary France against the Federalist and Aaron Burr's wing of the New York Democratic-Republicans in the 1790s also reflects Paine's positions. In short, there is nothing in the A Slave letter that conflicts with Paine's stated philosophical and political positions; on the contrary, the letter replicates his core political philosophy.

The most notable date A Slave emphasized was November 30, 1781, the same date as the letter, pegged as the start date for retroactive reparations for enslaved African Americans. The date November 30 also has parallels to Paine's biography: A Slave mentions Jefferson's 1781 publication of *Notes on the State of Virginia*, and November 30, 1774, was the date Paine first landed in Philadelphia. There is no other significant historical relevance of November 30 to the chronology and politics of the American Revolution. The author held Jefferson to account for his assertions in *Notes on the State of Virginia*, most notably the much-discussed section concerning slavery and race relations in which the future president rued slavery and predicted its eventual abolition, but discounted the possibility of a biracial America, and described people of African descent as biologically inferior to Euro-Americans. November 30 being both the day the Slave letter was written, and the day that Paine's struggles in America began, suggests a personal relationship between the author and Jefferson, which by 1808–1809 had grown strained and distant on Jefferson's part. Paine's letters to Jefferson reveal his frustration with Jefferson's perceived neglect of his old friend, a frustration that can be seen in the A Slave letter. Paine's admonishments to Jefferson—condescendingly calling him "Thomas" throughout—concerning the Virginian's stances on slavery and race relations boil over in the letter Paine possibly dictated from his deathbed. Further, in a letter to George Washington dated November 30, 1781 (the key date in A Slave's letter)—as concern over the newly independent United States bubbled over—Paine linked the sentiment of defending the character of the country with the contrary behavior toward him (see a quote from that letter below). His intent in 1781 of defending the character of America, surrenders to woe in the tyranny and brutality of slavery in this letter.

Paine revealed these feelings to Washington in detail:

> It is seven years, this day, since I arrived in America, and though I consider them as the most honorary time of my life, they have nevertheless been the most inconvenient and even distressing. From an anxiety to support, as far as laid in my power, the reputation of the Cause of America, as well as the Cause itself, I declined the customary profits which authors are entitled to, and I have always continued to do so; yet I never thought (if I thought at all on the matter), but that as I dealt generously and honorably by America, she would deal the same by me. But I have experienced the contrary—and it gives me much concern, not only on account of the inconvenience it has occasioned to me, but because it unpleasantly lessens my opinion of the character of a country which once appeared so fair, and it hurts my mind to see her so cold and inattentive to matters which affect her reputation.[23]

As for context, A Slave's letter is not written in Paine's hand, so an explanation is necessary for the method of its composition. In November 1808, Paine was practically bedridden. His correspondence had ended by July 1808, as far as it is extant, as his health began to seriously decline. Madame Bonneville was the primary caregiver to Paine at the end of life, and she moved him closer to her just after the letter was composed. Paine made his will at the end of December 1808, and while his physical health had declined precipitously, his mind remained active. Writing a long letter at this point would be taxing. As well, Jefferson knew Paine's handwriting, having exchanged correspondence with him many times, most of which had been in the last ten years. Both factors point to an acting secretary taking down Paine's words.

This brings Marguerite Brazier Bonneville into the story. There was a long relationship between the Bonneville family and Paine. Madame Bonneville was the wife of Nicholas, Paine's best friend in France. Paine and Nicholas Bonneville had worked together writing and producing publications supporting the French Revolution since 1791, and when Paine was released from prison, after staying at first with James Monroe, he moved in with the Bonneville family in Paris. The youngest child was named for him. Madame Bonneville and two children followed Paine to America in 1802, when Nicholas was imprisoned, and Paine returned their favors by looking after the three of them. He left his farm in New

Rochelle to the family, to use for the children's education. This is why Bonneville was at his bedside toward the end. Paine looked after her welfare, while not getting along with her, but he was paying a debt to his old friend back in France.

The handwriting of the A Slave letter is a major matter in determining authorship; previous examinations of the letter offer few clues connecting the handwriting to authorial identity. On initial investigation, it bore a resemblance to Bonneville's, and on closer examination, in comparing capitalization, the similarity became obvious: Bonneville's first language was French, with enough knowledge of English to teach French to English speakers. If English words were dictated, she would likely spell by sound than by experience. It is also possible that it was deliberately misspelled, to maintain the ruse of an uneducated slave writing it. But the former seems more likely.

We matched four obvious capital letters in Bonneville's letters (two to Jefferson, one a note) with the A Slave letter, and the matches are very similar:

> The longer she writes, the more the slant of the writing matches her letters. It starts out more vertical, then the right slant becomes noticeable half way through to the end. The B's are identical in all 4 documents: they never close at the bottom. In addition, several combinations of letters also matched the handwriting of M. Bonneville. Only two letters in Bonneville's handwriting are known, one to Jefferson, one a note to Paine on her arrival. But even in this short comparison, the similarities are evident.

The issue of Madame Bonneville contributing in some way to the text of the letter also needs to be dismissed, but even if she contributed some text, it would not alter Paine's dominance in the composition. There was enough text from Madame Bonneville's account of Paine's death and burial to create a file that produced the following test on the letter:

> A result of less than 20% for any author shows no presence in the text—this test showed less than 10% for Bonneville. Her role in the letter was primarily secretarial, even if Bonneville may have shared some of Paine's views of slavery, as well as political positions discussed above.

In summary, all three aspects of the text attribution methodology suggest that Paine verbally dictated the A Slave letter, and nothing excludes him. We know that Paine and Jefferson, for all their shared experiences as American "founders," and even in their shared philosophical antislavery views, viewed the prospect of an emancipated America very differently. Jefferson never wavered in his insistence that a biracial America consisting of free persons of color sharing real and imagined communities with whites was impossible; Paine never expressed such exclusionary concerns. Heightened by the cold shoulder Jefferson showed Paine at the end of his life, it was not in Paine's nature to ignore an injustice; and all the injustices, personal, political, and philosophical, boiled up in this last letter, leaving a mark that was signed, as usual, anonymously.

The authors believe that the TAP project offers compelling tools for exploring vexing questions of authorship, as in the case of the Lapidus Query. Paine's potential authorship of the letter by A Slave suggests new areas of research into Paine's thought, revealing a piece of antislavery writing heretofore unknown in Paine's well-known corpus of publications and letters. The absence of an extended, direct critique of chattel slavery and the slave trades from Paine's pen is regarded by historians as a curious omission from his oeuvre, but potentially no longer. The possibilities the TAP project augers for researchers engaged with anonymous writings—by one author or multiple authors—are suggested in this article, and the authors hope to stimulate research across disciplines through TAP software and methodology.

NOTES

1. Hereafter, "A Slave" will be rendered A Slave, sans quotation marks.

2. Thomas N. Baker, "'A Slave' Writes Thomas Jefferson," *William and Mary Quarterly*, 3rd ser., 68, no. 1 (January 2011), 127–54. In this chapter, I fully acknowledge Baker's key role in bringing the letter to researchers' attention as well as his thoughtful and suggestive analysis of internal clues in the letter that might hint at the author's identity. Baker also notes that, as his article went to press, he discovered a 2004 seminar paper that had also explored the letter (128n5). A scanned copy and transcription of the letter are available at Library of Congress, Founders Online, "To Thomas Jefferson from Pseudonym 'A Slave,'" November 30, 1808, https://founders.archives.gov/documents/Jefferson/99-01-02-9200.

3. The pioneering modern attempt to utilize computer technology to determine author attribution is Frederick Mosteller and David L. Wallace, *Applied Bayesian and Classical Inference: The Case of the "Federalist Papers"* (New York: Springer-Verlag, 1964). One recent project to apply computational analysis techniques to questions surrounding Paine's writings and authorship that deserves credit for forwarding research in these fields is Richard Forsyth and David Holmes, "The Writeprints of Man: A Stylometric Study of Lafayette's Hand in the *Rights of Man*," *DHQ: Digital Humanities Quarterly* 12, no. 1 (2018), http://www.digitalhumanities.org/dhq/vol/12/1/000371/000371.html. As the subtitle indicates, Forsyth and Holmes hypothesize that the Marquis de Lafayette, a friend of Paine's, was plagiarized by the latter in *Rights of Man*, particularly in several key passages that provided detailed chronological evidence and inside knowledge of French revolutionary government activities.

4. For an overview of the absence of Paine's writings on antislavery, see James V. Lynch, "The Limits of Revolutionary Radicalism: Tom Paine and Slavery," *Pennsylvania Journal of History and Biography* 123, no. 3 (July 1999): 177–99. Previous TAP research suggests that "African Slavery in America," an originally untitled article published in the *Pennsylvania Journal and Advertiser*, March 8, 1775, commonly attributed to Paine, was in fact not authored by him. Authorial attribution of "African Slavery in America" to Paine was claimed first in Moncure D. Conway, *The Life of Thomas Paine: With a History of His*

Literary, Political and Religious Career in America, France, and England (New York: G. P. Putnam's Sons, 1892), 1:50–53, without offering direct evidence of Paine's authorship.

5. Rosa María, Coyotl-Morales, Luis Villaseñor-Pineda, Manuel Montes-y-Gómez, and Paolo Rosso, "Authorship Attribution Using Word Sequences," in *Progress in Pattern Recognition: Image, Analysis and Applications*, Lecture Notes in Computer Science (Including Subseries Lecture Notes in Artificial Intelligence and Lecture Notes in Bioinformatics), vol. 4225 (Berlin: Springer, 2006), 844–53, https://doi.org/10.1007/11892755_87.

6. Efstathios Stamatatos, "Authorship Verification: A Review of Recent Advances," *Research in Computer Science* 123 (2016): 9–25; Efstathios Stamatatos, "A Survey of Modern Authorship Attribution Methods," *Journal of the American Society for Information Science and Technology* 60, no. 3 (2009): 538–56; Patrick Juola, "Authorship Attribution," *Foundations and Trends® in Information Retrieval* 1, no. 3 (2008): 233–334, https://doi.org/10.1561/1500000005.

7. Maciej Eder, Jan Rybicki, and Mike Kestemont, "Stylometry with R: A Package for Computational Text Analysis," *R Journal* 8 no. 1 (2016): 107–21, doi:10.32614/RJ-2016-007.

8. Lubomir Ivanov and Brandon Neilsen, "Consonance as a Stylistic Feature for Authorship Attribution of Historical Texts," in *Text, Speech, and Dialogue*, ed. K. Ekštein, Lecture Notes in Computer Science, vol 11697 (Berlin: Springer, 2019), 45–57; Lubomir Ivanov, "Comparing Assonance and Consonance for Authorship Attribution" (paper presented at the Digital Humanities Conference, Utrecht, The Netherlands, July 2019); Lubomir Ivanov, "Learning Patterns of Assonance for Authorship Attribution of Historical Texts" (paper presented at the Thirty-Second International FLAIRS Conference, Sarasota, Florida, May 2019), www.aaai.org; Lubomir Ivanov, Amanda Aebig, and Stephen Meerman, "Lexical Stress-Based Authorship Attribution with Accurate Pronunciation Patterns Selection" (paper presented at the 21st International TSD Conference, Brno, Czech Republic, September 2018), https://doi.org/10.1007/978-3-030-00794-2_7, 67–75; Lubomir Ivanov, "Using Alliteration in Authorship Attribution of Historical Texts," in *Text, Speech, and Dialogue*, ed. Petr Sojka, Aleš Horák, Ivan Kopeček, and Karel Pala, Lecture Notes in Computer Science, vol. 9924 (Berlin: Springer, 2016); Lubomir Ivanov and Smiljana Petrovic, "Using Lexical Stress in Authorship Attribution of Historical Texts," in *Text, Speech, and Dialogue*, ed. P. Král and V. Matoušek, Lecture Notes in Computer Science, vol. 9302 (Berlin: Springer, 2015), 105–13.

9. Mosteller and Wallace, *Applied Bayesian and Classical Inference*.

10. Gary Berton, Smiljana Petrovic, Lubomir Ivanov, and Robert Schiaffino, "Examining the Thomas Paine Corpus: Automated Computer Authorship Attribution Methodology Applied to Thomas Paine's Writings," in *New Directions in Thomas Paine Studies*, ed. Scott Cleary and Ivy Linton Stabell (New York: Palgrave Macmillan US, 2016), 31–47, https://doi.org/10.1057/9781137589996_3; Smiljana Petrovic, Gary Berton, Sean Campbell, and Lubomir Ivanov, "Attribution of 18th Century Political Writings Using Machine Learning," *Journal of Technologies in Society* 11, no. 3 (2015): 1–13, https://doi.org/10.18848/2381-9251/CGP/v11i03/56506.

11. Philip J. Boland, "Majority Systems and the Condorcet Jury Theorem," *Statistician* 38, no. 3 (1989): 181–89, https://doi.org/10.2307/2348873.

12. Petrovic, Ivan, Smiljana Petrovic, Ileana Palesi, and Anthony Calise, "Eliminating Sycophants to Improve Authorship Attribution" (paper presented at the Thirty-Second International FLAIRS Conference, Sarasota, Florida, May 2019), www.aaai.org.

13. Smiljana Petrovic, Ivan Petrovic, Ileana Palesi, and Anthony Calise, "Weighted Voting and Meta-Learning for Combining Authorship Attribution Methods," in *International Conference on Intelligent Data Engineering and Automated Learning (IDEAL 2018)* (Berlin: Springer, Cham), 328–35, https://doi.org/10.1007/978-3-030-03493-1_35.

14. Baker, "'A Slave,' Writes Thomas Jefferson," 129, provides a fuller account of Duane's book and his intentions.

15. Thomas Jefferson, *Notes on the State of Virginia* (1781), Query XVIII. Jefferson is (and was, in 1809) believed to have authored the Declaration's "Slave Trade Clause," a ringing (and hypocritical) condemnation of the British Empire's role in promoting American slavery and the transatlantic slave trade.

16. For a broader context of American radical democratic politics, and Paine's role as a figurehead, see Seth Cotlar, *Tom Paine's America: The Rise and Fall of Transatlantic Radicalism* (Charlottesville: University of Virginia Press, 2006).

17. Baker, "'A Slave' Writes Thomas Jefferson," 137n27. These include the possibility that the date related to Jefferson's drafting of *Notes on the State of Virginia*, and the possibility that A Slave mistakenly wrote November 30, 1781, instead of November 30, 1782, the latter being the date of a preliminary agreement between the United States and Great Britain to formally end the American War of Independence.

18. The concept of American slavery reparations began to take its modern form during the 1960s, but Ta-Nehisi Coates's "The Case for Reparations," *Atlantic*, June 2014, ignited a public controversy and has largely set the terms of debate for twenty-first-century public opinion.

19. Juola, "Authorship Attribution"; Petrovic et al., "Attribution of 18th Century Political Writings Using Machine Learning."

20. Baker, "'A Slave' Writes Thomas Jefferson," 134.

21. Julian P. Boyd reconstructed the original Declaration from Adams's copy; see *The Papers of Thomas Jefferson*, vol. 1, *1760–1776*, ed. Julian P. Boyd (Princeton, NJ: Princeton University Press, 1950), 243–47. This version contained the Slavery Clause, which was removed by Congress before ratification.

22. *Old Truths and Established Facts* was a 1792 pamphlet in support of abolishing the slave trade; Gregory Claeys supports Paine's participation in its writing. Claeys, *Thomas Paine: Social and Political Thought* (Boston: Unwin Hyman, 1989), 3. The *Crisis* issue published in London, March 12, 1776, #61, was Paine's composition, being mailed from Philadelphia after *Common Sense* was published—it tests very strongly for Paine.

23. Thomas Paine to George Washington, November 30, 1781, Washington Papers, Founders Online, Library of Congress, https://founders.archives.gov/?q=paine%20november%2030%201781&s=1111311111&sa=&r=2&sr=.

12

WHO STANDS IN THE DIGITAL SHADOWS?

"City of Refuge" at the Intersection of "Old" and "New" Media in the Age of the Digital Humanities

Marcus P. Nevius

Between 1763 and 1804, the Loyalist merchants William Aitchison and James Parker, based in Norfolk and Portsmouth in Virginia, kept an account book bound in green vellum. The book's faded covers are marked by splotches of black ink. Its edges bear evidence of the brass clasps and catches that once held its covers shut. Prepared by the staff of the University of Virginia Albert and Shirley Small Library's Special Collections, the account book's provenance offers a summary that highlights the book's potential utility to researchers. An undigitized, physical source, the account book is an important record of far-flung Atlantic World mercantile networks that linked the two merchants' locally produced goods to consigners in the Canary Islands, in Cuba, and in Honduras.

Yet if the account book is distinctive in the context of eighteenth-century sources, it is so because significant gaps in time break up the merchants' records. Aitchison and Parker's irregular use of the account book is most likely explained by the outbreak in the 1760s and early 1770s of protest, then war, within the British Empire. In 1776 Aitchison and Parker evacuated Virginia, never to return.[1] But before their departure—in a context of Loyalists who scattered to various Atlantic World locations—the merchants documented an unusual story.[2] Undated, it is the account of a "negroe man" whose name the merchants did not record. According to the legend, the man had lived in Norfolk in the two years prior to the merchants' entry in their account book. Before that, he lived by himself in the Great Dismal Swamp, the commonwealth's largest natural wetland, for nearly thirteen years. There, he subsisted by raising "Rice & other grain" and occupied himself producing "Chairs, Tables & musical instruments."[3] The man's story is

FIGURE 12.1. Cover of the William Aitchison and James Parker account book. Aitchison and Parker Account Book, 1763–1804, Accession #12992, Special Collections, University of Virginia Library, Charlottesville, VA.

situated midway in the account book, sandwiched between the merchants' summary notes of the swamp's geographic location near the top of the page and, at the bottom, the merchants' "query" about "200 Palantines put on shoar near the Capes" who froze to death where they were landed.

The merchants' notation of the Dismal Swamp's Black furniture maker reflects the challenges that characterize an effort to investigate the story of freedom-seeking people in Southside Virginia in the age of the American Revolution. The description of his time in the swamp was not his own; it was recorded by white merchants with a stake in maintaining transatlantic networks that were disrupted as the war took shape. Historians generally seek a broad, deep primary source base, held in physical archives, to inform the stories of the past. While such is the case for historical actors—William Aitchison and James Parker, for example—who enjoyed the privilege of literacy and prominence, enslaved people, barred from these privileges in all but a few cases, did not produce volumes of written records. Thus, the placement of the Black furniture maker's story in the Aitchison and Parker ledger has, to date, gone largely unnoticed. Neither his story nor the ledger feature in studies of the American Revolution, a subject shaped by studies of archives replete with military and personal correspondence; buttressed by an archive of Patriot papers that reported to broader publics the events of British and Patriot skirmishes, battles, and massacres; and supported further by postwar documents.

With these ideas as a premise, I aim in this chapter to consider the work of historical recovery at the core of research conducted in physical archives in our current age of digital archives. I do so informed by two key scholarly trends: new studies and projects in the Black digital humanities; and new historiographical directions in the history of the American revolutionary era. On the one hand, for African American history, digitization projects have over the last thirty years been a boon of sorts, making slave narratives and runaway advertisements widely available to historians and students. Such sources have been largely inaccessible to most scholars who, due to institutional constraints, have lacked the resources to facilitate travel to archives. On the other hand, broadened access risks obfuscating less traditionally mined archival sources—eighteenth- and nineteenth-century company account ledgers that assigned a speculative value to enslaved people's labor, for example—that remain undigitized as archival staff make deliberate choices to prioritize a broadening base of slave narratives and runaway advertisements over others, such as an obscure merchant ledger. Such decisions to prioritize primary sources that generate the broadest interest thus relegates to archival shadows the very notations in which the Dismal Swamp's Black furniture maker resides. Considering all of this, I suggest that careful study of the scholarship of the Black digital humanities yields important lessons for students and scholars with an interest in broadening traditional primary source bases in both African American history and the history of the American Revolution.

Finding a tale of a Black furniture maker in the pages of an account book might be seen as a surprise to some scholars of early African American history before the 1830s. The field is traditionally shaped by petitions and narratives authored by Prince Hall, Olaudah Equiano, Venture Smith, Phyllis Wheatley, Paul Cuffe, or the early organizers of free Black communities in Atlantic Coast ports from Boston to Charleston to New Orleans on the Gulf Coast, or the antebellum generation of freedom seekers who fled enslavement and whose experiences featured at the center of the era's rising generation of abolitionists.[4] However, in the last decade, new work defining "slavery's capitalism" in the first half of the nineteenth century signals a vanguard of change, astutely scrutinizing the history of women in the broad context of the domestic slave trade, and investigating the "technologies of capitalism" that reduced enslaved people to market abstractions.[5] New histories focused on liminal spaces in broader slave societies—the Great Dismal Swamp, for example—highlight the ways that slavery endured in Virginia by new mechanisms of expanded "slave hire" systems, even as slavery's critics, such as Frederick Douglass, made an example of the moral impropriety of the system.[6]

In the early 1990s, slave narratives were a core set of primary sources that incentivized the digitization of archival sources, placing African American history in the vanguard of digital history as the discipline took shape. Centrally, these efforts included projects such as the University of North Carolina Libraries' Documenting the American South (DocSouth)pilot launched in 1996, when it published on the internet six digitized, highly circulated slave narratives from the special collections of the university's libraries.[7] The DocSouth was a pioneer of sorts, anticipating similar projects including the University of Virginia Press's Rotunda Collections. Rotunda is a paid subscription database that, according to the latest information on its website, was "created for the publication of original digital scholarship" and "newly digitized critical and documentary editions in the humanities and social sciences."[8] Perhaps the most widely known digitization effort is the Trans-Atlantic Slave Trade Database, a trove of nearly thirty thousand Atlantic slave trade voyages launched in 1998 on CD-ROM.[9]

The imperative to digitize archival records, as the historian Sharon Block has observed, dates to the "digital turn" of the mid-1990s. This was a period marked by debates aimed at defining best practices for digitizing records to preserve and expand access to archives.[10] The historian Lara Putnam has cited the field-changing potential of the digital humanities as they benefit researchers and other audiences who gain increased access to far-flung archives no longer constrained by institutional structures defined by "political-territorial units."[11]

It is also important to note key lessons to be drawn from the scholarship of the Black digital humanities, emergent in the second decade of the twenty-first century. The historian Kim Gallon has highlighted the utility of a "digital episteme of humanity" that foregrounds the relationship between Black studies and the digitized humanities, particularly as digital projects generate a "technology of recovery" aimed at "recovering alternate constructions of humanity that have been historically excluded" from humanities fields predicated on Western cultural traditions.[12]

The Black digital humanities, as Gallon observes, compel scholars to consider key issues: the racialized foundation of humanities fields through the privileging of Western cultural traditions; the need to assess the previous use of digital tools, in hindsight, with an eye toward questioning how they might have been used if they had been developed to explore marginalized texts produced in the racialized humanities; and, importantly, the need to question how digital tool development might reflect the material circumstances of Blackness. Others, including the historians Vincent Brown and Jessica Marie Johnson, have offered similar advice for the utility of the Black digital humanities. Brown's cartographic narrative of the major slave revolt that began in St. Mary's Parish, Jamaica, in April 1760, for example, has highlighted the importance of new tools that present opportunities for historians. These new digital tools enhance researchers' abilities to evaluate high volumes of source materials (data); to reveal patterns previously unobservable with traditional research tools; to create multimedia graphics that illustrate the "contours of social life"; and to render data visualizations that become a form of digital or digitized storytelling. Nascent in Brown's study of death and power in the history of colonial Jamaica and transatlantic slavery, these insights are manifest in his study of the massive slave rebellion in Jamaica that erupted in 1760 during the Seven Years' War.[13] That the use of digital humanities tools must be attended by important cautions is an imperative imparted by the Black digital humanities. Johnson has interrogated conventional definitions of data—historical archival sources—to propose the stakes in historical data's "implicit claim to stability or objectivity." Noting that the historian's contextualization of sources informs the story one tells, Johnson reminds scholars that the archives of eighteenth- and nineteenth-century Atlantic slave trading provide more than just quantitative data for present-day digital humanities projects.[14]

Taken together, studies and projects in these trends—the Black digital humanities and new histories of the American Revolution—provide a sturdy foundation for asking some important questions. What are the benefits of digital humanities projects? Do they include preservation of degraded primary sources for posterity; leveling, as Putnam observes, the barriers of privilege that long rendered the archival sources inaccessible to broad audiences; opportunities for new epistemologies and technologies, as Gallon and Brown observe; or salient reminders that researchers, new and old, should attend primary sources with the work of historical recovery in mind, as Johnson advises? What emerges in the pages to follow is a brief, twofold story of my engagement with primary sources, informed in part by scholarship in the Black digital humanities. I aim in this short case study to reflect on my own approach to research undertaken in physical archives and in digitized archives.

Though the historical actors in this chapter remain largely cloaked in archival shadows, this project adds texture to our understanding of the American revolutionary era. It centers on key primary sources: the evidence of the Dismal Swamp's Black furniture maker recorded in the Aitchison and Parker account book; three runaway advertisements for as many as three men named Tom who fled to the Dismal Swamp in the late 1760s; and an appraisal list that records the names of fifty-four men, women, and children dispatched to the Dismal Swamp in the early 1760s. What is clear in my reading of this selection of primary records is that all of these people—the fifty-four men, women, and children at Dismal Plantation in 1764; the mysterious Black furniture maker; the three Toms; and countless others in their era—are part of a central population comprising those who sought refuge from enslavement in or near the Great Dismal Swamp.[15]

What is less clear in the Dismal Swamp's scant late eighteenth-century primary record might be understood by careful attention to the historical context of the American Revolution in Virginia. That the Black furniture maker lived in the Dismal Swamp for thirteen years, for example, cannot be corroborated by his own words. Yet he lived there at a time of great upheaval that would have been to his advantage in his flight. In the spring and summer months of 1775, the colony's last royal governor, John Murray, fourth earl of Dunmore, began to stoke fears of a massive slave rebellion. In April, six days before news of Lexington and Concord reached Williamsburg, local authorities had sternly criticized the governor's actions in the *Virginia Gazette*, blaming him for a series of slave conspiracy scares. Responding to a scare in Williamsburg, Dunmore ordered the removal of the gunpowder stored in the town's arsenal to HMS *Magdalen*, riding at anchor in the James River. By month's end, Dunmore made matters worse by threatening the House of Burgesses that unless they agreed to the gunpowder's removal, he would declare freedom to enslaved people and encourage them to set on Williamsburg to burn the town.[16]

Virginia's American Revolution thus began in a context of fears of slave revolts generated by rumors. These rumors were only exacerbated days later by the news that General Thomas Gage had also ordered the removal of weaponry from the New England arsenal in Concord, Massachusetts.[17] Throughout summer and early fall 1775, Dunmore, aboard the HMS *Otter* at anchor in the waters off Portsmouth and Norfolk, invited enslaved Virginians to take refuge with the flotilla. In early November, Dunmore gave orders to Captain Samuel Leslie and 130 men, including a significant number of freedom-seeking enslaved people, to attack Virginians who had compiled war matériel at Kemp's Landing, on the northeastern fringe of the Dismal Swamp. Though outnumbered, Leslie's force defeated the Virginians. Their victory emboldened Dunmore to raise the king's standard in Southside Virginia, signaling that he had declared the colony to be in rebellion. The victory also inspired Dunmore to issue the now-famous proclamation on November 14, based on two previous drafts, that created the "Ethiopian Regiment" and offered emancipation to enslaved people who joined the British forces. Though Virginians organized to hinder enslaved people from responding to Dunmore's proclamation, as word spread through the lower portion of Virginia, enslaved people sought out Dunmore's lines by the hundreds.[18] In early December, the Virginia colonel William Woodford's forces routed Dunmore's forces in the Dismal Swamp at Great Bridge. On December 14, Woodford's forces marched into Norfolk and on New Year's Day 1776, Dunmore's flotilla began a bombardment of the town. Ceding the burning town to the Virginians, Dunmore's flotilla began a slow retreat northward in the Chesapeake Bay, carrying with it several hundred enslaved people. Others escaped into the Dismal Swamp.

In late May 1776, Dunmore's flotilla landed on Gwynn's Island, a small sandbar situated between the mouths of the York and Rappahannock Rivers. Encamped on the island for several weeks to quarantine troops and freedom seekers stricken with smallpox, Dunmore's forces faced Virginians who began bombarding the island on July 9, the same day that New Yorkers pulled down King George III's statue in that city. To avoid Patriot cannon, Dunmore's forces evacuated the island, including several hundred Black people still alive in the midst of the smallpox epidemic, leaving many others to die from disease.[19] Dunmore, at the helm of the floating town, steered upstream from the Chesapeake Bay to the Potomac River. By early August, Dunmore's floating town reversed course to depart the Chesapeake region for good. Locals reported that at least one thousand enslaved Blacks left Virginia with Dunmore's fleet, and that at least one thousand more had died of smallpox, other diseases, and warfare.

It is likely that the Black furniture maker emerged into Norfolk from the swamp in the late months of 1775 or early months of 1776 during the war's disruptions. If he remained in town in Dunmore's wake, he would have witnessed more disruptions in the late months of the war. British forces returned to Virginia during several more sustained engagements beginning in May 1779 and ending in October 1781. In May 1779, British forces sacked two towns near Norfolk, pillaging in Portsmouth before burning Suffolk. In this latest round of fighting, at least five hundred enslaved people fled to British lines.[20] In October 1780, British forces again landed at Portsmouth, from which they pillaged Southside Virginia in activities that attracted runaway slaves to British lines.[21]

In April 1781, yet another British invasion force launched a campaign of pillage from Portsmouth. By May, news of this latest invasion force's disruptive activities had convinced General Charles Cornwallis, moving northward through North Carolina with seven thousand troops, to set Virginia as an objective. Cornwallis sent a detachment of five hundred soldiers to Hampton Roads by way of Edenton and the Great Dismal Swamp. On July 21, this detachment raided Dismal Plantation. Jacob Collee later reported that the soldiers made off with a debilitating amount of valuable property: draft steers, fifty head of cattle, two hundred barrels of corn, and the site's work tools. But most injurious to the Dismal Swamp Company's operations was the fact that the British soldiers induced more than half of Dismal Plantation's enslaved laborers to seek freedom behind British lines. By Collee's count, the group of freedom seekers numbered twenty-one men, one woman, and five children.[22] That October—with smallpox still raging in the ranks and among freedom seekers—Patriot columns, bolstered by French troops, besieged Cornwallis's forces who had holed up in Yorktown.[23] As the historian Robert G. Parkinson has observed, newspaper articles, exchanged in papers from Virginia northward to Massachusetts, described for contemporary audiences the "vast Concourse" of runaway slaves, thousands, who aided the British war effort at Yorktown.[24]

Many freedom seekers behind British lines at Yorktown evacuated North America with British troops and Loyalists in a broader context of slave flight and marronage. As Lathan A. Windley once observed, the *Virginia Gazette* published notices for more than 3,500 runaways between 1736 and 1783.[25] From 1775 to 1782, as the historian Cassandra Pybus has estimated, approximately 20,000 enslaved people fled to the British, 12,000 of whom fled from southern enslavers. Between 8,000 to 10,000 free African Americans evacuated North America with British Loyalists, about 2,000 of whom escaped from Virginia.[26] Countless other runaways chose refuge in the Dismal Swamp during the war.

Even before the war's disruptions inspired slave flight on a massive scale, in the late 1760s, several advertisements in the *Gazette* informed the Virginia colony's reading public of several Toms who had taken refuge in the Dismal Swamp. The historian Charles Royster cited a June 1768 advertisement for one Tom, a recent survivor of the Middle Passage, enslaved by Samuel Gist. Posted by John Augustine Washington acting as agent for the Dismal Swamp Company, the advertisement noted that Tom had fled more than a year earlier, in April 1767. Tom stood about five feet six and bore four "country marks" on each cheek. Such scarification appeared as distinctive patterns, displayed primarily on the face, that signified ethnic heritage in West Africa.[27]

In October 1768, John Mayo placed a different ad for a freedom seeker named Tom. It described this second Tom as "about 6 feet high" with a "roguish look" and noted that he had "lost part of one of his ears." In this second instance, Tom had been seen "in Nansemond and Norfolk counties" but was "supposed to be about the Dismal Swamp." The historical archaeologist Ted Maris-Wolf cited Tom in the 1768 advertisement as the "first documented runaway to be associated with the Dismal Swamp," explaining that it was unclear whether Tom (1768) had spent any time in the swamp before he was found in 1781 at a neighbor's house near the swamp.[28]

In April 1769, the *Virginia Gazette* ran a third advertisement for Tom, the second posted by Mayo. It described Tom as approximately six feet tall with one cropped ear, but Mayo added that he now believed Tom to be "about the Dismal Swamp. *or low down in North-Carolina*."[29] This third advertisement explicitly emphasized Mayo's suspicion that Tom might be found in the southern end of the swamp, and perhaps Mayo's implication that Tom might find support in flight from the historically independent small triracial communities that inhabited North Carolina near the Albemarle Sound.[30]

Two of the three advertisements for Tom found distribution in the first volume of Windley's work. A decade later, Tom Costa's 2005 Geography of Slavery in Virginia project digitized and made available the advertisements announcing Tom's flight.[31] Maris-Wolf notes the capture of one Tom in 1781, even after he had been aided in flight by local Quakers; an undigitized December 1783 letter now archived in the Dismal Swamp Land Company records at the David M. Rubenstein Rare Book and Manuscript Library at Duke University reveals the capture of the second Tom. In the letter, the company manager David Jameson wrote to John Driver, the Dismal Swamp Company's agent at Suffolk, to advise him of the company manager David Meade's directives regarding "an outlying fellow belonging to the D.S.Co. called Tom." Relaying Meade's instructions, Jameson explained that Tom needed "to be sold" out of North Carolina and directed Driver to arrange the sale.[32] No further evidence in the Rubenstein collection reveals Tom's fate after Meade and Jameson decided to sell him. Citing two undigitized 1784 letters revealing the point at which one Tom exits the primary record, Royster observed that cool winter weather might have compelled him to exit the swamp, leading to his capture.[33]

Notwithstanding the two Toms' fates, we can consider the context for the man in the Aitchison and Parker account book. If we take the initial year that the merchants began to keep the book, 1763, as a starting point, and place the man in the Dismal for as many as thirteen years, we have at a maximum the year 1776 as the last year that the man remained in the swamp. If we subtract two years to account for the time the legend observes that the man had recently spent in Norfolk, then we might assume that he emerged into town in about 1774, and certainly no later than 1776. The fact that the two merchants evacuated Virginia sometime after the burning of Norfolk in January 1776 provides yet another clue pointing to the man's emergence in Norfolk in about 1774. This triangulation of dates offers us one temporal context that—inspired by lessons about the work of historical recovery at the core of the Black digital humanities, and with the aim of restoring the humanity of the man at the legend's center—we might use to evaluate the story's veracity.

To deepen the contextualization of the Dismal Swamp as a refuge in the American revolutionary era as we move toward a conclusion in this chapter, I turn briefly to a third source: George Washington's copy of the Dismal Swamp Company's "Appraisement of Dismal Swamp Slaves."[34] In July 1764, the Dismal Swamp Company counted at Dismal Plantation fifty-four enslaved people: forty-three men, nine women, one boy, and one girl. Among this group were three men named Tom, whose appraised values differed, ranging from a little more than 52 British pounds to 75 British pounds. In the appraisal we confirm one Tom, advertised by John Augustine Washington, listed as enslaved by Samuel Gist.

The extant evidentiary record offers less to confirm if the two other men named Tom, advertised as in flight by John Mayo in 1768 and 1769, had both escaped from Dismal Plantation. Here, a careful comparison of the two advertisements reveals that they may have been the same person. In both advertisements, Mayo describes the Tom who escaped as about six feet tall, with one cropped ear, and taking refuge in the Dismal Swamp, with knowledge of two of the three Virginia counties—Nansemond and Norfolk—whose jurisdictions covered Dismal Swamp lands.

In truth, as Mayo's suspicion that Tom moved about Norfolk and Nansemond Counties suggests *before* the American Revolution's overt fighting reached Southside Virginia, Dismal Plantation's population was much less isolated from more traditionally known contexts for late eighteenth-century enslavement in the Old Dominion than some scholars have considered. Other freedom seekers documented in runaway advertisements were similarly suspected of moving across significant distances in their efforts to find freedom with their feet—authors of their own resistance stories and coauthors of the advertisements themselves, as the historian Antonio T. Bly has argued.[35] In November 1771, two thirty-year-olds enslaved by Nathaniel Burwell—Jack and Venus—fled from Warresqueak Bay. Nearly one month passed before Burwell posted an advertisement in the *Virginia Gazette*, articulating his suspicion that both Jack and Venus had fled *to* the Great Dismal Swamp, where both had labored for two years at Dismal Town Plantation.[36] There, both had been documented in the 1764 appraisal list.

Taken together with the Dismal Swamp Company's 1764 appraisal list, and the selection of runaway advertisements we have considered, the man who captured a fleeting moment of Aitchison and Parker's attention was not the only enslaved person of African descent to experience a life of liminal freedom in the mid-eighteenth-century Dismal Swamp. The two merchants recorded no evidence of the Dismal Swamp furniture maker's wares, though these were products that we might assume were of interest to the two men who sought marketable goods to consign to merchants in Havana or Tenerife or Honduras. Nor did the merchants provide any evidence of the swamp furniture maker's precise camp location within the swamp. The merchants also did not record further evidence of the man's subsistence activities, raising rice and grains for survival. Instead, the account book captures a series of decontextualized mercantile transactions from 1763 to 1804. What, then, explains such a peculiar placement for such a peculiar story about an African descendant, skilled in the practical work of making furniture and of growing grains? What explains the merchants' interest in the man such that they recorded it in an account book otherwise dedicated to transatlantic networks? Why do such questions about a single primary source matter in our present-day context of primary research on the American Revolution in a digitized age?

The answers to these questions, in part, demonstrate the utility of plumbing archival shadows for the stories of historical actors—enslaved, more often than not—whose lives mattered to contemporaries, if only in passing. In pursuing such research, it is important to carefully consider the context in which Aitchison and Parker likely regarded the Black furniture maker as more than just an enslaved human commodity. Informed by scholarship in the Black digital humanities, this perspective casts aside debate about the source's veracity to instead turn attention to a Black furniture maker obscured in the archival shadows cast by Aitchison and Parker's account book, the book's provenance notwithstanding. A core lesson of the Black digital humanities informs us that we must engage seriously the humanity of the enslaved people whose presence was recorded fleetingly in otherwise mundane archival records as a central aim of the method of verifying and contextualizing such sources in new histories and scholarship.

Moreover, the work of plumbing archival shadows must continue to happen in physical archives, even as digitized humanities projects expand access to archived materials. The limited reach of the Aitchison and Parker account book (or of other sources penned by the merchants) in the current digitized age is a function of its relative inaccessibility. To access the account book requires the privilege of a visit to the physical archives held the Albert and Shirley Small Special Collections of the University of Virginia's Alderman Library in Charlottesville. However, evidentiary fragments about the Dismal Swamp's enslaved population in the revolutionary era, such as the runaway advertisements for Tom(s), are today facilitated widely by a growing range of digitized humanities projects that have broadened access to previously hard to access primary sources through free access platforms, such as the Geography of Slavery in Virginia project. UVA Press's Rotunda project, a vital digitized research collection, provides access to a wide range of digitized papers from prominent actors of the revolutionary and early republican periods but only to researchers with the resources to pay a subscription fee. Available at disproportionate rates to researchers at the best-endowed universities and institutions, such resources are not readily available to researchers at historically disadvantaged institutions such as historically Black colleges and universities.

The example of the several Toms' flight *within* the Dismal Swamp and its immediate environs, of the Dismal Swamp's Black furniture maker, of Jack and Venus, and of the countless others who found refuge in the Dismal Swamp during the American Revolution and afterward, all highlight the challenges presented by the work of historical recovery. These men, women, and children hid from enslavers, and thus, from the instruments of the state by which their activities would appear in the extant primary sources on which we draw to write "history." So they remain obscured by archival shadows. Still, in this new era of studying the history of the American Revolution in the digital age, their stories nonetheless matter and inform in important ways our methods for uncovering new histories.

NOTES

1. John Selby, *The Revolution in Virginia, 1775–1783* (Charlottesville: University of Virginia Press, 2001), 67–68, 80–85. On James Parker, see Selby, 59–61; Woody Holton, *Forced Founders: Indians, Debtors, Slaves, and the Making of the American Revolution in Virginia* (Chapel Hill: University of North Carolina Press, 1999), 23–40; and Michael A. McDonnell, *The Politics of War: Race, Class, and Conflict in Revolutionary Virginia* (Chapel Hill: University of North Carolina Press, 2007), 169–70.

2. Maya Jasanoff, *Liberty's Exiles: American Loyalists in the Revolutionary World* (New York: Vintage, 2012); Maya Jasanoff, "The Other Side of Revolution: Loyalists in the British Empire," *William and Mary Quarterly*, 3rd ser., 65, no. 2 (April 2008): 205–32.

3. Account Book of William Aitchison and James Parker, 1763–1804, 51, Albert and Shirley Small Special Collections Library, University of Virginia, Charlottesville; Marcus P. Nevius, *City of Refuge: Slavery and Petit Marronage in the Great Dismal Swamp, 1763–1856* (Athens: University of Georgia Press, 2020), 20–22.

4. See, for example, Christopher James Bonner, *Remaking the Republic: Black Politics and the Creation of American Citizenship* (Philadelphia: University of Pennsylvania Press, 2020); Amrita Chakrabarti Myers, *Forging Freedom: Black Women and the Pursuit of Liberty in Antebellum Charleston* (Chapel Hill: University of North Carolina Press, 2011); and Leslie M. Alexander, *African or American? Black Identity and Political Activism in New York City, 1784–1861* (Urbana: University of Illinois Press, 2008). On Black resistance at the center of the early republic's abolition movement(s), see Manisha Sinha, *The Slave's Cause:*

A History of Abolition (New Haven, CT: Yale University Press, 2016); Manisha Sinha, "To 'Cast Just Obliquy' on Oppressors: Black Radicalism in the Age of Revolution," *William and Mary Quarterly* 64, no. 1 (January 2007): 149–60; and Benjamin Quarles, *Black Abolitionists* (New York: Oxford University Press, 1969).

5. See, for example, Alexandra J. Finley, *An Intimate Economy: Enslaved Women, Work, and America's Domestic Slave Trade* (Chapel Hill: University of North Carolina Press, 2020); Daina Ramey Berry, *The Price for a Pound of Their Flesh: The Value of the Enslaved, from Womb to Grave, in the Building of a Nation* (Boston: Beacon Press, 2017); Sven Beckert and Seth Rockman, eds., *Slavery's Capitalism: A New History of American Economic Development* (Philadelphia: University of Pennsylvania Press, 2016), 1–28; Calvin Schermerhorn, *The Business of Slavery and the Rise of American Capitalism, 1815–1860* (New Haven, CT: Yale University Press, 2015); Edward E. Baptist, *The Half Has Never Been Told: Slavery and the Making of American Capitalism* (New York: Basic Books, 2014); and Joshua D. Rothman, *Flush Times and Fever Dreams: A Story of Capitalism and Slavery in the Age of Jackson* (Athens: University of Georgia Press, 2012).

6. On marronage in the Great Dismal Swamp, see J. Brent Morris, *Dismal Freedom: A History of the Maroons of the Great Dismal Swamp* (Chapel Hill: University of North Carolina Press, 2022); J. Brent Morris, "'Running Servants and All Others': The Diverse and Elusive Maroons of the Great Dismal Swamp," in *Voices from within the Veil: African Americans and the Experience of Democracy*, ed. William Alexander, Cassandra Newby-Alexander, and Charles Ford (Newcastle-upon-Tyne: Cambridge Scholars, 2008), 85–112; Marcus P. Nevius, "New Histories of Marronage in the Anglo-Atlantic World and Early North America," *History Compass* 18, no. 5 (May 2020): 1–14; Kathryn Benjamin Golden, "Armed in the Great Swamp": Fear, Maroon Insurrection, and the Insurgent Ecology of the Great Dismal Swamp," *Journal of African American History* 106, no. 1 (Winter 2021): 1–26; Sylviane A. Diouf, *Slavery's Exiles: The Story of the American Maroons* (New York: New York University Press, 2014); Ted Maris-Wolf, "Hidden in Plain Sight: Maroon Life and Labor in Virginia's Dismal Swamp," *Slavery and Abolition* 34, no. 3 (2013): 446–64; and Megan Kate Nelson, "Hidden Away in the Woods and Swamps: Slavery, Fugitive Slaves, and Swamplands in the Southeastern Borderlands, 1739–1845," in *"We Shall Independent Be": African American Place Making and the Struggle to Claim Space in the United States*, ed. Angel David Nieves and Leslie M. Alexander (Boulder: University Press of Colorado, 2008), 251–72.

7. DocSouth dates back to 1991 and published on the internet its first slave narrative in 1995. The project was informed by broader efforts at the university to advocate the construction of a building to house what became the Sonja Haynes Stone Center for Black Culture and History. For more on DocSouth, see Patricia Buck Dominguez and Joe A. Hewitt, "A Public Good: Documenting the American South and Slave Narratives," *RBM: A Journal of Rare Books, Manuscripts, and Cultural Heritage* 8, no. 2 (2007): 106–23. My thanks to Maria Estorino, associate university librarian at the University of North Carolina at Chapel Hill Libraries, for providing crucial assistance with my attempt to historicize the development of the DocSouth.

8. Rotunda, accessed July 2, 2021, https://www.upress.virginia.edu/rotunda.

9. A decade later, David Eltis and David Richardson, in collaboration with scholars, students, and other researchers, moved the database to an online, open access website, hosted at Emory University, accompanied by a physical volume. See David Eltis and David Richardson, *Atlas of the Transatlantic Slave Trade* (New Haven, CT: Yale University Press, 2010). The Trans-Atlantic Slave Trade Database has been transferred to new hosts at Rice University.

10. Sharon Block, "#DigEarlyAm: Reflections on Digital Humanities and Early American Studies," *William and Mary Quarterly*, 3rd. ser., 76, no. 4 (October 2019): 611–48, 611, 612.

11. Lara Putnam, "The Transnational and the Text-Searchable: Digitized Sources and the Shadows They Cast," *American Historical Review* 121, no. 2 (April 2016): 377–402, 377.

12. Kim Gallon, "Making a Case for the Black Digital Humanities," in *Debates in the Digital Humanities 2016,* ed. Lauren F. Klein and Matthew K. Gold (Minneapolis: University of Minnesota Press, 2016), 42–49. Gallon's definition for Black studies builds, in part, on Alexander Weheliye's framing of the discipline as a "mode of knowledge production" for the "comparative study of black cultural and social experiences under white Eurocentric systems of power in the United States." See Alexander G. Weheliye, *Habea Viscus: Racializing Assemblages, Biopolitics, and Black Feminism Theories of the Human* (Durham, NC: Duke University Press, 2014).

13. Vincent Brown, "Mapping a Slave Revolt: Visualizing Spatial History through the Archives of Slavery," *Social Text* 33, no. 4 (December 2015): 134–41; Vincent Brown, Slave Revolt in Jamaica, 1760–1761: A Cartographic Narrative, revolt.axismaps.com; Vincent Brown, *Tacky's Revolt: The Story of an Atlantic Slave War* (Cambridge, MA: Belknap Press of Harvard University Press, 2020); Vincent Brown, *The Reaper's Garden: Death and Power in the World of Atlantic Slavery* (Cambridge, MA: Harvard University Press, 2010).

14. Jessica Marie Johnson, "Markup Bodies: Black [Life] Studies and Slavery [Death] Studies at the Digital Crossroads," *Social Text* 36, no. 4 (137) (December 2018): 57–79.

15. Nevius, *City of Refuge*, 26–28.

16. *Virginia Gazette* (Dixon and Hunter), April 22, 1775; Robert G. Parkinson, *The Common Cause: Creating Race and Nation in the American Revolution* (Chapel Hill: University of North Carolina Press, 2016), 82–85; Holton, *Forced Founders*, 140–48; Selby, *Revolution in Virginia*, 1–7; James Corbett David, *Dunmore's New World: The Extraordinary Life of a Royal Governor in Revolutionary America with Jacobites, Counterfeiters, Land Schemes, Shipwrecks, Scalping, Indian Politics, Runaway Slaves, and Two Illegal Royal Weddings* (Charlottesville: University of Virginia Press, 2013), 122–23.

17. Alan Taylor, *The Internal Enemy: Slavery and War in Virginia, 1772–1832* (New York: W. W. Norton, 2014); Sylvia R. Frey, *Water from the Rock: Black Resistance in a Revolutionary Age* (Princeton, NJ: Princeton University Press, 1991).

18. Parkinson, *Common Cause*, 142–57. On the previous drafts of Dunmore's proclamation, see McDonnell, *Politics of War*, 131–32; Selby, *Revolution in Virginia*, 64; Holton, *Forced Founders*, 155. For a list of fourteen newspapers that exchanged reports on Dunmore's efforts to encourage slave flight, see Parkinson, *Common Cause*, 143n60. For the classic story of Dunmore's efforts to raise the "Ethiopian Regiment," see Benjamin Quarles, *The Negro in the American Revolution* (Chapel Hill: University of North Carolina Press, 1961), 19–32.

19. Parkinson, *Common Cause*, 163, 245–49, 277–83.

20. "Return of Persons Who Came off from Virginia with General Matthew in the Fleet of the 24 May 1779," microfilm reel 28, item no. 10325, British Headquarters (Sir Guy Carleton) Papers, Library of Congress, Washington, DC, 1957; Parkinson, *Common Cause*, 460–61; Selby, *Revolution in Virginia*, 204–6, 220–25; McDonnell, *Politics of War*, 343–44, 398–434.

21. Parkinson, *Common Cause*, 507–11; Selby, *Revolution in Virginia*, 220–25.

22. Cassandra Pybus, *Epic Journeys of Freedom: Runaway Slaves of the American Revolution and Their Global Quest for Liberty* (Boston: Beacon Press, 2006), 47; Charles Royster, *The Fabulous History of the Dismal Swamp Company: A Story of George Washington's Times* (New York: Vintage Books, 1999), 271–72.

23. Elizabeth A. Fenn, *Pox Americana: The Great Smallpox Epidemic of 1775–82* (New York: Hill and Wang, 2001), 129–33; Philip Ranlet, "The British, Slaves, and Smallpox in Revolutionary Virginia," *Journal of Negro History* 84, no. 3 (1999): 217–26.

24. Parkinson, *Common Cause*, 512–26.

25. Lathan A. Windley, *A Profile of Runaway Slaves in Virginia and South Carolina from 1730 through 1787* (New York: Garland, 1995), 287, 291–92.

26. Cassandra Pybus, "Jefferson's Faulty Math: The Question of Slave Defections in the American Revolution," *William and Mary Quarterly*, 3rd ser., 62, no. 2 (April 2005): 243–64. Figures quoted on 261.

27. *Virginia Gazette*, June 23, October 6, 1768; Royster, *Fabulous History*, 147. For more on country marks, see Michael A. Gomez, *Exchanging Our Country Marks: The Transformation of African Identities in the Colonial and Antebellum South* (Chapel Hill: University of North Carolina Press, 1998), esp. 1–16.

28. Edward D. Maris-Wolf, "Between Slavery and Freedom: African Americans in the Great Dismal Swamp, 1763–1863" (MA thesis, College of William and Mary, 2002), 51–54. See also Daniel O. Sayers, *A Desolate Place for a Defiant People: The Archaeology of Maroons, Indigenous Americans, and Enslaved Laborers in the Great Dismal Swamp* (Gainesville: University Press of Florida, 2014), 88–89.

29. *Virginia Gazette*, April 13, 1769, emphasis added.

30. Noeleen McIlvenna, *Early American Rebels: Pursuing Democracy from Maryland to Carolina, 1640–1700* (Chapel Hill: University of North Carolina Press, 2020); Noeleen McIlvenna, *A Very Mutinous People: The Struggle for North Carolina, 1660–1713* (Chapel Hill: University of North Carolina Press, 2009); Arwin D. Smallwood, "A History of Native American and African Relations from 1502 to 1900," *Negro History Bulletin* 62, no. 2/3 (April–September 1999): 18–31; Arwin D. Smallwood, "A History of Three Cultures: Indian Woods, North Carolina 1585 to 1995" (PhD diss., Ohio State University, 1998).

31. Tom Costa, The Geography of Slavery in Virginia, http://www2.vcdh.virginia.edu/gos/.

32. David Jameson to John Driver, December 5, 1783, copy of letter, box 1, folder 2, Dismal Swamp Land Company Records, David M. Rubenstein Rare Book and Manuscript Library, Duke University; Nevius, *City of Refuge*, 26–29.

33. Jacob Collee to David Jameson, December 26, 1784; David Jameson to Jacob Collee, December 30, 1784, copy of letter, box 1, folder 2, Dismal Swamp Land Company Records, David M. Rubenstein Rare Book and Manuscript Library, Duke University; Royster, *Fabulous History*, 290.

34. "Appraisement of Dismal Swamp Slaves," George Washington copy, July 4, 1764, accessed June 2020, University of Virginia Press Rotunda American History Collection, https://rotunda.upress.virginia.edu/founders/GEWN-02-07-02-0191.

35. Antonio T. Bly, "'Indubitable Signs': Reading Silence as Text in New England Runaway Slave Advertisements," *Slavery and Abolition* 42, no. 2 (2021): 240–68.

36. *Virginia Gazette*, December 5, 1771.

Part IV
ECHOES IN THE PRESENT

13
MEDIA LITERACY IN REVOLUTIONARY AMERICA

Jordan E. Taylor

"I thank God *there are no free schools,* nor *printing,*" Governor William Berkeley of Virginia famously wrote in 1671, "for *learning* has brought disobedience, and heresy, and sects into the world, and *printing* has divulged them, and libels against the best governments. God keep us from both!"[1] To British officials in colonial North America, the best colonist was an ignorant one. They believed that farmers, artisans, and laborers should leave the great questions of the day to the wealthy and wise. Though Berkeley thanked heaven for preserving the settlers' ignorance, it was also a matter for men. For many years, British imperial policy had discouraged the free and open flow of information, in part by limiting printers from divulging these heresies.

But even the weight of empire could not dam up these exchanges for long. A print revolution in eighteenth-century North America saw newspapers, books, and other media spread rapidly across Anglo-America, causing a massive expansion in the volume of information available to ordinary people. A century after Berkeley, this democratization of information helped bring about the American Revolution, as colonists questioned the received wisdom of the mighty. And the demands of republican self-government created by the American Revolution convinced many former colonists that their fellow citizens needed to be aware of matters beyond their personal experience. If people could not access information about the broader world, they could not debate the issues, petition leaders, or elect the best representatives. To ensure that the voters of the United States would be well-informed—as historians have shown—political leaders attempted to mobilize institutions to provide news, facts, ideas, and knowledge to a broad citizenry. The most important of these institutions were public schools, the press, and the post office.² As the periodical press expanded rapidly in the aftermath of the American Revolution, the US Congress decided in 1792 to set particularly low postage rates for newspapers, hoping that this would encourage the spread of useful information. With the passage of this legislation, the North Carolina congressman John Steele explained, "If the people hereafter remain uninform'd it must be their own fault."³ For him, ensuring that the people were well-informed was a quantitative issue: when information was scarce, ignorance was forgivable, but with greater access to news, the people would be responsible for identifying truth and shunning falsehoods.⁴

In fact, Steele had approached the problem almost exactly backward. Like many before and after him, he had confused information with knowledge. Tragically, a greater volume of news and information would not equip people to participate meaningfully in politics but would, instead, ultimately provoke greater confusion and misunderstanding. Even as information became more abundant than ever before in the late eighteenth century, many observers believed that they were experiencing an unprecedented epidemic of error. Almost every important report seemed to come accompanied by another claiming the opposite. "The many contradictory reports which are continually flitting through America," noted the Philadelphia printer Benjamin Franklin Bache, "point out the importance of authentic information." He wished for "some means" of "procuring intelligence which could be relied on, frequently and regularly from Europe. At the present crisis this is peculiarly necessary, the situation of transatlantic politics is now uncommonly interesting to Americans."⁵ The printer Joseph Dennie put it more succinctly: "Truth seems to fly from curiosity."⁶ Indeed, there was much to be curious about. The late eighteenth century was a time of important events at home and abroad. Revolutions, declarations, wars, speeches, trade disruptions, and transnational movements occupied the attention of Americans. The world seemed to tilt toward some great conclusion, perhaps a global regeneration or maybe the rebirth of despotism, and the news seemed more important than ever. In this context, a false report might enflame a mob. It might also elect a senator.

The rapid circulation of falsehoods alarmed many elites. By the time they were considering a new national government, the delegates at the Constitutional Convention worried about the peoples' capacity for discerning truth. Roger Sherman of Connecticut, for example, opposed popular elections, arguing instead for state legislatures to choose representatives. "The people," he argued, "should have as little to do as may be about the government. They want information, and are constantly liable to be misled." The Massachusetts delegate Elbridge Gerry concurred, noting that the people were "daily misled into the most baneful measures and opinions by the false reports circulated by designing men, and which no one on the spot can refute." For Gerry, this was at the root of the "excess of democracy" that the Constitutional Convention aimed to redress.[7]

Allergic to nuanced models of causality, Americans usually attributed false news to the intentional efforts of malicious deceivers.[8] Indeed, this was the most basic tenet of early American media literacy. Contradictory accounts of the same event seemed to indicate that at least one person, somewhere, had aimed to dupe others for selfish purposes. According to one revolutionary essayist, while it was true that the "People when well informed never act wrong," they were nevertheless frequently misled "by the artful and vile Insinuations of designing Men."[9] The presence of so much factual discrepancy during the Age of Revolutions, therefore, lent itself to conspiratorial thinking. One lie might be the work of a single ill-intended individual, but piles of contradiction pointed toward a pack of unscrupulous mediators. It was often irresistible, then, for those working to separate truth from falsehood to imagine themselves to be heroic detectives unmasking a conspiracy of deception. What we call "media literacy" today, Americans in the Age of Revolutions viewed as the detection of traitors, fraudsters, and schemers. Of course, falsehoods resulted as often from accidental misunderstandings as from intentional deception, but Americans rarely considered that possibility.

Historians have long puzzled over why the peoples of revolutionary America believed so many things that we know, in retrospect, to have been untrue. Most famously, for example, Patriot leaders in the 1760s and 1770s falsely claimed that the leadership of Parliament was conspiring to enslave them by stripping away their "English liberties." If this was untrue, though, why did they accept it? In the early twentieth century, as the World Wars induced fears of propaganda, a group of scholars subsequently known as the "Progressive" historians suggested that Patriot leaders did not really believe these absurd things, but were only sharing these exaggerations and falsehoods in order to mobilize others.[10] By the middle of the twentieth century, however, another group of historians suggested that the ideologies and politics of the late eighteenth century made colonists more suspicious (they sometimes used the word "paranoid") of these falsehoods.[11] There is another way of thinking about this problem. To a large extent, the inhabitants of revolutionary America were simply absorbing the information they encountered in ways that aligned with the era's expectations for practical media literacy. But those standards happened to be defective, in ways that mirror the ineffective media literacy of our digital present. Understanding where eighteenth-century media literacy erred can help us better navigate the informational landscape of the twenty-first century. Instead of evaluating news as the revolutionary generation did, the denizens of the digital age would do better to take advantage of resources that were unavailable in early America and embrace habits of verification through fact-checking.

In their efforts to detect deceivers, eighteenth-century observers focused primarily on the internal markers of a piece of news—its dates, its stated source, and the social status of its utterer. This was not a very effective approach. Frustratingly, more than two centuries later, Americans are still engaging with information in many of the same ways that served the peoples of revolutionary America so poorly. In the late twentieth and early twenty-first centuries, as in the print revolution centuries before, new technologies and media caused the volume of information available to Americans to increase exponentially.[12] This digital revolution led millions of Americans to hope that democratic networking and communications tools—including social media platforms such as Twitter and fact repositories such as Wikipedia—would liberate people around the world by allowing them access to more information and more unmediated forms of communications. In 2008 the existence of such market-driven platforms led one "futurist" to proclaim "Death to the Gatekeepers."[13] For these techno-utopians, so-called experts had shut out ordinary people from important information for too long, and digital networking promised to break down such barriers. They believed, as Steele did in 1792, that easy access to information would promote greater knowledge about public affairs.

Yet, for every expert sharing meaningful forms of knowledge in newly accessible, nonhierarchical digital spaces, there are many more charlatans, hoaxers, and propagandists competing for attention. As in the late eighteenth century, falsehoods and contradictions have mounted. Less than a decade after some called for the death of the gatekeepers, others mourned the "Death of Expertise."[14] Today, few would agree that access to information makes for an informed citizenry. The digital tools of the twenty-first century have produced a simultaneous explosion of information and ignorance. The internet provides a fertile ground not only for useful political information but also for conspiratorial thinking, "fake news," and deceptive advertising. Though conspiracy theories have been a constant force in American history, the digital age has made unfounded accounts of the world, from QAnon to certain versions of "Russia-gate," more accessible and more adaptable, mutating rapidly to meet changing circumstances and popular attitudes.[15] The result is that Americans have never been so well-informed or so misinformed. While techno-utopians once asked how the digital age will improve society, technologists are now more likely to ask if a digitally abetted flood of information will destroy global democracy.

At the core of these concerns is a fear that Americans lack media literacy. To learn to navigate the twisted pathways of the internet, some argue that Americans must be trained to recognize "bias" and rely on unbiased sources. Yet, as the fittingly named media literacy organization AllSides puts it, "There is no such thing as completely unbiased news. We're all biased, making it impossible to write or curate perfectly objective news."[16] This cynical view, in which all sources are suspect, unintentionally repeats the mistakes of the eighteenth century, when Americans frequently misattributed falsehoods to the wiles of deceptive enemies. Moreover, asking users to personally evaluate information mediators produces greater polarization of information since their judgments of the degree of a source's "bias" will inevitably mirror their own preexisting perceptions.[17] Asking others to see the world through the lens of "bias" is inviting them to climb into filter bubbles and ignore uncongenial news. This emphasis on bias invites a dangerous epistemic nihilism. If all sources are biased, what is the point in upholding the concept of truth at all?

Other critics have insisted that meaningful media literacy requires more than exercising suspicion. Instead, they insist that identifying false news and discerning truth requires information consumers to build productive habits and exercise self-discipline to resist the urge to automatically label uncongenial news as false. Effective media literacy requires that the internet's itinerants work to actively unearth valuable evidence that can verify or disprove discrete claims about the world. Using a search engine in a thoughtful way, seeking previous work performed by professional fact-checkers, or even taking a well-directed peek at Wikipedia, ordinary users can fact-check the information they encounter in a few minutes or less.[18] If digital tools offer countless pathways to falsehood, they also offer relatively easy access to powerful practices of verification. With these tools, a fact-checking habit is far more effective for combating false news than dismissing "biased" news sources.[19]

Whereas Americans today take these fact-checking tools for granted, their predecessors in the late eighteenth century would have envied them. Throughout the late eighteenth century, across rapidly changing political and social contexts, their methods for ferreting out falsehoods changed little; they were consistently difficult, slow, uncertain, and ineffective. Because the most important news of the day usually originated across forbidding oceans, which took weeks to cross, it was not possible to verify reports as they arrived in North America. Instead, infrastructural limitations restricted the observers of revolutionary America to relying on several futile techniques for determining the truth value of the news they received: they evaluated the timeline of a piece of news, counseled patience, linked truthfulness with the status of mediators, and dismissed biased sources.

The most straightforward of these media literacy tools available in revolutionary America involved closely examining the dates attached to a piece of news. Early American newspaper reports often carried elaborate headings full of details that today's observers might consider extraneous. Take one example: "New York, Dec. 2. A letter from a respectable gentleman, dated Amsterdam, September 8, 1786, brought by capt. Baas, who arrived at Charleston, S. C. the 13th ult."[20]

Modern readers can take it for granted that the paper had gathered its information according to professional norms, but eighteenth-century readers could make no such assumptions. This lengthy heading served as a chain of evidence, tracing the letter from Amsterdam in September to Charleston in November to New York in early December. If one link in this chain proved to be weak, it might cast doubt on the authenticity of the letter.

Anyone forging a letter or report would need to produce such a chain of evidence but could easily make a mistake in the process. News that arrived with an error or absence in the chain was automatically suspect. During the American Revolution, an enterprising writer in the Patriot *Boston Evening-Post*, for example, claimed to have exposed a 1774 letter from Philadelphia as a forgery on this basis: "I knew from the palpable Falshoods contained in it that it must have been forged, but looking at the date every one must be convinced of this." The letter had been dated May 17 but referenced news from Boston on May 13, which created a timeline that seemed unlikely to this author: "How the Post could carry in 4 days this account to Philadelphia, is beyond my comprehension." Information that moved at a suspiciously quick pace must have been invented. Because the letter offered bad news for Patriots, he concluded that it was "one Instance among many of the lying tricks of our Enemies."[21]

Similarly, when evaluating an account, a reader might compare it to news from the same place that had arrived "of a later date," or, in the terminology of the time, the "freshest advices." If the later report contradicted the earlier account, observers often settled for the former. Challenging an account of "great disturbances" in Guadeloupe described by a letter dated September 25, 1791, a writer in a Boston newspaper noted the arrival of a letter dated October 2 that did not mention such disturbances. This author chided newspaper editors for circulating the original account, arguing that in matters "which may affect the commerce of our country, the Editors of newspapers ought to be peculiarly cautious."[22] Likewise, in 1794 Bache dismissed a report of the French revolutionary wars from the *London Gazette*, and instead passed along "accounts from French papers," which were "4 or 5 days later from the great scene of action in Europe, than any intelligence before published here."[23] A rabidly prorevolutionary Francophile, Bache was suspicious of the British accounts to begin with and may have used their dates to wave them away while boosting news originating in France.

Another technique involved withholding judgment about doubtful information and patiently awaiting more news. Observers assumed that while two conflicting pieces of news created uncertainty, ten reports about the same event would likely skew in the direction of the truth. As a result, they often noted if a piece of news was as-yet unconfirmed. Some criticized those who published "premature" accounts without supporting evidence. At the end of the American

Revolutionary War, for example, the Rhode Island printer John Carter noted that some ships had arrived with an account of the destruction of a British fleet. But while he noted that he was happy to "communicate glad Tidings," he held back, believing it was "equally his Duty to contradict premature Intelligence, when proved to be such."[24] In some cases, the lack of follow-up confirmation, rather than an explicit contradiction, was enough to lead observers to discount news. In 1793 Susanna Dillwyn of Philadelphia wrote to her father describing a rumor about the seizure of a ship carrying her aunt and uncle. When she first heard the news, she explained, it was "generally believed throughout the Town," while today "the circumstance is mention'd in the papers, but nobody seems to have heard anything new, so that I think it may possibly be a false rumor."[25] The nature of the early modern information economy ensured that reports of significant events would usually arrive through multiple conveyances and multiple media. A piece of news that stood alone was suspect.

Compared to other strategies, waiting for more news was reasonably effective. But caution for confirmation sometimes resulted from partisan or personal motives. When the Massachusetts lawyer Dwight Foster heard reports of the American army's defeat at Ticonderoga in 1775, he wrote to his father that "we hope however that this is false and want to hear it confirmed ere we credit it."[26] Foster's expression of his "hope" and "want" regarding this news indicates that his preferences were guiding his use of this media literacy strategy. This was not unusual. The historian Matthew Rainbow Hale has noted that in the early 1790s members of the Francophobic Federalist Party in the United States were particularly likely to urge observers to wait patiently for news about the French Revolution as the rush of reports threatened to overwhelm them.[27] With war, terror, and revolution spreading through Europe, Federalist commentators regularly argued that until the "storm is blown over . . . certainty cannot be attained."[28] This caution was not a feature of Federalism but of partisanship more generally. Partisans of all stripes counseled caution when news arrived that they perceived as harmful to their political cause. In 1798, as reports from France of what became known as the X. Y. Z. Affair seemed to run against Republican interests, Republican mediators urged Americans to wait for the French government's response before "we form our judgement."[29] This tool of media literacy was easily refashioned into a partisan weapon.

Another approach that Americans used to analyze information was to interrogate the social status of the news bearer. Many associated the authenticity of a letter or oral report with the wealth and reputation of the person who had articulated it. Because North Americans generally viewed false news as the product of intention, rather than error, they assumed that respectable gentlemen would not stake their reputations on deception. Eighteenth-century American elites inhabited an honor culture that valued above all else a gentleman's reputation for truthfulness.[30] As a result, newspaper headings regularly offered extracts of letters from gentlemen of "reputation," "undoubted veracity," "great political information," or "a very respectable [mercantile] house."[31] Others were introduced as being from a "respectable citizen," an "unquestionable authority," or a "high authority."[32] This rhetoric drew on a long-standing habit in the Western world of attributing higher truth values to claims made by elite, educated, white men.[33] Not yet recognized as a logical fallacy, appeals to authority held great sway in convincing many readers.

Unsurprisingly, though, some observers were more status-conscious than others. During the 1790s, those allied with the elitist Federalist Party went out of their way to note the status of their information sources. Federalists were particularly likely to rely on ship captains and merchants, while emphasizing their reliability. When Alexander Young and Thomas Minns, the Federalist printers of the *Massachusetts Mercury*, received news from a Captain Bacon, who had just returned from western France, about a French attack on the Spanish city of Bilboa in late 1794, they noted that they were relying on the word of a "gentleman of veracity, and intelligence." Yet they also pointed out that Bacon did not speak French, and therefore was just passing on what others told him had been printed in the French newspapers.[34] Some Republicans would have scoffed at this. Indeed, in the summer of 1791, Bache had mocked the idea of relying on ship captains who did not speak the local language. It was absurd, he argued, that other Americans should trust the word of captains "who generally not speaking the language of the inhabitants, and being the greatest part of their time on board their vessels, which, from the nature of the harbor, are anchored out at some distance from the town, have not many opportunities of gathering information."[35] For Bache, status and character were not as important as a person's direct experience with a news source.

Indeed, during the 1790s members of the populist Republican Party often vocally rejected the idea of evaluating information based on the status of an information source. In fact, the Republican printer Philip Freneau addressed this point directly in his Philadelphia *National Gazette*. He complained that some people had asked him about the character or status of some of his correspondents: "whether he be a foreigner, or home born, or well-born . . . a man of property, or a no property man?" He concluded by asking "such inquisitive persons" to "mind your own business."[36] For Freneau, a correspondent's social status had little to do with the value of the information they shared. Likewise, the Republican printer Thomas Adams dismissed the truth value of a 1794 letter from France even though, as he noted, the author was "well known in this town." Adams might have easily used the author's status to assert that the letter's news was authentic, but instead he cast doubt on it by asserting that he would "not undertake to determine ... whether its very important contents are true."[37]

Judging news based on the identity of its bearer often backfired. Indeed, the Boston preacher Jedidiah Morse unleashed one of the great conspiracy theories of the eighteenth century based largely on the apparent credibility of a Scottish academic by the name of John Robison.[38] Convinced that a secretive group known as the Illuminati had kindled the French Revolution and was plotting against Christendom in Europe and the United States, Morse initiated what became known as the "New England Illuminati Scare." When challenged about the absurdity of these claims, Federalist supporters inevitably pointed out that Robison was a "gentleman of character and station."[39] By any standard of the time, this was true. Nevertheless, his most important claims proved to be totally unfounded, to the embarrassment of Morse and his friends. Even important men often got it wrong.

While some of these tools were more useful than others, the most destructive, counterproductive, and widespread media literacy strategy in the late eighteenth century was the application of the concept of "bias" to news. While this was not a new idea, in the late eighteenth century Americans increasingly asserted that the mediators involved in the production of news were unfair, untrustworthy, or otherwise corrupt. The events of the late 1760s and early 1770s led some American colonists to worry extensively about the biases of information mediators. In a series of incidents well known to historians of revolutionary America, colonists became concerned that two successive governors of Massachusetts, Francis Bernard and Thomas Hutchinson, were misrepresenting the state of affairs in their colony in order to beg for troops and make themselves seem more important in the eyes of their superiors. The publication of two collections of intercepted letters from Bernard, Hutchinson, and others contained little evidence of dissimulation but nevertheless convinced many Patriots that British officials were biased against them and were providing an exaggerated account of events.[40] When Britain sent troops to Boston to enforce order, Patriot leaders blamed their arrival on the (apparent) falsehoods propagated by these men.[41]

By 1770, as the Tory ministry led by Lord North ascended to power in London, Patriot leaders began to worry about a new kind of misrepresentation and a new kind of bias. They focused on the relationship between the British press and the ministry, arguing that newspapers from London reported whatever aided the ministry, regardless of whether or not it was true. British opposition papers seeded this notion, complaining that North was working to "deceive both Parliament and the world" through subsidies to the ministerial press.[42] This was not without foundation. The ministry did indeed publish a propaganda outlet in the *London Gazette* and subsidized a number of loyal newspapers.[43] By the middle of the decade, though, Patriot colonists in North America had become nearly obsessed with the idea that the London "ministerial press" was biased and unreliable. In 1776, for example, the Philadelphia Patriot printer John Dunlap passed along an account of troop movements with the loaded heading "From the London *(or lying)* Gazette of May 3."[44] Even when the *Gazette* surprised Patriots by publishing materials that ran against its interests, printers nevertheless returned to the bias framework. In 1780 the *Maryland Journal* printer Mary Katherine Goddard exclaimed, "The London Gazette gives a melancholy Picture of conquering America!"[45] If the ministry's biased mouthpiece was making such an admission, Goddard suggested, then it must be true. For many, the politics of a news source had become the sole criterion through which they judged a report's authenticity.

As war approached, a few Loyalists suggested that the Patriot colonists' grievances were founded on their imbalanced diet of news. Because they only read letters and newspapers from their allies in England, they pressed, of course the Patriots imagined the ministry to be their inveterate enemies. In 1774 an essayist signing off as "Mercator" wrote to James Rivington's Loyalist paper alleging that his rival paper, run by the Patriot printer James Holt, was little more than a calamity of confirmation bias: "With respect to foreign intelligence, those paragraphs" that disparaged "the Ministry and the Parliament, and tend to widen the breach between Great Britain and the Colonies, are industriously selected."[46] While Holt denied the charge, he had indeed reprinted considerably less material from London ministry papers and more from opposition papers than the city's other printers.[47] A few months later, a Boston Loyalist with the pseudonym "An Observer" mocked their neighbors' "implicit credulity" toward news that pleased them. Instead of hearing from "both sides," they charged, Patriots had simply dropped their subscriptions to Loyalist papers, preferring to "listen with greediness to one side only."[48] The Patriots' defective media literacy and insufficient understanding of bias, these Loyalists charged, had led them to the brink of war. And perhaps it had. By the mid-1770s, leading Patriots believed many things about the internal politics of London that turned out to be false.[49] If they had taken the *London Gazette* and other "ministerial" papers into account, Patriots might have arrived at a fuller, or at least different, picture of London's politics.

Americans continued to accuse British newspapers of bias after the American Revolutionary War. In the 1780s and 1790s, though, they seldom distinguished between the politics of different British newspapers. In 1786 the New York printer Francis Childs took note of British newspapers' "reproaches, invectives, and indignities" toward the new nation. He suggested that Americans should "be no longer indebted for our intelligence to channels so corrupt."[50] Why should American newspapers continue to reprint so much of their material from British papers that had exhibited such a clear anti-American bias during the war? This rhetoric ramped up with the onset of the French Revolution as Americans wondered why they should trust reports appearing in London newspapers about Britain's ancient enemy. One Rhode Island essayist explained his thought process for dismissing the legitimacy of British newspapers: "If the English or their rulers are embittered toward France by repeated losses, if they apprehend danger from the progress of French politics to their own internal safety, they will sometimes, to cover their losses and secure that safety, disguise the truth in a robe of their own making, or conceal her wholly from our eyes."[51] Because Britain had an interest in hiding truth about the French Revolution from its people, many commentators argued, its newspapers could not be trusted.[52]

During the late eighteenth century, Americans used a promiscuous mix of these four techniques—attending to dates, waiting for more information, assessing the status of a mediator, and dissecting the "biased" politics of a source—to judge the authenticity of any given piece of news. Why weren't they effective? In evaluating information sources, these techniques led observers to appeal mostly to their intuition. "Bias" and "status" are, after all, cousins to "common sense," "critical thinking," and other slippery subjectivities. They empower individuals to determine the truth value of a piece of information on their own, rather than relying on a broader community of expertise and knowledge. Instinctive interpretations of news lead to what social scientists today call motivated reasoning: the tendency for an individual to subject uncongenial information to greater scrutiny than information that supports their preexisting beliefs.

Indeed, eighteenth-century observers tended to wield these tools unevenly. They scrutinized news they disliked while asking fewer questions about news they approved of. They rarely questioned the "bias" of news taken from a friendly information source, any more than they questioned the status of a bearer of good news. Motivated reasoning exerts a powerful counterweight against media literacy strategies centered around an untrained individual's heroic capacity for discerning truth. Professional fact-checkers, though, argue that the first and most important question to ask is not "Is this a trustworthy mediator?" but rather "Can this be verified using trusted techniques?"[53] The most effective advocates for media literacy in the twenty-first century tend to focus more on developing an individual's ability to tap into a globally networked community of expertise than on developing an individual's capacity to detect nonsense through critical thinking.

The use of faulty media literacy tools likely contributed to the highly polarized politics of revolutionary America. Relying on their own sense of a piece of news or a mediator's trustworthiness led observers to accept news that confirmed their expectations while rejecting nearly everything else. This pattern is most clearly evident in the era's newspapers. In the late 1770s and early 1780s, Loyalist newspaper printers shared news from the British ministry's *London Gazette* at about twice the rate of Patriot presses. Patriot papers, in turn, shared reprints from friendly radical and opposition papers three to four times as often as Loyalist papers. During the early 1790s, partisan newspapers also found themselves in similar information silos. From 1793 through 1795, when news from revolutionary France was particularly controversial, Francophilic Republican newspapers cited news from French presses significantly more often than their anti-French Federalist counterparts.[54] The era's media literacy techniques produced something like the political echo chambers of today's media environment.

Americans past and present have struggled to make sense of the torrent of information unleashed by successive media revolutions—first a print revolution, then a digital revolution. As in revolutionary America, false news runs rampant today. But there are crucial differences between the informational dynamics of these two eras. In the late eighteenth century, much of the news Americans cared most about was simply unverifiable. Slow communications across prohibitive distances prevented anything that resembled fact-checking. Because it took weeks for news from Europe or the Caribbean to arrive, it was not possible for observers to seek confirmation for the transnational news that most occupied them. They could do little more than analyze what they received, and so "media literacy" tools focused on evaluating the credibility of a source according to its internal markers. Today, though, it is by comparison fairly easy to verify a piece of news by relying on experts who obey ethical standards and professional norms regarding the verification of evidence. Entire professions—with painful deterrents against fabrication—exist to sift through information, evaluate its authenticity, and share their findings with anyone who will listen. Because of this work, amateur fact-checkers can easily punch a claim into a search engine, run a reverse image search, or go "upstream" to determine if a cited source really says what a Facebook post claims it does. Proficient amateur fact-checkers can dart from tweet to tweet like a hummingbird in a garden. Though hardly perfect, simple fact-checking protocols can quickly answer many digital media users' questions about the authenticity of any given piece of news.[55]

Yet the lazy logic of "bias" remains far more popular in the United States today than these near-miraculous digital tools of verification. Researchers have shown that news consumers in the United States are more likely to interrogate the credibility and bias of an information mediator rather than fact-check the evidence presented in a news item, even though the latter course of action is significantly more effective.[56] Despite massive shifts in media technologies and cultures since the late eighteenth century, Americans have retained the same defective strategies for navigating the news. Modern militaries do not fight each other with flintlock muskets and bayonets, yet Americans today fight disinformation with tools developed in response to the print revolution. It is time that the inheritors of the digital revolution leave behind early modern methods of evaluating the truthfulness of the news and instead take up the best tools available in the digital age. While fact-checking tools were unavailable to eighteenth-century observers, Americans in the digital age have no excuses.

NOTES

1. William Waller Hening, ed., *The Statutes at Large . . . of Virginia* (New York: R. & W. & G. Bartow, 1810–32), 2:517.

2. Richard D. Brown, *The Strength of a People: The Idea of an Informed Citizenry in America, 1650–1870* (Chapel Hill: University of North Carolina Press, 1996), esp. chaps. 2–3; Richard R. John, *Spreading the News: The American Postal System from Franklin to Morse* (Cambridge, MA: Harvard University Press, 1998); Johann Neem, *Democracy's Schools: The Rise of Public Education in America* (Baltimore: Johns Hopkins University Press, 2017).

3. Noble E. Cunningham and Dorothy H. Cappel, eds., *Circular Letters of Congressmen to Their Constituents, 1789–1829* (Chapel Hill: University of North Carolina Press, 1978), 1:9.

4. On information and republican politics, see Sophia Rosenfeld, *Democracy and Truth: A Short History* (Philadelphia: University of Pennsylvania Press, 2019).

5. Benjamin Franklin Bache, "Philadelphia," *General Advertiser* (Philadelphia), July 26, 1793.

6. Joseph Dennie, *Prospectus of a New Weekly Paper . . . The Port Folio* (Philadelphia: Joseph Dennie and Asbury Dickins, 1800).

7. [James Madison], *Journal of the Constitutional Convention*, ed. E. H. Scott (Chicago: Scott, Foresman, 1893), 78.

8. Gordon S. Wood, "Conspiracy and the Paranoid Style: Causality and Deceit in the Eighteenth Century," *William and Mary Quarterly* 39, no. 3 (July 1982): 401–41.

9. "Mr. Draper," *Boston News-Letter*, March 19, 1772.

10. Among the most prominent works within this body of scholarship are John C. Miller, *Samuel Adams, Pioneer in Propaganda* (Boston: Little Brown, 1936); Philip Davidson, *Propaganda and the American Revolution, 1763–1783* (Chapel Hill: University of North Carolina Press, 1941); and Arthur M. Schlesinger, *Prelude to Independence: The Newspaper War on Britain, 1764–1776* (New York: Alfred A. Knopf, 1958).

11. For a historiographical discussion of this transition, see Gordon S. Wood, "Rhetoric and Reality in the American Revolution," *William and Mary Quarterly* 23, no. 1 (January 1966): 3–32.

12. Martin Gurri, *The Revolt of the Public and the Crisis of Authority in the New Millennium* (San Francisco: Stripe Press, 2018), chap. 1.

13. Thomas Frey, "Death to the Gatekeepers," *Futurist Speaker*, April 3, 2008, https://futuristspeaker.com/business-trends/death-to-the-gatekeepers/.

14. Thomas M. Nichols, *The Death of Expertise: The Campaign against Established Knowledge and Why It Matters* (New York: Oxford University Press, 2017).

15. On conspiracy theories in twentieth-century America, see Kathryn S. Olmsted, *Real Enemies: Conspiracy Theories and American Democracy, World War I to 9/11* (Oxford: Oxford University Press, 2009). On the widespread nature of conspiratorial thought in modern America, see Joseph E. Uscinski and Joseph M. Parent, *American Conspiracy Theories* (Oxford: Oxford University Press, 2014), chaps. 4–5.

250 CHAPTER 13

16. "Recognizing Bias," Newseum, https://newseumed.org/curated-stack/recognizing-bias; "Media Bias," AllSides, https://www.allsides.com/topics/media-bias-media-watch.

17. Mike Caulfield, "Cynicism, Not Gullibility, Will Kill Our Humanity," https://hapgood.us/2018/11/27/cynicism-not-gullibility-will-kill-our-humanity/.

18. Mike Caulfield, *Web Literacy for Student Fact Checkers* (Pressbooks, 2017), https://webliteracy.pressbooks.com.

19. Sam Wineburg and Sarah McGrew, "Lateral Reading and the Nature of Expertise: Reading Less and Learning More When Evaluating Digital Information," *Teachers College Record* 121, no. 11 (2019): 1–40.

20. New York heading, *Pennsylvania Evening Herald*, December 2, 1786.

21. "Messrs Fleets," *Boston Evening-Post*, May 30, 1774. The original letter appears in "Extract of a Letter from a Gentleman in Philadelphia to His Friend in Boston, dated May 17, 1774," *Boston News-Letter*, May 26, 1774. See also "Rio de la Plata," *Mercantile Advertiser* (New York), August 23, 1802.

22. Boston heading, *Cumberland Gazette* (Portland, OR), November 28, 1791.

23. Philadelphia heading, *General Advertiser* (Philadelphia), July 19, 1794.

24. Providence heading, *Providence (RI) Gazette*, March 3, 1781.

25. Susanna Dillwyn to William Dillwyn, April 27, 1793, box 2, folder 8: 1793 January–April, Dillwyn and Emlen Family Correspondence, Library Company of Philadelphia (Philadelphia, PA).

26. Dwight Foster to Jedidiah Foster, October 9, 1775, box 1, folder: Dwight Foster, 1774–1775, Dwight Foster Papers, Massachusetts Historical Society (Boston, MA).

27. Matthew Rainbow Hale, "On Their Tiptoes: Political Time and Newspapers during the Advent of the Radicalized French Revolution, circa 1792–1793," *Journal of the Early Republic* 29, no. 2 (Summer 2009): 203–4.

28. "For the United States Chronicle," *United States Chronicle* (Providence, RI), August 15, 1793.

29. *Independent Chronicle* (Boston, MA), April 16, 1798. In the late 1790s, as Republican observers faced an onslaught of Franco-American diplomatic problems, they often displayed more patience for news than their Federalist counterparts. In a 2009 essay, Matthew Rainbow Hale found the opposite during the years 1792–1793. This suggests that, in this regard at least, American partisans' understanding of political time was rather malleable depending on the circumstances. Hale, "On Their Tiptoes," 191–218.

30. Joanne B. Freeman, *Affairs of Honor: National Politics in the New Republic* (New Haven, CT: Yale University Press, 2001).

31. See, respectively, *Catskill Packet* (New York), May 20, 1793; *Columbian Centinel* (Boston), January 29, 1794; *Oracle of the Day* (Portsmouth, NH), July 15, 1794; and *Daily Advertiser* (New York), May 7, 1793.

32. *Gazette of the United States* (Philadelphia), August 25, 1790; *Columbian Centinel* (Boston), May 2, 1795; *Columbian Herald, Or, the Southern Star* (Charleston, SC), November 9, 1793. See also *Columbian Centinel* (Boston), April 6, 1793; and *American Apollo* (Boston), April 5, 1793.

33. Steven Shapin, *A Social History of Truth: Civility and Science in Seventeenth-Century England* (Chicago: University of Chicago Press, 1995), chap. 3.

34. *Massachusetts Mercury* (Boston), December 16, 1794.

35. Philadelphia heading, *General Advertiser* (Philadelphia), August 4, 1791.

36. *National Gazette* (Philadelphia), June 18, 1792.

37. *New Hampshire Gazette* (Portsmouth), March 29, 1794.

38. Jordan E. Taylor, "The Literati and the Illuminati: Atlantic Knowledge Networks and Augustin Barruel's Conspiracy Theories in the United States, 1794–1800," *Mémoires du livre—Studies in Book Culture* 11, no. 1 (Autumn 2019): 1–36.

39. David Tappan, *A Discourse Delivered in the Chapel of Harvard College, June 19, 1798* (Boston: Printed by Manning and Loring, 1798), 16.

40. *Letters to the Ministry from Governor Bernard, General Gage, and Commodore Hood* (Boston: Edes and Gill, 1769); *The Representations of Governor Hutchinson and Others, Contained in Certain Letters Transmitted to England* (Boston: Edes and Gill, 1773).

41. *A Short Narrative of the Horrid Massacre in Boston, Perpetrated in the Evening of the Fifth Day of March, 1770* (Boston: Edes and Gill, 1770), 3–5, 7, 8.

42. *Morning Post* (London), September 5, 1776. See also "To the Printer," London heading, *New York Journal*, May 12, 1774.

43. See Arthur Aspinall, *Politics and the Press, c. 1780–1850* (London: Home and Van Thal, 1949); and Lucy Werkmeister, *The London Daily Press, 1772–1792* (Lincoln: University of Nebraska Press, 1963), 4.

44. *Pennsylvania Packet* (Philadelphia), August 5, 1776. See a similar heading in *Massachusetts Spy* (Boston), December 24, 1778. On the term's use in London, see Solomon Lutnick, *American Revolution and the British Press, 1775–1783* (Columbia: University of Missouri Press, 1967), 22. On the term's use against Loyalist papers, see *Massachusetts Spy* (Boston), February 16, 1775; "Boston, June 9" heading, *Connecticut Gazette* (New Haven), June 13, 1777; *New Jersey Gazette* (Burlington), June 24, 1778; *Continental Journal* (Boston), November 5, 1778, September 14, 1780.

45. *Maryland Journal* (Baltimore), February 29, 1780.

46. *Rivington's New-York Gazetteer*, August 11, 1774.

47. *New-York Journal*, August 18, 1774.

48. *Boston News-Letter*, February 16, 1775.

49. Paul Langford, "British Correspondence in the Colonial Press, 1763–1775: A Study in Anglo-American Misunderstanding before the American Revolution," in *The Press and the American Revolution*, ed. Bernard Bailyn and John B. Hench (Worcester, MA: American Antiquarian Society, 1981), 279–301.

50. New York heading, *Daily Advertiser* (New York), August 16, 1786.

51. "For the United States Chronicle," *United States Chronicle* (Providence, RI), August 15, 1793.

52. For more detail on this, see Jordan E. Taylor, "The Reign of Error: North American Information Politics and the French Revolution, 1789–1795," *Journal of the Early Republic* 39, no. 3 (Fall 2019): 437–66.

53. Lucas Graves, "Anatomy of a Fact Check: Objective Practice and the Contested Epistemology of Fact Checking," *Communication, Culture & Critique* 10, no. 3 (2017), 518–37.

54. Jordan E. Taylor, *Misinformation Nation: Foreign News and the Politics of Truth in Revolutionary America* (Baltimore: Johns Hopkins University Press, 2022), 75–76, 168–69.

55. Caulfield, *Web Literacy*.

56. Melissa Tully, Emily K. Vraga, and Anne-Bennett Smithson, "News Media Literacy, Perceptions of Bias, and Interpretation of News," *Journalism* 21, no. 2 (2020): 209–26.

14

"A BUSY, BUSTLING, DISPUTATIOUS TONE"

News Anxiety in the Age of Revolutions and Today

Joseph M. Adelman

Set in the Catskills region of the Hudson River in upstate New York, "Rip Van Winkle" is a story familiar to many Americans. The title character, a henpecked husband, wanders into the mountains in search of solitude, where he meets oddly dressed Dutchmen playing a game that looks like ninepins and sounds like peals of thunder. They offer him some of their drink and he soon falls asleep. Awaking the next morning to find his gun rusted and his dog vanished, Rip meanders back to town and finds it transformed: rather than a single night, twenty years have passed. After several minutes of confusion—and a near mob reaction to his proclamation of loyalty to King George III—his identity is revealed, his now-grown daughter takes him in, and he resumes his life of loafing and leisure.[1]

Washington Irving used the short story as a vehicle to play with the concept of time and its passage, folding past, present, and future on top of one another for both his readers and characters.[2] To guide his readers through the uneven passage of time, Irving places forms of news consumption at the center of his descriptions of the pre- and postrevolutionary Hudson village in which Van Winkle lived. When the story opens in the colonial era, Rip Van Winkle wiles away his days at the village inn, marked by a "rubicund portrait of his majesty George the Third." The gathered men would talk with the inn's owner, turning to news only when "by chance an old newspaper fell into their hands from some passing traveler."[3] The news, Irving implied, occurred far away and filtered out to the countryside slowly, with time for locals to ruminate. When Rip Van Winkle returns from the mountains two decades later, he is stunned at the disappearance of the village he once knew. First among the changes he notices is the viciousness of politics and the pace of news. The village inn has been replaced by "The Union Hotel," and the sign depicting George III has been altered to show instead "General Washington" in a Continental uniform. The "character of the people" suddenly had a "busy, bustling, disputatious tone" that stunned Van Winkle.[4] Nonetheless, he quickly resumed his place among the village elders, and slowly caught up with their discussions.

Over the two centuries since the story's first publication, Rip Van Winkle has assumed the role of a stock character of sorts, a metaphor for people who have been absent for long stretches of time (and occasionally for men, the facial hair growth that accompanies such an absence). At the beginning of the COVID-19 pandemic in early 2020, for example, several news stories featured the character in their depiction of how people were experiencing the isolation of quarantine or the illness itself. One man earned a *New York Times* headline with a Van Winkle reference because he entered a seventy-five-day silent meditation retreat in northern Vermont just days before much of the country went into a lockdown.[5] Another, a doctor in Seattle who was in intensive care early in the crisis, described himself as feeling "like Rip van Winkle, waking up . . . and realizing that the world had shut down."[6]

The comparison is apt. In identifying the rapid pace of news and politics as a marker of change, Irving touched on an aspect of life that would have been immediately familiar to his readers in 1819. The American Revolution brought massive changes to the business of news. The number of newspapers increased exponentially in the first decades of the republic. Printers and editors adopted more openly partisan positions, setting aside the tradition of outward neutrality that characterized the colonial press. To some scholars, in fact, the distinctive feature of the age was a tendency toward conspiratorial thinking.[7] Changes between the 1990s and 2020s feel eerily similar. The news media landscape of the early 1990s was certainly not free of controversy, but Americans then experienced the end of an era of relative stability. The bulk of national news appeared on the major broadcast networks, CNN, a few newspapers with broad distribution, wire services, and across several popular weekly newsmagazines. Local news coverage appeared in regional newspapers as well as on local radio and television stations. Publications of a more partisan slant existed, of course, but entry into the market was costly and placed some limits on reach. Thirty years later, anyone can start a free website, open free social media accounts, and build a news audience with little more than time, effort, and a dream. Events become national news within moments as individuals offer commentary and stream videos live from the scene, such as the Boston Marathon bombing in 2013, the 2017 white supremacist marches in Charlottesville, or the January 6, 2021, insurrection against the US Capitol. The information flows unfiltered into feeds, and individuals and journalists often struggle to make sense of what they are seeing, hearing, and reading.[8]

The technology may have changed, but the feeling of being overwhelmed echoes across time. In both eras, the paramount challenge for news readers and consumers is how to manage information within a decentralized and less than fully regulated media ecosystem. Rather than focus on conspiratorial fears, therefore, I suggest that the issue was (and is) about information management. It is therefore worth examining what we can draw from how Americans during the revolutionary era managed their anxieties about the news—and how our study of that era shapes and limits what we can see. This chapter offers three lessons. First, communications infrastructure matters. During the revolutionary era, news producers—primarily the printers and editors of newspapers—managed sophisticated networks of connections through a variety of communications media to get news from place to place. Second, Americans sought out reliable and credible flows of news. That included both issues of physical geography, such as the quality of roads or protections against the elements, and a range of reading techniques to determine trustworthiness. Finally, the digital tools that are crucial to researching the revolutionary era and processing contemporary news have introduced a set of biases and limitations that shape the context of news. These lessons will not unlock the secrets of avoiding news anxiety in the twenty-first century, but they can clarify what is happening and how to manage it.

A Revolutionary Infrastructure

As practices for circulating news and information around the British Atlantic world coalesced in the eighteenth century, newspaper printers, editors, and readers concurrently developed ways of managing their expectations for the accuracy and timeliness of the news. Because of the integration of life around the Atlantic Basin—both within the British Empire and among the many peoples whose lives were affected by movement across the ocean—the speed, accuracy, and efficiency with which news traveled had implications for military, political, economic, and social affairs.

Patterns for how to deal with news in the British Atlantic first cohered in the early eighteenth century in London and radiated outward. The capital boasted a bustling market for news, with dozens of newspapers among other publications. Demand for news was so high, in fact, that theatrical productions began to feature a stock character known as "Quidnunc" (which translates literally from Latin to "What now?"), who was obsessed with learning the latest news at the expense of the rest of his life. In financial markets, speculators traded on the timing and content of new information from transatlantic ships. And readers—or at least the white male readers of middling or higher economic status—extended the conversation from the pages of newspapers to discussions in public spaces such as taverns and coffeehouses on both sides of the Atlantic.[9]

Anxiety about the circulation of information, its pace, and the potential for it to be lost pervaded the letters and literature of the late eighteenth-century Anglo-Atlantic world. When sending letters across the Atlantic, for example, writers regularly cataloged the correspondence they sent and received over the previous few months because they could not guarantee the arrival of any individual letter, and they often arrived in batches rather than at the same pace they were sent. In literature, the popularity of the epistolary form of novel (that is, one written primarily as an exchange of letters) meant that thousands of readers could encounter the same anxiety about waiting for personal news as characters traveled within the world of the books.[10] That the plots of epistolary novels often employed the same ripped-from-the-headlines technique as such modern television dramas as the *Law & Order* franchise only reinforces the sense that the news was important.

At the same time, news producers worked hard to manage the flow of information and maintain a steady supply of news for their publications and readers. Colonial newspapers were largely compilations of texts from a variety of sources, including oral sources (people who walked into the printer's office with news), handwritten sources such as letters, and printed sources, most especially newspapers from other towns. They included not only paragraphs of news but also local advertisements and notices as well as the occasional woodcut or poem. To produce a newspaper on a weekly schedule, therefore, required printers to constantly be in search of information from whatever sources were available. To do that, printers and editors relied on the imperial postal system, and after 1775 the Continental Post Office, to ensure delivery of both correspondence and newspapers from other towns and across the Atlantic.[11]

Communications technology placed clear limits on the speed at which news circulated. Obviously, colonists and others had no concept that news could travel faster than by horse on land or a well-outfitted sailing ship at sea, but they nonetheless chafed at the days and weeks it took to circulate information from one place to another (and back). By the 1760s, though, the timing had become mostly regularized. A trip from Britain to the Atlantic coast could take six to ten weeks depending on the exact destination, and the trip east about four to eight weeks thanks to the Gulf Stream.[12] Travel between Philadelphia and Boston along the post road was relatively reliable. Using the time lags for paragraphs reprinted from other newspapers as a guide, news could travel between Philadelphia and New York within two or three days and between New York and Boston within a week. Further south, overland travel was somewhat slower and more difficult—most information that circulated between those northern ports and Charleston in South Carolina, for example, arrived much more quickly by ship. The circulation of the Declaration of Independence, even in the midst of a war, offers a useful case study of how news circulated. News of independence (and usually a printing of the Declaration's text) appeared within six days after July 4 in New York, twelve in Boston, sixteen days in Williamsburg, but took nearly a month to reach Charleston.[13]

In addition to travel times, the geographical design of the postal system and imperial communications network also had an outsized impact on the circulation of news. Until 1775 the British Empire operated the post office in the North American colonies, and by law any revenue it generated went to support the Crown. There were rules in place to facilitate the circulation of "exchange papers," or individual copies of newspapers sent between printers, but the system operated primarily along the Atlantic seaboard, with a few inland routes in New England, New York, and Pennsylvania. Because of the costs of sending mail, most of the post's users were relatively well-to-do or well-connected—merchants, government officials, and others for whom official communications circuits were vital. Away from the coast, official correspondence often traveled via the British army, which by the 1760s had established a series of roads and pathways in the North American interior.[14] Most others who wanted to share news and communicate sought out informal channels, whether by sending letters via a friend, unofficial post riders, or some other method.

The scale of twenty-first-century communications notwithstanding, contemporary media bear a strong family resemblance to their revolutionary-era predecessors. In some distinctive ways, the news business is reverting to what it looked like in the late eighteenth century: highly decentralized and networked, with a strong interest in aggregation and collection from a variety of sources. There are two significant and interrelated differences. The first is the scale of capital involved. Obviously, there were no media megacorporations during the revolutionary era, whereas now many media outlets, whether local, national, or international in their scope, are owned by large conglomerates. The second relates to the epistemology of news. In the revolutionary era, nearly all printers and editors claimed that their newspapers were unbiased and open forums for all to express their political opinions, even if in practice editors often published what we would describe as a partisan take on the news.[15] It is practically a cliché at this point to say that the twenty-first-century media environment has fractured into a set of partisan niches. That is true and not-quite-true at the same time. What is occurring now is a retreat from a mid-twentieth-century ideal of objectivity. That position, now sometimes described as "The View from Nowhere" by media scholars, was based on the claim through much of the twentieth century that mainstream news organizations reported on events from outside the fray with no interests of their own.[16] In reality, their editorial decisions largely reflected the viewpoints of economically secure white men. Everyone else, the ideal implied, was compromised by their identity. In recent decades, new journalism outfits have turned that idea against traditional media, suggesting that they are unwilling to share essential truths. These alternative media outlets do so because they see themselves not as bystanders observing the news from a distance but instead as activists who must participate in the news creation process.[17]

Beyond formal news organizations, an even more robust informal political economy of news has grown in the first decades of the twenty-first century. The growth of social media platforms—Facebook, Twitter, Instagram, Snapchat, to name just the most prominent at this writing—owes in part to the ways they facilitate news circulation. That democratization of access to the public sphere has opened space for many Americans—especially those who are not white, cisgendered men—to tell more diverse stories about American politics, society, and culture. At the same time, that opening has operated to the detriment of society. As in the eighteenth century, access is often anonymous (or pseudonymous), but unlike the revolutionary era, influence can accrue broadly based on a number of factors, not least of which is entertainment value. Claims of "fake news" about the 2016 and 2020 elections and the COVID-19 pandemic, the QAnon conspiracy theory, and other media phenomena have all been fueled by access to social media. To the extent that social media corporations have devoted attention to these issues, they have tended to argue for a libertarian hands-off approach to content, which allows them to disclaim responsibility at the same time as they fuel engagement with the most vitriolic and vicious material posted on their platforms.[18]

In both eras, then, the issue of infrastructure had an outsized impact on how people consumed and experienced news. Information was and often is only partial owing to the necessarily limited point of view of any individual reporter or correspondent. How one understood an event shifted, sometimes dramatically, in response to new information and conditions. To the extent that differences in infrastructure between the two eras can be drawn, they are clearest in terms of amount and pace of information. Those in the revolutionary era often worried that they received too little information too slowly, while today news consumers feel overwhelmed by massive (if somehow still incomplete) dumps of information arriving at a rapid-fire pace.[19]

In Search of Reliability and Credibility

The structure of news created a baseline for the pace of news, but Americans in the revolutionary era nonetheless found other reasons to worry. People worried not just about time lags in general, but also specifically about delays because of weather and travel. The delays were often regional in character. During New England winters, newspapers would sometimes go one or more weeks without receiving any news from faraway towns and cities as snow and ice impeded the travel of post riders. Newspaper editors would occasionally post notices to that effect, such as one that William Goddard printed in the *Pennsylvania Chronicle* in January 1768. He begged his readers' pardon for the "Want of that Variety of fresh Intilligence [sic] which they might wish to see" and promised that "no Pains shall be spared to procure, from Time to Time, every Thing that may tend to their *Benefit* and *Satisfaction*."[20] Further south, especially in the Caribbean, hurricane season posed a threat to travel during the late summer and fall months, which meant that far more ships arrived in those ports from Europe during the winter and spring months.[21]

Readers and editors both fretted about the credibility of the information they were receiving. Colonists, like their counterparts in Britain, negotiated a sophisticated system to determine whether to trust another person based on relationships, social networks and status, and personal reputation. People knew that newspapers provided an incomplete picture; they promised the "freshest advices," but information could therefore change over time for particular events. In the early months of the Revolutionary War, colonists in Virginia and the Carolinas feared an outbreak of rebellions by enslaved people. These rumors grew because of the threat by the earl of Dunmore to offer freedom to men enslaved by Patriots should they run away to enlist in the British army. Once started, the rumors spread through the colonies, fueled by newspaper exchanges and including speculation that the rumors themselves might spark a revolt.[22]

Printers and editors played a significant role in mediating that process for their readers. In 1781, for example, news reached the Continental Army camp in northern New Jersey that the British and French had engaged in a massive naval battle in the Caribbean. Shepard Kollock, the printer of the *New Jersey Journal*, included a brief paragraph on the battle. After printing it, Kollock sent one copy north to Worcester, where Isaiah Thomas printed the *Massachusetts Spy*. His copy of that newspaper survives in the American Antiquarian Society (which Thomas founded later in his life), so we know that on that copy he marked the paragraph along the side for reprinting, but added four words by hand: "this account wants confirmation." While there is no evidence to help us understand why Thomas did that, those words then appear in the *Spy*'s version of the story, casting doubt on the accuracy of the underlying account.[23] What had been a straightforward news account, in other words, became just a rumor. Anyone who read the *Spy* would have no idea that it was Thomas's decision to downgrade the quality of this information. No conspiracy was to be found, besides the editorial process occurring inside dozens of printing offices.

In the twenty-first century, what seems like a glut of information at the same time feels like no information at all. As the poet Claudia Rankine put it in the *New York Times Book Review*, "We scramble in the drought of information / held back by inside traders."[24] She was writing about the confluence of the COVID-19 pandemic and the protests in the wake of the murder of George Floyd in Minneapolis, but her observation readily applies more broadly. In recent decades, the decentralization of news has created an illusion of reliable circulation and at the same time raised the level of difficulty in determining the credibility of a news source. After any major event—from an election to the World Series to a celebrity death—information floods social media outlets. Within a short time, metacommentary about the event can overwhelm the actual information. Either way, the flood itself can cause anxiety about the news.[25] However, anxiety about the news often pushes people to seek out more information—and perhaps ironically, often information that is more terrifying than reassuring.[26] In other words, many would feel a sense of recognition in Rip Van Winkle's bewilderment at the world he returned to.

At the same time, producers of news face increasing challenges to engage with news consumers. Because consumers have so many different media, genres, and forms to choose from, producers now see a struggle for attention as the dominant issue in thinking about news flows.[27] Ironically, many Americans express disgust at the "bias" of the news they read, even as the demand for news from a partisan standpoint remains strong, if not increasing.[28] All of these factors contribute to the anxiety Americans face about getting accurate and reliable information, whether about politics, pandemics, even the weather.

The Bias Inherent in the System

The advent of digital databases for primary source material transformed research into the Age of Revolutions, just as the shift of news to online platforms has forced a reimagining of how news circulates in the early twenty-first century. In each case, digital formats have democratized access overall, but with significant caveats. Because of the expense of digitizing materials, institutions have required funding assistance, which means that substantial amounts of material exist only behind gated paywalls. As a consequence, researchers and readers of news need to develop sophisticated skill sets in order to navigate not only the content they read but also the tools that structure their encounters with news.

The most important databases for news from the era of the American Revolution are *America's Historical Newspapers* and *America's Historical Imprints*. Both are owned by the Readex Corporation, a subsidiary of NewsBank, a major publisher of online academic databases. They developed in tandem through a partnership dating to the 1950s between Readex and the American Antiquarian Society (AAS), the leading collector of North American imprints published before 1800. In their first digital iterations, users accessed early American texts on microfilm (for newspapers) and microfiche (for other imprints). The microfiche collection used *American Bibliography* by the longtime AAS librarian Charles Evans as its organizing principle—in fact, most pre-1800 imprints are still cited according to their "Evans number."[29] The resulting microforms and microfilms allowed for a much broader group of scholars who could not necessarily travel to Worcester for lengthy stays at AAS to engage in research in early American printing. However, those scholars still needed access to a library whose funding permitted the acquisition of large and expensive microfilm and microfiche readers in addition to the reels and cards.

The physical cards and reels remained in use into the early twenty-first century (I used them for research on my undergraduate thesis on the American Revolution in 2001 and 2002). By then, however, AAS and Readex were already at work on a digital interface that would reproduce the images as scans and make them text searchable through optical character recognition (OCR). When "Evans Digital" first appeared in 2002, reviewers extolled the possibilities for the study of early American history, literature, and related fields. "The Evans Digital encourages us to make connections that would have been exceedingly difficult (if not impossible) before," wrote Cathy Davidson in a review for *Commonplace*. "It will facilitate new areas of research that could not have been accomplished in a lifetime spent only in the library or bent over a microform reader."[30] In the ensuing two decades, scholars have used the databases for all manner of research, including extensive study of print culture and the book trades as well as nearly every subfield in early American studies. Readex's databases now encompass about a dozen series each for newspapers and imprints spanning the seventeenth through the nineteenth centuries. And they are not the only available sources. *Chronicling America*, developed in cooperation between the Library of Congress and the National Endowment for the Humanities, includes millions of pages of newspapers from the revolutionary era to the twentieth century. Other private companies, such as Accessible Archives, have digitized runs of specific newspapers, and nonprofit institutions such as Colonial Williamsburg have also placed newspaper issues online.

It would be difficult to quantify how many scholars have made use of these tools in their research because they are ubiquitous—though that is not obvious in citations because scholars often neglect to specify the format in which they encounter a source.[31] Nonetheless, scholars have reimagined the world of revolutionary America. We now know a great deal more about the role of advertising on the Revolution, the ways political toasts circulated to celebrate figures from George Washington to Simón Bolívar, the ways transatlantic political relationships developed, and how Americans employed racialized and racist terms and ideas in service of the Patriot cause.[32] My own research on the circulation of news would have been next to impossible to complete within a plausible time frame without access to *America's Historical Newspapers*. Finding paragraphs reprinted in some number of the several dozen newspapers would have been a monumental task with only printed volumes. The existence of the database—and the ability to search the OCR text for the repetition of phrases across publications—facilitates a variety of reinterpretations of classic questions. Take the case of *Common Sense*, for example. First published as a pamphlet in Philadelphia in January 1776, scholars have long known that Thomas Paine's most famous work was reprinted in a total of twenty-five editions in thirteen American towns that year, with additional editions appearing in London. As Trish Loughran has shown, its publication was far more concentrated in Philadelphia, and far more related to internal printing trade disputes in that city, than is widely assumed. Nonetheless, we know that there was discussion of the pamphlet throughout the colonies—Congress distributed copies, and the text reached as far as 1,000 miles inland in Kentucky.[33] Newspapers can help us further gauge that spread. I searched for mentions of *Common Sense* in 1776 using *America's Historical Newspapers*, from which I found that at least 23 newspapers referred to the pamphlet a minimum of 137 times. Of those, 56 were advertisements (some of them repeats from one week to the next), 55 were in the form of responses, 15 were excerpts of the pamphlet, and 13 contained news items related to the pamphlet or Paine. Spread across the colonies, that indicates a fair bit of discussion.

But go back and reread the sentence outlining the numbers and you will see already some of the limits of the databases. References appeared *a minimum of* 137 times in *at least* twenty-three newspapers. Why did I have to qualify that accounting? First, the database does not include every issue of every newspaper—only those that survive and for which Readex has access to a copy to digitize (whether from AAS or another archive). Second, researchers see different results based on the subscription access they have. If you log in from the AAS reading room or a large research university, you get to see results from all series. I was not able to do that, which means that, though early series are relatively comprehensive for the revolutionary era, there may have been additional but paywalled results. Third, the OCR is not perfect. It is, I should emphasize, amazing considering the constraints of teaching a machine to read eighteenth-century print with variable spellings, different characters, and the occasional ink smudge. But if the underlying text of the search does not scan accurately, the result is lost. As a consequence, the precision of an analysis can sometimes obscure its comprehensiveness, even when we have reasonable confidence of the phenomenon it shows.

Furthermore, the ease of access and the encouragement to locate materials through searching rather than browsing has reshaped both research about the Age of Revolutions and how people engage with news in the early twenty-first century. We now encounter individual news stories far more frequently disaggregated from the rest of the news. Today, that involves clicking on a link on social media or, sometimes, on a news website. But one infrequently sees two stories set side by side. To get to the next one, you have to click again. In a similar way, the design of nearly all historical research databases (not just Readex) encourages users to search for keywords, subject terms, or full text for individual articles or paragraphs within a newspaper. Search results produce disembodied snippets of text divorced from the context in which they appear (and sometimes "snipped" in ways that obscure the meaning of what a reader is looking for).[34] It is actually much more difficult to simply read a run of a newspaper by clicking through page after page. In applying twenty-first-century user experience to eighteenth-century sources, in other words, the databases introduce barriers to interpreting sources within their historical context.

Today, we all feel like Rip Van Winkle, taken aback by the pace and scale of news. I first proposed writing this chapter in 2019, before the global COVID-19 pandemic, the Black Lives Matter protests that followed the murder of George Floyd, the 2020 presidential election, and several otherwise tumultuous years for news consumption. Yet, even when I thought of this topic, anxiety about the news was an ever-present fact of life for many Americans. Over the past forty years, the pace of the news cycle has accelerated from an era in which people received newspapers once or twice a day, checked in on the day's events on an evening network newscast, and perhaps caught an update or two on news radio. Now those same news organizations post stories to their websites as events are occurring and do so in competition with multiple cable news channels, news websites, not to mention social media sites like YouTube, Facebook, Twitter, Instagram, WhatsApp, and more.

One of the key components that makes the system of news operate today is the algorithm, a machine-learning process that predicts what you are likely to want to click on and read when you open Google, Facebook, Twitter, or Instagram. People are therefore likely only to get a small slice of the complete picture. And those algorithms are held privately by corporations acting in their own interests rather than conceptualizing themselves as acting for the public interest—sometimes to the detriment of factual information.[35] Even more troubling, significant evidence exists that algorithms introduce and reinforce gender and racial biases. These manifest online through search results that privilege particular perspectives or assume that searches for terms about people who are Black or female (or both) offer results that presume the worst stereotypes about them.[36] For the revolutionary era, similar biases make historical research on non-white people more difficult. One must read sources against the grain or work at cross-purposes to the intent of database designers in order to extract evidence about Black people, especially women.[37] To a certain extent, databases constructed around full-text searching are bound by the limits of historical language. But even they nonetheless have significant work to do to address these shortcomings.

It seems a strange coincidence that Americans reading the news today face many of the same challenges not only compared to their revolutionary-era forebears but also to scholars researching the Age of Revolutions. The comparison bears out across the lessons offered here. The pathways and platforms that facilitate communications are vital to the functioning of mass media and therefore also crucial objects of study. The design of a system has wide-ranging consequences, from placing editorial power in the hands of artisans or ordinary Twitter users to defining in a quite literal sense where information can travel. Readers need skill and time to evaluate fragmentary information fed by algorithms and sometimes lacking context or a broader narrative—what we today call "media literacy." At a basic level, researching the eighteenth century underscores the importance of attending to the context in which one is reading. Understanding how the news business is structured, how sites update their stories, what issues captivate reporters and editors—all of these skills clarify the world. Finally, we need to examine and understand the tools that provide access to the news from whichever century. Any tool—even with the best intention of developers—introduces a barrier between source and reader. News databases in particular must work to overcome both historical and contemporary bias about gender and race, and users must account for any shortcomings. In the end, however, there is some solace in knowing that we have been here before. Americans have undergone societal upheaval before, and reacted to it in much the same way, by attempting to embrace as much information as possible. One hopes that, like old Rip, we too can resume a steady place in the world before too long.

NOTES

1. Washington Irving, "Rip Van Winkle: A Posthumous Writing of Diedrich Knickerbocker," in *The Sketch-Book of Geoffrey Crayon, Gent.*, Early American Imprints, ser. 2, no. 48355 (New York: Printed by C. S. Van Winkle, 1819), 59–94.

2. Michelle R. Sizemore, "'Changing by Enchantment': Temporal Convergence, Early National Comparisons, and Washington Irving's *Sketchbook*," *Studies in American Fiction* 40, no. 2 (2013): 157–83; Jeffrey Insko, "Diedrich Knickerbocker, Regular Bred Historian," *Early American Literature* 43, no. 3 (November 2008): 605–41; Andrew Burstein,

The Original Knickerbocker: The Life of Washington Irving (New York: Basic Books, 2007), 125–32, 336–41.

3. Irving, "Rip Van Winkle," 67.

4. Irving, "Rip Van Winkle," 82.

5. Ellen Barry, "A Rip Van Winkle for 2020 Emerges after 75 Days," *New York Times*, June 2, 2020.

6. "'I Felt Like Rip Van Winkle': One of the 1st U.S. Doctors with COVID-19 Is Back Home," *All Things Considered*, NPR, April 13, 2020, https://www.npr.org/sections/coronavirus-live-updates/2020/04/13/833734452/-i-felt-like-rip-van-winkle-one-of-first-u-s-doctors-with-covid-19-is-back-home.

7. Gordon S. Wood, "Conspiracy and the Paranoid Style: Causality and Deceit in the Eighteenth Century," *William and Mary Quarterly*, 3rd ser., 39, no. 3 (1982): 402–41.

8. Jordan E. Taylor, *Misinformation Nation: Foreign News and the Politics of Truth in Revolutionary America* (Baltimore: Johns Hopkins University Press, 2022); Jeffrey L. Pasley, *"The Tyranny of Printers": Newspaper Politics in the Early American Republic*, Jeffersonian America (Charlottesville: University Press of Virginia, 2001); "The Breaking News Consumer's Handbook," On the Media, WNYC Studios, September 20, 2013, https://www.wnycstudios.org/podcasts/otm/articles/breaking-news-consumers-handbook-pdf.

9. Uriel Heyd, *Reading Newspapers: Press and Public in Eighteenth-Century Britain and America*, SVEC 2012 (Oxford: Voltaire Foundation, 2012), 195–230; Will Slauter, "Forward-Looking Statements: News and Speculation in the Age of the American Revolution," *Journal of Modern History* 81, no. 4 (2009): 759–92; Paul Starr, *The Creation of the Media: Political Origins of Modern Communications* (New York: Basic Books, 2004), chap. 2.

10. Hannah W. Foster, *The Coquette*, ed. Cathy N. Davidson, Early American Women Writers (New York: Oxford University Press, 1986); Cathy N. Davidson, *Revolution and the Word: The Rise of the Novel in America*, expanded ed. (New York: Oxford University Press, 2004); Rachael Scarborough King, *Writing to the World: Letters and the Origins of Modern Print Genres* (Baltimore: Johns Hopkins University Press, 2018), chap. 3.

11. Joseph M. Adelman, *Revolutionary Networks: The Business and Politics of Printing the News, 1763–1789*, Studies in Early American Economy and Society (Baltimore: Johns Hopkins University Press, 2019), 4–11, 147–50, 178–79; Robert G. Parkinson, *The Common Cause: Creating Race and Nation in the American Revolution* (Chapel Hill: published by the Omohundro Institute of Early American History and Culture and the University of North Carolina Press, 2016), 63–73.

12. Ian K. Steele, *The English Atlantic, 1675–1740: An Exploration of Communication and Community* (New York: Oxford University Press, 1986); Joyce E. Chaplin, "Knowing the Ocean: Benjamin Franklin and the Circulation of Atlantic Knowledge," in *Science and Empire in the Atlantic World*, ed. James Delbourgo and Nicholas Dew (New York: Routledge, 2008), 73–96.

13. "When and How Did the Colonies Find Out about the Declaration?," Declaration Resources Project, accessed August 19, 2020, https://declaration.fas.harvard.edu/resources/when-how.

14. A. Zuercher Reichardt, "War for the Interior: Imperial Conflict and the Formation of North American and Transatlantic Communications Infrastructure, 1727–1774" (PhD diss., Yale University, 2017).

15. Adelman, *Revolutionary Networks*, 45–49.

16. Atossa Araxia Abrahamian, "View from Nowhere: Is It the Press's Job to Create a Community That Transcends Borders?," *Columbia Journalism Review* (Summer 2019), https://www.cjr.org/special_report/view-from-nowhere.php/; Jay Rosen, "The View from

Nowhere: Questions and Answers," PressThink, November 10, 2010, https://pressthink.org/2010/11/the-view-from-nowhere-questions-and-answers/.

17. Claire Bond Potter, *Political Junkies: From Talk Radio to Twitter, How Alternative Media Hooked Us on Politics and Broke Our Democracy* (New York: Basic Books, 2020).

18. Siva Vaidhyanathan, *Antisocial Media: How Facebook Disconnects Us and Undermines Democracy* (New York: Oxford University Press, 2018); Kate Starbird, "Disinformation Campaigns Are Murky Blends of Truth, Lies and Sincere Beliefs—Lessons from the Pandemic," The Conversation, July 23, 2020, http://theconversation.com/disinformation-campaigns-are-murky-blends-of-truth-lies-and-sincere-beliefs-lessons-from-the-pandemic-140677; Matthew Hannah, "QAnon and the Information Dark Age," First Monday, January 15, 2021, https://doi.org/10.5210/fm.v26i2.10868; Melissa Gira Grant, "Americans Anonymous," *New Republic* 251, no. 9 (September 2020): 4–5.

19. At the same time, it is important to remember that "information overload" has a long history in American and European societies. See Ann Blair, *Too Much to Know: Managing Scholarly Information before the Modern Age* (New Haven, CT: Yale University Press, 2010).

20. *Pennsylvania Chronicle*, January 4, 1768; Jordan E. Taylor, "Now Is the Winter of Our Dull Content: Seasonality and the Atlantic Communications Frontier in Eighteenth-Century New England," *New England Quarterly* 95, no. 1 (March 2022): 8–38.

21. Julius S. Scott, *The Common Wind: Afro-American Currents in the Age of the Haitian Revolution* (London: Verso, 2018), 85–86.

22. Parkinson, *Common Cause*, 106–10, 141–76.

23. *New-Jersey Journal* (Chatham), July 11, 1781; *Massachusetts Spy* (Worcester), July 26, 1781. See also Joseph M. Adelman, "'Meer Mechanics' No More: How Printers Shaped Information in the Revolutionary Age," *Age of Revolutions* (blog), September 11, 2017, https://ageofrevolutions.com/2017/09/11/meer-mechanics-no-more-how-printers-shaped-information-in-the-revolutionary-age/.

24. Claudia Rankine, "'Weather,'" *New York Times*, June 15, 2020, Book Review section, https://www.nytimes.com/2020/06/15/books/review/claudia-rankine-weather-poem-coronavirus.html.

25. The most common iteration of this phenomenon is when a celebrity's name (especially one who is elderly) trends on social media. In very little time, the top search results for the name are comments about the celebrity not being dead rather than whatever it was that caused their name to trend in the first place. This is often accompanied by a GIF of Denzel Washington looking relieved.

26. Bethany Albertson and Shana Kushner Gadarian, *Anxious Politics: Democratic Citizenship in a Threatening World* (Cambridge: Cambridge University Press, 2015).

27. Zeynep Tufekci, *Twitter and Tear Gas: The Power and Fragility of Networked Protest* (New Haven, CT: Yale University Press, 2017), 28–48; Vaidhyanathan, *Antisocial Media*, 77–105.

28. "American Views 2020: Trust, Media and Democracy," Knight Foundation, 2020, https://knightfoundation.org/reports/american-views-2020-trust-media-and-democracy/.

29. Charles Evans, Clifford Kenyon Shipton, and Roger P. Bristol, *American Bibliography: A Chronological Dictionary of All Books, Pamphlets, and Periodical Publications Printed in the United States of America from the Genesis of Printing in 1639 down to and Including the Year 1820: With Bibliographical and Biographical Notes*, 14 vols. (Chicago: privately printed for the author by Blakely Press, 1903); Marcus McCorison, "Into the Unknown in 1955—AAS and Readex," *Proceedings of the American Antiquarian Society* 115, part 2 (October 2005): 279–88; Philip F. Gura, *The American Antiquarian Society, 1812–2012: A Bicentennial History*, rev. ed. (Worcester, MA: American Antiquarian Society, 2013).

30. Cathy N. Davidson, "From Movable Type to Searchable Text," *Commonplace*, April 2003, http://commonplace.online/article/movable-type-searchable-text/.

31. Sharon Block, "#DigEarlyAm: Reflections on Digital Humanities and Early American Studies," *William and Mary Quarterly*, 3rd ser., 76, no. 4 (2019): 611–48.

32. Carl Robert Keyes, "History Prints, Newspaper Advertisements, and Cultivating Citizen Consumers: Patriotism and Partisanship in Marketing Campaigns in the Era of the Revolution," *American Periodicals* 24, no. 2 (2014): 145–85; T. H. Breen, *The Marketplace of Revolution: How Consumer Politics Shaped American Independence* (New York: Oxford University Press, 2004); David Waldstreicher, *In the Midst of Perpetual Fetes: The Making of American Nationalism, 1776–1820* (Chapel Hill: published by the Omohundro Institute of Early American History and Culture and the University of North Carolina Press, 1997); Caitlin Fitz, *Our Sister Republics: The United States in an Age of American Revolutions* (New York: Liveright, 2016); Matthew Rainbow Hale, "On Their Tiptoes: Political Time and Newspapers during the Advent of the Radicalized French Revolution, circa 1792–1793," *Journal of the Early Republic* 29, no. 2 (Spring 2009): 191–218; Jordan E. Taylor, "The Reign of Error: North American Information Politics and the French Revolution, 1789–1795," *Journal of the Early Republic* 39, no. 3 (2019): 437–66; Sharon Block, *Colonial Complexions: Race and Bodies in Eighteenth-Century America*, Early American Studies (Philadelphia: University of Pennsylvania Press, 2018); Parkinson, *Common Cause*.

33. Thomas R. Adams, *American Independence: The Growth of an Idea; A Bibliographical Study of the American Political Pamphlets Printed between 1764 and 1776 Dealing with the Dispute between Great Britain and Her Colonies* (Providence, RI: Brown University Press, 1965), xi, 164–72; Trish Loughran, *The Republic in Print: Print Culture in the Age of U.S. Nation Building, 1770–1870* (New York: Columbia University Press, 2007), chap. 2; Parkinson, *Common Cause*, 189–95; Nicholas Cresswell, *A Man Apart: The Journal of Nicholas Cresswell, 1774–1781*, ed. Harold B. Gill and George M. Curtis (Lanham, MD: Lexington Books, 2009), 111–12.

34. Lara Putnam, "The Transnational and the Text-Searchable: Digitized Sources and the Shadows They Cast," *American Historical Review* 121, no. 2 (April 2016): 377–402.

35. Tufekci, *Twitter and Tear Gas*, 136–38; Vaidhyanathan, *Antisocial Media*, 175–95.

36. Safiya Umoja Noble, *Algorithms of Oppression: How Search Engines Reinforce Racism* (New York: New York University Press, 2018).

37. Jessica Marie Johnson, "Markup Bodies: Black [Life] Studies and Slavery [Death] Studies at the Digital Crossroads," *Social Text* 36, no. 4 (137) (2018): 57–79; Marisa J. Fuentes, *Dispossessed Lives: Enslaved Women, Violence, and the Archive* (Philadelphia: University of Pennsylvania Press, 2016); Block, *Colonial Complexions*.

15

COPYRIGHT AND HISTORICAL DANGERS OF LICENSING REGIMES IN THE DIGITAL AGE

Kyle K. Courtney

Technology has long driven the advancement of copyright law. However, technology, in all its forms, from the movable-type printing press to the social media platforms, has also advanced concerns around the control and censorship of works using these same distributive technologies. In all this time, licensing has been an instrument that can enhance control over or supersede copyright and, in some cases, help suppress free access to copyrighted works.

Government control of printing press technology was a function of emerging "copyright,"[1] a word yet to be developed in the sixteenth century but also part of England's developing law and policy related to trade regulation. Although Italy and Germany claimed some earlier attempts at establishing the "right to print" under exclusive licenses granted by governmental authority,[2] one of the first English examples of using licensing as a form of both copyright and censorship emerged in the early sixteenth century. This early copyright was a function of the English government's concerns around subversive expression, and it was especially useful in controlling the new technological means by which free expression could be widely distributed: the printing press. These concerns are similarly echoed in the modern technological struggle with publication, access, and copyright, triggered by the integration of the internet into every aspect of the creation and dissemination of copyrighted works in our culture. The same old battles are being fought in this modern landscape as they were in the past.

Copyright and censorship are complex topics in their own right. However, in their development, one was not entirely the product of the other. In fact, many historians have noted that the development of censorship and copyright could have been established independently but were intertwined as a result of the intersection of the rising need for government control, new trade and commerce considerations, and the development of distributive technology.[3] As printing press technology became more dispersed, new licensing laws developed hand in hand with regulations designed to control the publishing system and the presses themselves, and this created the unique conditions for modern copyright to develop.

In the modern context, copyright and licensing have been similarly weaponized to control use and dissemination of materials, whether for purposes of censorship or to deny ownership rights afforded to creators, users, or cultural institutions. As with the printing press, much of this work is tied directly to technology that enhances dissemination and distribution. Whether it is traditional licensing terms, prepublication licensing requirements, upload filters, DMCA (Digital Millennium Copyright Act) takedown requests, or the latest challenge to CDA (Communications Decency Act) Section 230 immunity, the end result is less access to materials, less preservation of works, and harm to the public. Examination of these current examples in US law, compared with their historical English and British predecessors, with a focus on the late eighteenth century, will reveal the continual, and sometimes repetitive, struggle over copyright, access, and technology.

Old Dog, New Tricks: Control of Copyright and Modern Licensing

The publishing landscape has undoubtedly changed over the centuries and, despite the foundational role copyright has played in the dissemination of works, there has been a continual struggle for understanding and control over copyrights. In the United States, for example, the constitutional purpose of copyright is to promote the progress of art and science.[4] This 225-year-old law has always sought a balance between the needs of the creator to reap rewards from the limited economic monopoly of their work and the benefit to the public, once the limited economic monopoly on those rights expired. Even as far back as 1783, a committee of the Continental Congress stated that "nothing is more properly a man's own than the fruit of his study, and that the protection and security of literary property would greatly tend to encourage genius and to promote useful discoveries."[5] These words are very much in line with modern thoughts surrounding creative labor, which view the creation of art, photos, music, books, and scholarship as "fruit[s] of . . . study."[6]

Typically, a copyright law grants the creator a set of certain rights. These rights give creators the right to copy, modify, display, perform, and create other works modified from the original. These are typically referred to as the exclusive rights, or in common parlance, the "bundle" of rights. The creator automatically gains these bundled exclusive rights as long as the work is fixed and creative; generally, no registration or other formality is required.[7]

But it was not always this simple. In fact, the very notion of authors controlling their rights in their own creations was an idea that had to overcome vast legal and political hurdles, sometimes in the form of restrictive licensing agreements. Through these government or private licenses, the publisher (or "rightsholder," since the publisher has acquired the copyright of the author) controls the rights in the author's work by virtue of their monopolistic control over the technology used for dissemination. In order to see the work published, the authors handed over the rights to their creations to the owner of the printing presses. Licensing serving as the foil to authors' rights is a very old tactic, and one that authors and creators are still struggling with to this day. In fact, this modern incarnation of "the old licensing trick" may be much more harmful to the creators, their communities, and public access to the cultural record.

This modern publishing environment features the same historical use of licensing agreements that frequently strips the creator of any rights. As the copyright holder, the author, until or unless copyright is transferred, is the sole possessor and decision maker about the use of this exclusive "bundle" of rights—the right to copy, distribute, perform, display, and create derivative works.[8] Each of these rights is unique and can be transferred in whole or in part. For example, an author could transfer her right to distribute her essay to an individual or organization. Thereafter, under the law, that individual or organization could distribute that essay as if they were the original author.

However, such decisions to transfer outright some or all of the bundle of copyright could prevent authors from exercising their rights. For example, if an author transferred all of her exclusive copyrights in one agreement—as is often the case with traditional publishing contracts—she might not be able to make copies, place the work online, or share the work with colleagues. The law would view the new rightsholder as the original author for legal purposes. Often, especially in the digital era, the ability to access, use, share, or disseminate through normal communication channels are greatly restricted by licensing agreements with the authors.[9]

The way in which others might use and access the material is also greatly restricted once this copyright is licensed to a publisher or rightsholder. In the modern context, this user-focused restrictive licensing culture is out of control. This has never been clearer than during the COVID-19 pandemic when, for example, hundreds of millions of books and media that were purchased by libraries, archives, and other cultural institutions had become inaccessible due to COVID-19 closures.[10] While the authors of these "trapped" works may have desired to grant access to their copyrighted works to the world, the licensing agreements they had signed shifted that decision-making authority to the publishers. As a result, the publishers—the owners of the copyright under the law—set their own limited terms for access.[11]

Further, licenses have also prevented users from accessing and enjoying these copyrighted works. Licenses are written in a way that often denies some of the "ownership" rights that are afforded to purchasers. For example, when a user purchases a book, they can make use of the book in various ways: they can lend the book to others, sell the book on eBay, give it away at a yard sale, and other actions—all without permission of the rightsholder. However, when a user purchases an e-book, they are not actually "buying" the work—they are merely renting it. The language of the license reads more like a lease in that it restricts the "ownership" and potential downstream uses of the copyrighted work. In this way, there is no real "purchase," just a lease—with terms that can even deny the user from accessing the work subject to the whim of the licensor at any time. The licensor controls all the uses and makes determinations on where, when, and how the user may even access or use the work.[12]

From where did this methodology to separate both the creator and users from their rights via licensing and technology emerge? That moment may have been when the printing press became the newest form of technology to enhance access and distribution of copyrighted works. When William Caxton brought the printing press to England in 1476, it began a technological revolution that produced two specific new routes in law and publication: the path to the creation of a new property right—copyright—given to a select few, and a secondary specialized system of control in the form of licenses and other printing privileges, regulations, and customs.

This second route, one of control, helped implement some of the English government's earliest attempts to censor materials, especially in the printed works that ran counter to the religious and political affinities of the government. For nearly two hundred years, this new property "copyright" interest, emerging from economic and trade principles, did not come out of a legislative or judicial process—it was controlled by those select few who served, and feared punishment from, the government. The new copyright interest gave those select few who controlled the presses a national monopoly via a specialized and restrictive license.

There have been dozens of scholarly works examining the development of regulations, proclamations, royal patents, and licensing systems employed by the English government.[13] Here, we cover some of the most important legal highlights and look in depth at the licensing schemes of control that have, in many ways, repeated themselves in our modern technological environment, emerging from the current state of copyright law and technology-based control of expression.

Note that it was not until the eighteenth century that copyright truly started to look like the modern version well represented now in statutes, common law, and regulations. But it is worth noting that the methodologies used in these early licensing acts were the earliest attempts to comingle copyright law with the means to control access to information via technology.

Prepublication Licensing

In the late sixteenth century, there was a shift in the use of single proclamations to ban particular books and their printing to a fully formalized prepublication licensing system. This methodology endured in England for over 150 years, from Henry VIII to William of Orange. It was influential in understanding how the government uses copyright as a tool of censorship to impede access to new distributive technology like the printing press and, later, how that system was rejected by the English legislature. This earlier history previewed and enabled the important policy and law discussions that would emerge in the early United States, post-Revolution.

In 1538 King Henry VIII was arguably the first to attempt to establish a preemptive licensing system for printed materials in England. The king made a proclamation that transferred all responsibility for licensing, and thereby permission to publish works, to his Privy Council. The system was designed so that all new books had to be approved by the council before publication. In fact, from 1538 through to 1641, the Crown exercised almost unlimited authority over the world's newest technology. These powers over the printing press were expansive, well beyond the modern licensing schemes that we see surrounding most of our media today.[14] They also included the regulation of officially approved presses and allowed search, confiscation, destruction, and potentially—after the investigation was complete—imprisonment for those operating outside the proper authorization. The powers were shared by groups that were given special authority under the English Crown to conduct such operations, including the Court of Star Chamber ("Star Chamber"), and later the Worshipful Company of Stationers ("Stationers").

Beyond Henry VIII's 1538 proclamation, the Star Chamber promulgated similar declarations asserting this nascent technological control in 1556, 1585, 1623, and 1687.[15] These declarations regulated the number of printing presses and the manner of printing throughout England. They required licensing for printing and prohibited any publication or importation of unlicensed books.

While these declarations looked like a familiar protective economic trade regulation, the effects were much more profound due to the special nature of the Star Chamber's power.

The Star Chamber was a specialized court comprising judges and privy councilors that emerged from the historical role of an advisory council which was used to supplement the judiciary.[16] However, these advisory councils were in strict service of the king or, on occasion, subject to their own interests. The history of the Star Chamber reads like an early historical lesson covering the need for "checks and balances" in government and the doctrine of separation of powers.

The role of the Star Chamber was expanded under Henry VIII to help enforce laws because, according to many, the courts were corrupt and inefficient. Later, Charles I used the Star Chamber to enforce unpopular political and religious policies, and as a result, it became the finest symbol of oppression and censorship, serving as an intergovernmental rival to the traditional courts and legislature.[17]

The Star Chamber also represented the first move in creating real licensing restrictions. In 1556, by a decree of the Star Chamber, it was forbidden, among other things, to print contrary "to any ordinance, prohibition, or commandment in any of the statutes or laws of the realm; or in any injunction, letters patent, or ordinances set forth or to be set forth by the Queen's grant, commission, or authority."[18] By a related decree in 1585, every book was *required to be licensed*, and all persons were prohibited from *printing* "any book, work, or copy against the form or meaning of any restraint contained in any statute or laws of this realm, or in any injunction made by her Majesty or her Privy Council; or against the true intent and meaning of any letters-patent, commissions, or prohibitions under the great seal; or contrary to any allowed ordinance set down for the good government of the Stationers' Company."[19] Within a year, the Star Chamber had built on this licensing-only mandate and passed their infamous Star Chamber Decree of 1586. This may have been one of the most effective and extensive licensing systems created to date. As a result of this decree, any use of the new printing technology could only take place in London, Oxford, or Cambridge. Additionally, it limited the number of presses allowed, and all permitted presses were required to report to the Stationers' Company.[20]

In 1623 a proclamation was issued to enforce this earlier decree.[21] The proclamation noted that the 1585 decree had been evaded, among other ways, "by *printing* beyond sea such allowed books, works, or writings as have been imprinted within the realm by such to whom the sole printing thereof, by letters-patent or lawful ordinance or authority, doth appertain." In 1637 the Star Chamber again decreed that "no person is to print or import (printed abroad) any book or copy which the Company of Stationers, or any other person, hath or shall, by any letters-patent, order or entrance in their register book, or otherwise, have the right, privilege, authority or allowance, solely to print."[22]

While these decrees were made under the guise of early trade regulations, designed to protect the market for English booksellers and printers, it set an important process and precedent for greater control over nascent "copyright" and censorship of publications via a licensing regime. Although the Star Chamber was abolished in 1640, the decrees drew up the blueprint for a system of continued control over the printing press technology via mandatory licensing, and now under the control of another deputized governmental agency, the Stationers' Company.

Stationer's Company and Printing Press Control

To understand how modern publisher and technological platforms learned to assert their control over digital works via licensing agreements in this era, it is critical to examine the origins of the Stationer's Company. Mass production and distribution of technology like the printing press had obvious potential as dangers to the government—it could allow the spread of ideas that were contrary to the government's principles. However, because of the nature of this new technology, production and distribution of works required considerable investment for the many operational and overhead costs. A printer had to raise capital to purchase a printing press, hire and pay skilled labor, and buy printing supplies—not to mention the time to prepare the works themselves. Further, after spending all this money on setting up the printing press shop, the printer then had to sell the works or make deals with other booksellers. As a result, to return any profit and stabilize any financial risk, printing had to be performed at scale. When the number of printing presses grew and the trade became more competitive, it was foreseeable that a group of printing press owners might combine forces to address these risks and work more efficiently together.

However, as discussed above, the English government, having been the grantor of various rights to print via royal proclamation in the past, was also aware of the new role that these technological and economic organizations might play related to distribution of the printed texts. If the government could find a means of effective control of this printing technology to curb the dangers of dissemination of unauthorized texts on a vast scale, then they could merge both interests—economics and printing control—in one single authority. By adding a privileged economic incentive for only the *authorized* printers, the government could regulate trade and control the free flow of information.

In this environment emerged the first real powerbroker in the world of copyright and censorship: the Stationers' Company. While Queen Mary issued many of her own proclamations banning the printing and distribution of certain works, her greatest historical role in copyright was when she traded an economic monopoly on the printing of books to gain an efficient royal censorship system simply by granting the Stationers a charter in 1557.[23]

However, like the current social media corporations emerging from the early internet, the history of the Stationers truly predates their charter. In 1403 the mayor and aldermen of London granted a petition by a new guild—a group of writers, illuminators, bookbinders, and booksellers. As early as 1542, the Stationers had already requested their own charter. It is interesting to note that this 1542 request for incorporation was refused, but it was not the end of the attempt to gain all-encompassing control over the book trade by government consent.[24]

The term "stationer" could have been applied to any member of the book trade at the time, including the printer, bookseller, or bookbinder. The word "stationer" actually derived from the term for a member of the book trade who was in a fixed, stationary position in a stall or shop that was working to meet these new demands for books. At this point, book production had moved out from the monasteries, where books were copied by hand to serve the educational needs of the church or government.[25] Later, education access expanded, and children needed schoolbooks or primers, as did students at major university centers like Cambridge and Oxford. The slow hand-copied production by the monasteries could hardly meet this new educational demand. The shops were set up as a result.[26]

The "stationer" at a shop or stall would accept orders for a copy to be made. In the shop that handled this trade, there were several craftsmen working in association: the scribe or writer, the illustrator, the person who prepared the books, the binder, and the business agent who took the orders. This guild of workers in the shop was the origin of the Stationers' Company.[27]

At the time, the Stationers and the government were eager to control the importation of books, although for very different reasons. The printers and booksellers wanted to eliminate sources of competition from foreign printers, or eliminate any competitive, unregulated, or illicit presses. The government wanted to eliminate any sources of treason. Both needs became more critical as the sociopolitical tides continued to shift in England. It was in the interest of both parties that the printers and booksellers should be organized in a single body that would be the channel to achieve both the Stationers' and the government's ambitions.

According to the preamble of the charter granted in 1557, Queen Mary incorporated the Stationers to provide a remedy against seditious and heretical material that was being printed and distributed.[28] These materials not only threatened her control, but the government also believed that they were created to set her subjects against her reign and that they challenged the religious authority of the Catholic Church. The situation was ripe for a new regulation, via charter, to control this risk of subversion stemming from a relatively new technology that enhanced mass distribution.

The charter itself is a well-crafted work and created a monumental partnership between the Stationers and the government. It is arguably one of the most notorious public-private partnerships in copyright history. The government effectively created an agency for censorship, imbuing it with all the powers of the government in the guise of a standard business charter. These charters were common, and frequently granted to other trade guilds and companies to promote trade and the economy. This is not to imply that the charter did not have an economic effect; it was still a fairly complete and total economic monopoly over the printing business in England. However, it certainly served to create an effective and willing partner to the government's censorship campaign.

The charter granted the Stationers the exclusive right "of printing any book or any thing for sale or traffic" within England and other English dominions. The charter also stated that the Stationers could develop "ordinances, provisions and statutes whenever it shall seem to them to be opportune and fit."[29] The Stationers wasted little time with these new powers. They immediately began to create their own licensing system to further their control and, in effect, further the government's censorship goals. In 1559 the Stationers drafted an ordinance which required that "Every book or thing [is] *to be allowed by the Stationers* before it is printed."[30] This idea of authorization, as expressed in the particular language, emerged from an early custom to simply document their day-to-day work with new books. However, the recording took on a new meaning as a form of explicit permission to print in their official register. In 1562, when the Stationers finally agreed and issued its ordinances, this critical provision was repeated.

These new ordinances—utilizing the phrase "allowed by the Stationers"—had a significant impact on both sides of the developing relationship between copyright and censorship. Certainly, the previous proclamations, decrees, and laws regulating the printing press mandated that a work was required to undergo a review by some authority for purposes of censorship.[31] However, here, the use of this phrase also indicated the particular right to print a particular book. This is the emergence of the concept of "copy-right"—the work was previously unpublished, and the author had given the Stationer the right to make copies of the work. As some scholars have suggested, this is "precisely the mechanism which the crown needed for enforcement of its own control of the process, since no book was 'allowed' (in the Stationer's sense) until it had been 'allowed' or licensed by the royal censors."[32]

Although the company's charter had been granted by Mary, Elizabeth I also confirmed the charter in November 1559, without any change, and for substantially the same reasons it had originally been granted: so that the Stationers might aid the government in controlling the printing presses. Merely because there was a shift in the monarchy did not mean that the new government wanted to assert any less control over this printing technology.

In effect, the Stationers were granted powers that we would define today as "enforcement." These provisions were typically reserved for the government and those tasked with enforcing the laws created by the government. However, the Stationers were a private entity given these powers to investigate, seize, and burn books, and even destroy illicit presses. These powers were only furthered by their own internal regulations that maintained order in the profession via a new "copy-right," which determined who had the right to print a work.

A member acquired a right to copy by registering the title of a book in the Stationer's register. The legal title of that work was assigned in that moment of registration, and the assignee then had the sole authority to print it. The right was fully transferable; it could be sold, assigned, or given to an heir. This new right, then, was a *perpetual copyright for the publishers*, and not for the authors. Further, this early copyright for the publishers was also supported by other decrees, ordinances, and statutes that equally promoted censorship while affirming the official policy of granting the Stationer's total control over the printing press functions.

It is interesting to note that copyright in these early stages was not concerned with the rights of authors. The right granted was an exclusive right to print for the publisher. Copyright, as a rule, was created and limited to the Stationers, and the Stationers continuously fought to maintain this perpetual right through petitions, legislation, influence, and policy. They had the support of the government with respect to the government's lack of concern that the publishers claimed the author's writings as perpetual property, but merely that the copyright was one of many effective tools of censorship. So copyright was and remained, for a time, fully alienated from authorship. Authors simply could not enjoy any copyright protection. Later, based on the shifting priorities of the government and the book trade, the authors' rights would play a crucial role in altering the Stationer's mission—freeing copyright from its role as a form of mandatory censorship.

The Licensing Act

Before the Stationers lost their epic printing monopoly from the government,[33] they had enjoyed a period of near total control of the book trade. The Licensing Act of 1662—titled "An Act for Preventing Abuses in Printing Seditious, Treasonable, and Unlicensed Books and Pamphlets, and for Regulating of Printing and Printing Presses"[34]—was an explicit legislative codification of the Stationers' copyright. The act moved printing press authority from the Crown's "executive order" model to a fully formed legislative statute as passed by Parliament.[35]

Although its execution was periodic, lasting from 1662 to 1679, and adapted again between 1685 and 1695, the Licensing Act represents the best example of government censorship advantageously linked with copyright and licensing.

The Licensing Act stated plainly:

> Whereas the well-government and regulating of printers and printing-presses is [a] matter of publick care, and of great concernment, especially considering, that by the general licentiousness of the late times, many evil disposed persons have been encouraged to print and sell heretical, schismatical, blasphemous, seditious and treasonable books, pamphlets and papers, and still do continue such their unlawful and exorbitant practice, to the high dishonour of Almighty God, the indangering the peace of these kingdoms, and raising a disaffection to his most excellent Majesty and his government: . . . (2) for prevention whereof, no surer means can be advised, than by reducing and limiting the number of printing-presses, and by ordering and settling the said art or mystery of printing by act of parliament, in [a] manner as herein after is expressed.

Perhaps some of its impact is lost in the language of the previous centuries, but the disposition is clear: under the Licensing Act, the Stationers were given the power to determine which materials could be made accessible to the public. Again, much like the Star Chamber's licensees from the earlier era, this is nearly a perfect legal construct: a small guild of enforcers, lining their own pockets with an economic monopoly and harboring a protected legal status because they were acting on behalf of the government's censorship program, now legitimized by Parliament. The outcome was control of public access to ideas that are embedded in the copyrighted materials themselves—whether a play, poem, textbook, or other creative work.

We see that the Licensing Act's focus was on *the printing* of seditious and treasonable illegal, not the *writing* of such material. The focus on the new technology's potential for wide distribution of works using the printing presses was tantamount to any other related concerns. Perhaps because the Stationers and the Licensing Act simply ignored their potential copyright, the authors were similarly ignored for creating the illegal material. It could be surmised that this was because the potential use of the distributive technology was the subject of greater concern by the government and the Stationers. As discussed later in this chapter, that concern is similarly reflected hundreds of years later by a movement in copyright and licensing law, shifting liability from the authors and creators and toward the owners of this distributive technology.

Statute of Anne and Anglo-American Copyright

Because the Licensing Act was a strategic combination of copyright and censorship, when there was a political shift in the law to promote free expression, the Licensing Act became out of step with the new liberties afforded under new English rule. The Glorious Revolution of 1688, which ensured Protestant succession to the English Crown, influenced the idea that there should be a legal split between copyright and censorship. Parliament enacted a new Bill of Rights (1689), the Toleration Act (also 1689), and other laws that strengthened certain freedoms and rights of English citizens. As a result, there was less concern about the printing presses' publication of religious, educational, and other potentially controversial materials. Now, for the concept of copyright to survive, it had to be separated from its censorship roots. A related aspect was that the Licensing Act was not merely preventing the Stationers from printing prohibited works, it was also, through government controls over the technology itself, preventing the public from greater access to copyrighted works.

In this divorce from its origins, the Licensing Act may have divulged its true purpose: the Stationers' total economic monopoly of the book trade at the cost of freedom of expression and rights for authors and users. Under the Stationers' regulations, an author could not make a copy of their work without seeking permission from the Stationers' Company. However, even if a book was accepted for printing by the Stationers' Company, the author did not see any profits in the sale of any copies. Authors were paid only once for their work and the Stationers held all the rights to collect profits from the sale of the copies. Authors did not, therefore, have any real economic incentive to continue to create works, plays, textbooks, drawings, and other creative works if they saw no other profit than a one-time payment.[36] If this practice continued, it could cause the modern notion of intellectual "brain drain," a departure of educated people from writing books or, worse, departing England altogether for other places that rewarded an author's creativity and innovation.

In 1695 the House of Lords let the Licensing Act expire without a renewal or other law to take its place. This obviously caused the Stationers great concern. They were not willing to sacrifice their lucrative economic monopoly simply because the government had changed its mind on censorship and controlling the printing press. However, it is apparent that members of Parliament were perhaps tired of the Stationers' monopoly because, for over a decade, Parliament refused to pass a new statute regulating the printing presses, as was proposed by the Stationers.

Eventually, a draft of a new copyright law—the origin of the Statute of Anne—was created to counter the economic privileges found in the Licensing Act and to promote free expression. In effect, this new copyright statute would no longer be used by the government as a method for censorship. Parliament's passage of the Statute of Anne remains the defining moment in Anglo-American copyright history of transforming a publisher's private law copyright into a benefit to the public.

However, it is interesting to note that the Statute of Anne actually originated from the Stationers' repeated attempts to renew their powers under the Licensing Act, which had lapsed in 1695. At first, the Stationers lobbied for legislative authority that matched the previous Licensing Act. Their arguments ranged from process, progress, and economic factors, but as might be imagined, they emphasized control and censorship as the most critical reason to maintain legislative authority.[37] For sixteen years, these arguments were ignored. One opponent of the Stationer's request to extend the Licensing Act said that the previous law "subjects all Learning and true Information to the arbitrary Will and Pleasure of a mercenary, and perhaps ignorant, Licenser; destroys the Properties of Authors in their Copies; and sets up many Monopolies."[38]

Stationers made a strategic assessment and changed their arguments. They quickly lobbied for protection of authors' rights in the new law and framed the new scope of copyright as promoting and supporting learning. This is not to say they gave up the battle for protecting their economic interests; as some scholars have pointed out, their attempts were mainly to secure some continued version of the publisher's right, and indirectly protect some of the authors' interests.[39] However, this new lobbying, combined with some broad support from other members of Parliament and some authors, led to the Statute of Anne's passage. The new statute even started with the phrase "An Act for the Encouragement of Learning."[40]

For the first time in history, this new law gave authors a rudimentary economic incentive to create new works: authors had some control of their own works and the copies made of them, via a limited economic monopoly—not unlike our modern understanding of copyright. The Statute of Anne attempted to capture the first balance between authors' rights and the public benefit in copyright, shuttling works into the public domain when the rights expired. This temporary economic right was designed to be just enough incentive for authors to continue to create new works. Of course, when the rights expired (after fourteen years) and the work entered the public domain, then anyone could print, share, and disseminate the work thereafter without permission. This encapsulated the cycle of copyright: creation, limited economic control, and expiration, with the hope that further works could be created using what lapsed into the public domain.

However, as many historians have noted, the language of the Statute of Anne left much to be desired.[41] While Parliament focused on the author's role to advance science and encourage learning, the law may have done nothing more than safeguard "the control of production by a few wealthy capitalists ... [and] the continued dominance of English publishing by a few London firms."[42] As Professor Oren Bracha has stated, "the only entitlement conferred on authors and their assignees [under the new Act] was 'the sole Liberty of Printing and Reprinting' a book for the prescribed term. The statute did not create a new concept of 'ownership' of a copy."[43] There was no overnight revolution in authors' rights. Authors continued to use the system with which they were familiar, handing away any rights to their works, with the Stationers continuing to claim copyright in the formal registration of the books. In fact, authors had to wait for a court case, some sixty years after the passage of the Statute of Anne, to gain any clarification of their rights.[44] "Copyright," as we understand it in the modern meaning, still had a long way to go before it incorporated a concept of true authors' rights to and ownership of their own works.

Nevertheless, the Statute of Anne served as a launching pad for a "grand compromise" between all competing interests in the realm of early copyright policies. Lessons learned from its passage included securing economic rights for authors; breaking any total monopoly over authorized technology, like the printing presses; and securing the public interest by limiting copyright to a specific time period or, as the statute stated, encouraging learning by creating a path for all works to enter the public domain.

This compromise was well represented in the debates surrounding the adoption of the Statute of Anne into US law. However, these concepts enshrined in the British law were not immediately recognized in the American colonies. Certainly, the members of the United States Constitutional Convention were aware of the ideas of control and censorship as the United States emerged from English rule. But in the early days of the colonies, "promoting the progress of science and useful arts" through copying was strictly forbidden.[45] Sir William Berkeley, Charles I's royal governor in Virginia, stated in 1671, "I thank God there are no free schools nor printing, and I hope we shall not have these hundred years; for learning has brought disobedience, and heresy, and sects into the world, *and printing has divulged them*, and libels against the best government. God keep us from both."[46] Sir Berkeley was perhaps clairvoyant; one hundred years later, the Constitutional Convention debated that very subject. America's founding elite, much like their peers in English Parliament, no doubt had the same concerns over printing press control and laws like the Licensing Act of 1662 because they ran counter to the laws and policies that were developing in early America, including freedom of the press and freedom of speech.

Even in the earliest day of the American colonies, there was a vision emerging that this new technology, the printing press, should be freed from the permissive-based control that existed under the Licensing Act of 1662. The act certainly haunted the early printers in the American colonies as they were still technically governed by it under British law. One of the most famous of all colonial printers was William Bradford. And in a fascinating appeal to Governor John Blackwell of Pennsylvania in 1689, Bradford expressed his vision for the importance of permission-free printing, clearly in direct violation of the goals of the Licensing Act. Bradford stated "It is my imploy, my trade and calling, and that by which I get my living, to print; and if I may not print ... I cannot live.... If I print one thing today, and the contrary party bring me another tomorrow, to contradict it, I cannot say that I shall not print it. Printing is a manufacture of the nation, and therefore ought rather be encouraged than suppressed."[47] Apparently, Governor Blackwell did not agree, and Bradford had to post a bond and pledge to never publish without permission again.[48] However, these sentences belied something that was to become revolutionary—a free press that unbound the tangle of copyright, licensing, and censorship.

Yet any control of technological innovation, as manifested by the printing press, would continue to serve the cause of censorship, not copyright, if the British laws continued to govern the colonies. Eventually, this was partially remedied. The Founders drafted the US Constitution's Copyright Clause, "to promote the Progress of Science and useful Arts, by securing for limited Times to Authors and Inventors the exclusive Right to their respective Writings and Discoveries," to equally balance the concern over access and the distribution of controversial materials and the rights of authors. They created an exclusive right as originated, but not necessarily well-practiced, in the Statute of Anne. This was the essence of what is known in the law as a "content-neutral law," wherein the focus is not on any prepublication license or regime that intertwines censorship and copyright but rather a system that allows authors to assert some control over their own works. It represented a continuing shift in the adoption of the lesson from the English struggles with copyright; the focus of the law was no longer on controlling the distributive technology—rather, it was focused on the author's right and the benefit to the public. It also incentivized free expression, a concept that the Founders discussed in the debates surrounding the creation of the First Amendment.

In 1790, pursuant to their constitutional authority, Congress passed, and President George Washington signed, the first copyright law in the United States. The law—like the US Constitution's Copyright Clause and the Statute of Anne before it—was focused on the public benefit rationale for copyright protection. Its language stated it was "An Act for the encouragement of learning, by securing the copies of maps, charts, and books, to the authors and proprietors of such copies, during the times therein mentioned."[49] It clearly emphasized and mirrored some of the exact statutory language found in the Statute of Anne. Further, the language featured the same balance that the English had devised with the Statute of Anne: an incentive of a limited economic monopoly granted to authors over their works, followed by the expiration of those rights, whereby the works dropped into the public domain. This bargain for authors, featuring some initial control, was consolidated along with the public good—a limited economic benefit granting the creator a temporary monopoly and the expiration of that exclusivity as the works drop into the public domain. In this sense, US copyright was, in Thomas Macaulay's words, a "tax on readers for the purpose of giving a bounty to writers"[50]—a bounty designed to encourage new creation.[51]

Since then, copyright law has been updated many times. Currently, copyright lasts for the entire lifetime of the creator, plus an additional seventy years after the creator has died. Typically, copyright law grants the creator a set of certain rights. These rights, codified in the United States Code at 17 U.S.C. § 106, grant creators the right to copy, modify, display, perform, and create other works modified from the original. These are typically referred to as the exclusive rights or, in common parlance, the "bundle" of rights. The creator automatically has control over these bundled exclusive rights as long as the work is fixed and creative; no registration or other formality is required.

Licensing and Ownership

As can be seen from the historical record, licensing is a separate area of law but nonetheless inextricably tied to copyright, as licenses are often used to transfer copyright. Licenses are most often granted within the context of a contractual relationship. Often the same words used to create the license are also contained in the same instrument that memorializes a contract between two parties. The best way to view a license is a "contract not to sue." A license, then, is a legal interest created by a rightsholder granting some privileges to a non-rightsholder.

Attempts at utilizing a preemptive "Licensing Act"–like license have been made in the recent past, especially with regards to publication, purchase, and distribution of print works. In the early twentieth century, in a case called *Bobbs-Merrill v. Strauss*, the publisher of a popular novel, *The Castaway*, placed an attempted license, in the form of a notice, after the title page that read, "The price of this book at retail is $1 net. No dealer is licensed to sell it at a less [sic] price, and a sale at a less [sic] price will be treated as an infringement of the copyright."⁵² Disregarding the notice, Isidore and Nathan Strauss, owners and operators of Macy's department stores, had purchased a large number of these books and were selling them at a discount for eighty-nine cents. This, according to publisher, was illegal. Adhering to its notice placed on the book's front page, the publisher sued the Strausses for copyright infringement.

The publisher argued that the notice inside the cover of every book printed and sold constituted a binding *license* on purchasers. Further, this notice attempted to prevent consumers and retailers from reselling the purchased book at the price of their choice. In endorsing a new doctrine called first sale, the US Supreme Court rejected the publisher's argument and held that since there was no agreement or contract between the Strauss brothers and Bobbs-Merrill, the publishing company could not impose any post-sale restrictions downstream, such as a reduced resale price. According to the court, copyright law did not create the right to limit, via a simple notice, a person's ability to resell any purchased copyrighted work, such as a book.⁵³ Later, the US Supreme Court described this licensing attempt at controlling the downstream market as "hateful to the law from Lord Coke's day to ours, because it is obnoxious to the public interest."⁵⁴

The copyright owners' rights, the court said, were first exercised, and then exhausted, upon the first sale of a copy. This exhaustion occurred when Strauss bought the books to eventually sell in the Macy's stores. Under the law, Strauss and any subsequent purchasers who bought the book for a discount were both free to sell their copies at whatever price they set, whether higher or lower than the copyright owner preferred (or, additionally, either party could give the work away for free). Rightsholders could not interfere with these actions, nor did a person have to seek permission in order to lend, sell, or give the work away downstream in the secondary market.⁵⁵

Shortly after this ruling, Congress adopted its first iteration of the first sale doctrine in the Copyright Act of 1909. Later, the Copyright Act of 1976 clarified the doctrine in the present-day Section 109. Now, much like fair use, first sale is a statutory exception to copyright. Any owner of a copy of a legally acquired work may "sell or otherwise dispose of the possession of that copy" without the permission of the original copyright owner.⁵⁶

This policy permitted the development of a critical part of the copyright economy: the secondary market. We probably are more familiar with the secondary market as represented in the "used" stores. This protection of the "used" market for copyrighted works supports consumers' ability to do what they will with legally purchased copies. Used bookstores, used music stores—even libraries⁵⁷— could not exist if the sellers retained some right to individual copies or used restrictive licensing to control this secondary market.

However, the current digital space has seen an enormous restriction in the ability for first sale to operate. This certainly threatens the secondary market, users' rights, and the cultural record. Once again, licensing is the problem. In fact, modern licensing limits have achieved precisely what the Stationers publishers sought in 1662 and *Bobbs-Merrill* in 1908: downstream control via license to destroy secondary markets. That is because licensing is not actually a "purchase."

Licenses are not the equivalent of a purchase under the law. Although e-media vendors may use the word "buy it now" or "purchase" or "for sale," according to the fine print, it is not a "purchase" for which the first sale doctrine applies. When an individual agrees to the terms of these licenses, at best they are merely *renting or leasing* temporary access to these copyrighted works. Depending on the terms of the license, the individual may have to renew or pay for access over and over. As a result, these licenses are preventing anyone—consumers, users, libraries, and schools—from owning e-books, instead forcing them to stream movies, music, and other copyrighted works. All these works are more akin to temporary rentals under the license, offering only a limited, nonexclusive, nontransferable right to access.

Additionally, in some licenses, the terms are so limited that individuals and institutions are restricted from the very uses, exceptions, or exemptions made legal in the Copyright Act. For example, a common media license might state: "Except as explicitly authorized in these Terms of Use, you agree not to archive, download, reproduce, distribute, modify, display, perform, publish, license, create derivative works from, offer for sale." As read, this license takes away nearly all the copyright exceptions for the public user. In this way, the license serves the same purpose as the Licensing Act and all the other limiting laws, decrees, and regulations of the previous eras—it is limiting access, control, and use of the copyrighted works.

The Modern De Facto Licensing Acts: Digital Millennium Copyright Act, Section 230 Reform, and Royal Executive Orders

Predating the Stationers and the Licensing Act, one of the first ventures into government control of printing technology was the appointment of one person, via a license, proclamation, or printing privilege, to oversee the printing of certain books or materials for the Crown. In 1504 Henry VII appointed William Facques as the first royal printer, granting him the exclusive right to print the official documents of the government. Later, in 1518, a second royal printer was appointed for producing religious sermons.[58] These special royal printing privileges are the first example of government press control, except they were controlling an individual person who was the beneficiary of the licensing scheme. Later, the essence of this method would be adopted into a similar privilege system, via broader licensing schemes given to an entire guild of printers, the Stationers. In many ways, there is a near parallel in these predecessor royal proclamations and decrees with the current state of reform efforts surrounding the Digital Millennium Copyright Act (DMCA) and the Communications Decency Act (CDA) Section 230.

Additionally, the origins of the Licensing Act were subject-specific, limited to censoring materials that were objectionable to the government. Topics such as religion, political opinions, government criticism, and related commentaries were the target. If the work was not about a banned topic, the economic process moved forward, and the Stationers printed the work. However, the strategy behind modern laws and preventative regulations in the DMCA extends beyond the approval by the government of the material's subject. The technological platform *itself* has an adopted process (rules, regulations, or policies) that give rise to a burdensome censorship procedure which can be wielded by powerful private media organizations that fear criticism or subversion as much as any government.

This is where the danger of licensing-only culture reemerges. Having a well-defined and well-policed licensing market, designed to generate income, can serve as an efficient smokescreen for using the DMCA copyright takedown process or a Section 230 reform agenda as a tool of censorship. These takedowns are done in the name of market economics and rarely, if ever, invoke censorship. However, the real end goal is clearly censorship—and the data indicates that works are being censored at a disturbing level.[59] As an interrelated law governing some uses of copyrighted materials, CDA Section 230 generally provides immunity for website platforms for hosting third-party created content.[60]

On their face, both the historical laws and the modern reform efforts appear as a form of censorship of a creator's copyrighted work that is distributed via particular technology—the printing press in the former, social media platforms in the latter. However, before we can examine how this call for DMCA and CDA reform is aberrantly moving us back to the goals of pre–Statute of Anne licensing regimes, we must look at the origins of liability that had attached to the distribution of a copyrighted work through technology in the modern era.

Both the DMCA and CDA represent the unique problems created by the growth of the internet under copyright law. The scale at which a user could distribute works through an online service grew exponentially. For these growing internet platform companies, it would be impossible to scrutinize each file that was uploaded and spread on their platform. Inevitably, some of these files might contain material that infringed the copyright of third parties or worse, did reputational harm or other damage.

For example, a person or entity who publishes a disparaging or defamatory statement by a third party generally bears the same liability for the statement as if they had created it. A book, journal, or newspaper publisher can be held liable for anything that appears within its publication. The theory behind this "publisher" liability centers around control: the publisher has the knowledge, opportunity, and ability to exercise editorial control over the content of its publications.

On the other hand, "distributor" liability developed as a much more limited policy concern. For example, libraries and bookstores are generally not held liable for the content of the materials they distribute. Much like the modern internet platform company, it would be impossible for libraries and bookstores to read every publication before they distributed it to the public. If this were to happen, then free speech and the ability to distribute information would be harmed. Libraries and bookstores would start limiting holdings and regulating their sales as part of a risk-mitigating strategy, leading, inevitably, toward self-censorship. In addition, it would be difficult for distributors to know whether a work, comment, or expression is actually defamation. Libraries and bookstores do not have time to check the veracity of each publication. To hold them liable for the actions of third parties would be prejudicial. However, as the internet became more and more common as a means of communication and commerce, the question remained: Does "publisher" or "distributor" liability apply to these platforms and online services?

Under the DMCA, the law asked similar questions. Since 1998 the DMCA in the United States and its European counterpart, the E-commerce Directive (2000/31/EC), have provided a safe harbor program for online service providers. This safe harbor prevents platforms from incurring liability for copyright infringement carried out by their users. These laws were created as a response to fears around digital piracy and copyright infringement on the internet. The DMCA provided internet service providers (ISPs) with more certainty regarding liability for copyright infringement by third parties and increased copyright owners' ability to protect their rights on the internet in the form of notice-and-takedown procedures. As is often the case with copyright legislation, this was a balancing act, and the US Congress had been investigating how to maintain copyright's principles while allowing for the internet to operate at its fullest capacity since the 1990s.[61] Congress's interest was also in allowing the new technology to work the way it was designed: increasing the free flow of information and access to knowledge, often in the form of copyrighted materials. Congress decided that absent legislative intervention, copyright law would fail to protect copyrighted works in a manner that would "make digital networks safe places to disseminate and exploit" copyrighted works.[62]

Additionally, many ISPs faced uncertain liability, as there were several conflicting cases on the issue of ISP copyright infringement liability that had the potential to hamper the expanding speed and capacity of the internet. Congress recognized that ISPs were experiencing a high level of risk for copyright infringement because the technology required that they process, cache, or host third-party copyrighted materials that were either created by or posted by their users.[63] As a result, Congress created Section 512, a specific section in the DMCA: a safe harbor law that would shield ISPs from copyright liability via a multifaceted notice-and-takedown system.

The system was designed to benefit those internet companies for whom the business model was based on hosting and distributing copyrighted materials. For example, YouTube, which relies on its users to upload materials for access on their platform, might inadvertently host infringing copyrighted content that has been posted by users. Under the DMCA, YouTube would not be held liable for the actions of its users if it has a DMCA policy and follows the note-and-takedown provisions in Section 512. YouTube operates within the law, and avoids potential massive copyright infringement liability, if it creates and deploys this policy and expeditiously responds to rightsholders' takedown notices.

Section 512 of the DMCA was meant to be another specific arrangement for distributive technology, balancing the needs of tech companies, rightsholders, and users. ISPs develop policies for their platforms to gain protection from liability, copyright owners have a system to police copyright infringement, and users maintain the ability to upload work and respond to takedown requests—all without having to go to court. Section 512—much like the CDA Section 230 covered below—is another liability-limiting federal law that helped generate all the ISPs, modern platforms, and social media companies in use today. However, much like the modern reading of the Statute of Anne (see previous section on scholars' criticisms of it), there is much to be desired from the employment of Section 512 by publishers and other rightsholders that, in some cases, are using this copyright instrument as a means of censorship. Arguably, the law actually aids censorship because the Section 512 rules clearly state that an ISP must remove a copyrighted work immediately upon receipt of a takedown request, regardless of any potential liability or viable defenses.[64] This certainly hearkens back to the government-sponsored licensing regimes created for the Stationers, where censorship was disguised through a copyright-adjacent system of technological control.

If looked at carefully, it is clear how legitimate copyrighted materials, often expressions of the author or creator, are subject to "at whim" censorship from internet platforms by virtue of this Section 512 law. If a rightsholder does not like what the author might be doing—criticizing, commenting, or creating using a copyrighted work—the rightsholder can merely send a takedown via the platform's DMCA process, and the work will automatically come down. This is a "shoot first, ask questions later" system. And while it is true that the Section 512 process allows investigation after the takedown, the process can be burdensome.[65]

Further, since the DMCA law that controls these rules is created by the government—with specific takedown regulations deployed by the social media platforms—the analogy to the English government, the Star Chamber, and the Stationers' relationship surrounding censorship and control is obvious. Like these organizations and their licensing decrees of old, this modern DMCA note-and-takedown system accomplishes all parties' goals of control without any real oversight by any judicial authority. Again, the means of access to technology and using its distributive properties is subject to the "modern" Star Chamber or Stationers' total authority.

CDA Section 230, another safe harbor–style law, was born out of an early internet decision from 1995, called *Stratton Oakmont, Inc. v. Prodigy Services*.[66] Much like the early printing presses, early online services (like Prodigy Services) granted access to the internet and provided forums (or virtual bulletin boards) for publication of various copyrighted materials by third-party creators. In *Stratton*, a third party posted allegedly defamatory statements on a bulletin board discussion group called "Money Talk." This bulletin board was operated by Prodigy. The user claimed that a securities investment banking firm had committed criminal and fraudulent acts in connection with an initial public offering of stock and was a "cult of brokers who either lie for a living or get fired."[67]

In a beautiful analogy that hearkened back to early case law, the Supreme Court of New York evaluated whether Prodigy itself was merely a *distributor* of the posted comments or exercised sufficient control over the bulletin board to render it a *publisher*. Prodigy had content guidelines and had hired staff to police users to follow the guidelines. Additionally, Prodigy had an automatic software screening program that filtered out offensive language. Reviewing these facts, the court found that, although it claimed a role as a distributor, Prodigy exercised substantial *control* over the bulletin board and therefore was not a *mere distributor* entitled to special protection under defamation law. Therefore, Prodigy was liable for the defamatory statements.

In this case, the court punished this early online provider for making a "conscious choice" to regulate the content of its bulletin boards.[68] This presented a problem: all future companies using the distributive technology of the internet thereafter could potentially be exposing their entire company to a greater risk of liability for having an active community with well-developed policies and guidelines. There was also some fear that this case left little to no incentive for companies to even attempt to create any basic guidelines and policies for safe online communities. From the vantage point of many new startup online services in 1995, including AOL, Yahoo!, and others, it might not be worth the risk of allowing creators to author statements, create works, or share ideas using their technology in open discussion communities, considering the potential for enormous damages resulting from defamation claims.

All these fears emerged because the law was highly dependent on the older interpretation of a distributor's editorial control of the material—now posted on websites or online forums—very much like the early publishers and printing press technology. This raised concerns even in the halls of Congress, where some lawmakers pondered whether internet companies would stop monitoring any content to avoid potential liability. Further, Congress found that because these discussion boards and other forms of early social media had the capacity for exponential economic development in the United States, it would be against economic policy to allow the *Stratton* decision to stifle innovation in the digital space by requiring these providers to police everything that occurred on their platforms.[69]

To help continue development of the internet, preserve "the vibrant and competitive free market . . . unfettered by Federal or State regulation," and remove "disincentives for the development and utilization of blocking and filtering technologies," Congress passed CDA Section 230, which solidified a "safe harbor," or immunity from lawsuits, in this space for service providers. Much like "Good Samaritan" provisions in other parts of US law, Section 230 grants interactive computer services, including social media platforms, safe harbor protection from liability arising from user-generated content. For such a powerful provision, the key portion of Section 230 is relatively brief and consists of only twenty-six words: "No provider or user of an interactive computer service shall be treated as the publisher or speaker of any information provided by another information content provider."[70] The provision also grants platforms the capability to restrict "material that the provider or user considers obscene, lewd, lascivious, filthy, excessively violent, harassing, or otherwise objectionable, whether or not such material is constitutionally protected,"[71] without the potential of civil liability.

For decades, courts have interpreted CDA Section 230 to consistently provide broad immunity for online providers of all types. Investors also viewed the CDA as a risk-mitigating factor when investing in online social media companies; as a result, the companies grew and innovated using those investments. However, as of the time of this writing (2023), the same legal strategies of the past—as far back as the seventeenth and eighteenth centuries—are attempting to shift the liability back on those who control the technology: the social media platforms. This shift may have consequences that chill innovation, the economy, and copyright expression.

In the past several years, content restriction by social media companies has risen to the forefront of the debate about the role those companies should play in combating "misinformation" and "harmful speech" on the internet. During these debates, the Trump administration, the Biden campaign, and others have repeatedly called for the repeal of Section 230. The Trump administration even went as far as to say that Section 230 is a "serious threat to our National Security & Election Integrity."[72] This messaging surrounding reform reached a fevered pitch after Twitter and Facebook began labeling Trump's posts with fact checks. The Trump administration moved the federal government into action. First, the administration claimed that the companies that controlled this distributive technology had a conservative bias and therefore were censoring conservative viewpoints. Many conservative lawmakers, on both the state and federal levels, introduced bills and other administrative acts to repeal Section 230.[73] Trump wrote and signed *Executive Order 13925: Preventing Online Censorship*, which included a provision asking the secretary of commerce and the attorney general to file a petition for rulemaking with the Federal Communications Commission (FCC) requesting that the FCC rewrite and clarify Section 230.[74] Even the Department of Justice sent a letter to Congress proposing amendments to change Section 230.[75] The FCC chairman announced that the commission would work on clarifying the meaning of Section 230.[76]

Returning to the DMCA, a similar reform pattern is working its way into the law and, like Section 230 reforms, its effects could be detrimental to the copyright ecosystem. Under the current DMCA notice-and-takedown regime, the rightsholders must do much of the work. A rightsholder must find and notify platforms about the specific instances of infringement. These requests go into the system designed by the platforms, which then follows rules that govern the receipt and processing of any takedown requests. As recently edified in a ten-year-long DMCA case on fair use, the law requires that rightsholders consider some of the many exceptions and limitations to copyright, including fair use, before sending a takedown request.[77] This aspect of the law—protecting fair use and allowing counternotices to challenge the takedown request—protects First Amendment freedoms enshrined in copyright law, including news reporting, journalism, commentary, and criticism.

Over time, the DMCA takedown system has developed to meet the demands of the internet. For example, more than 500 hours of content is uploaded to YouTube every minute and more than 100 million photos and videos are uploaded to Instagram every day.[78] These numbers are staggering. For internet platforms to maintain their limited liability under the DMCA, it was inevitable that technology was going to have to develop to ease the burden of dealing with a potentially endless amount of takedown requests. While the DMCA does not require the platforms to actively monitor their own websites for infringing content, in order keep up with the exponential uploads, many companies have deployed an automated filtering system. This acts as a "prefilter" to find infringing content before the rightsholders check themselves. These automated filter systems examine the user's uploaded digital files, looking for particular characteristics of copyrighted works by using an algorithm. If there is a match, the video is flagged, taken down, and the system sends a notification to the copyright holder. The YouTube Content ID system is arguably the most famous of these automated filters as it was first introduced in 2007 and has been continually built up since then. However, in addition to YouTube, other large platforms have also developed automated content filters that enable the detection of copyrighted files at the time of user upload.[79]

This is not only the case in the United States. Other parts of the world are considering adoption of prepublication licenses or "content checks" as a means of controlling potential copyright infringement on major internet platforms. Like the printing press that emerged from Europe, one of the threats to copyright is also emerging in the form of the relatively new EU Copyright Directive, which significantly alters this balanced system in favor of mandatory licensing or permission. Directive 2019/790 on Copyright in the Digital Single Market ("EU Copyright Directive") was adopted in 2019 and is designed to be implemented across all EU member states through their national laws.[80] Some of the objectives stated in the EU Copyright Directive are to reform the law so that creators and rightsholders are treated more fairly within the system. Additionally, rightsholders could exert greater control over the distribution of their content and capture any revenue derived from the use.

Article 17 of the EU Copyright Directive was written to meet these goals with an extremely specific change: the platforms *must* have rightsholders' permission to distribute the content—even if users had uploaded the content. Under these rules, the platforms will be considered to have committed copyright infringement by making illegal user-uploaded content available. This shift is why Article 17 was one of the most controversial provisions in the EU Copyright Directive. It significantly altered the existing copyright safe harbor system by distinguishing specific rules for a newly defined category of internet platform companies: online content-sharing service providers (OCSSP). An OCSSP is "a provider of an information society service of which the main or one of the main purposes is to store and give the public access to a large amount of copyright-protected works uploaded by its users, which it organizes and promotes for profit-making purposes."[81] For example, YouTube and Instagram are OCSSPs and therefore are subject to the rules in Article 17.[82]

In a dramatic shift, Article 17 requires that OCSSPs obtain a *license* from a rightsholder for *all works uploaded by their users*.[83] In many cases, this license should be acquired prior to the user uploading any content. The OCSSPs are expected to negotiate a license or permission with the major rightsholders for *potentially* infringing uploads by users. Additionally, the OCSSPs are required to make "best efforts" to ensure they are not distributing unauthorized content by either obtaining that licensed authorization, preventing distribution of the content, or removing the content entirely. In cases where the rightsholder of the uploaded content is unknown, OCSSPs like YouTube and Instagram must prove their diligent search efforts for a rightsholder in order to request a license or permission.

Again, it is evident how a government—here the EU—in its role as a supranational organization pushes a licensing agenda for rightsholders with this law. Article 17 provides rightsholders with powerful language that will inevitably result in license terms that clearly favor the rightsholders in negotiations. This may be an advantage in terms of access, price, or other restrictions that benefit the rightsholder.

It is also interesting to note that this is not only an EU problem. For foreign platforms and companies, Article 17 will apply to any OCSSP hosting copyright-infringing user-uploaded content targeted at the European market, regardless of the jurisdiction governing the OCSSP. For example, an OCSSP with its legal principal place of business in the United States doing business in EU member states would be subject to Article 17. Therefore, much like the spread of the copyright ideas emerging from the Statute of Anne as they made their way to the United States, it may come as no surprise that the US government has taken note of this type of Article 17 copyright system.

Congress commenced a series of hearings in 2020 on the safe harbor rules. The US Copyright Office issued a study that recommended several significant changes to existing safe harbor rules. In a fairly pro-rightsholder-oriented document, the Copyright Office did conclude that the DMCA notice-and-takedown regime is "unbalanced" and Congress may need to "fine-tune" the safe harbor section.[84] However, in its deliberations over this fine tuning, it appears that the Copyright Office did not consider the millions of internet users and creators who rely on these internet platforms every day.

The controversy over Article 17 is simple: the law could force platforms with user-uploaded content to automatically check the uploads for copyright infringement before distributing them online. This automatic process to deploy an upload filter was, as in the case of the Content ID system developed by YouTube, a voluntary action. However, it is clear that Article 17 could predictably lead to platforms implementing such filters just to lower the massive liability risk that could result from the vast volume of copyrighted content that is uploaded every day. How could a platform, under Article 17, otherwise negotiate for all the licenses, permissions, fees, and due diligent searching without some sort of technological filter? Common sense would dictate that it might be the only way to avoid liability under the EU system.

In all of these actions—from Section 230 reform, through the specter of the EU Copyright Directive's Article 17—the modern strategy was the same as that of Henry VIII's or Queen Mary's laws: shifting the law's focus from those creating the "wicked" works, to those who were in control of the technology where those works were either distributed or censored. These early royal proclamations were created to control any opposing ideological printed content, including broadsides, pamphlets, and books.[85] Much like the equivalent executive orders issued by state and federal authorities in the modern era that remove a technology platform's safe harbor because of alleged bias or in the name of efficiency, the original "Royal executive orders" were often a result of the struggles facing whichever monarch had contemporary control of the government.

For example, the Treason Act, announced under the reign of Henry VIII, punished those who "slanderously and maliciously *publish* and pronounce, by express *writing* or words, that the King our Sovereign Lord should be Heretick, Schismatick, Tyrant, Infidel, or Usurper of the Crown."[86] If someone published a work in violation of this act, the penalty was death. This law, which clearly impacted any free expression regarding religion, was the result of Henry's infamous battle with the Catholic Church in Rome.[87] Putting aside the punishment—death, which is thankfully not part of today's publication laws—what we see in this early law is a shifting liability with a focus on who controls the distributive technology. These laws similarly move the potential liability away from those who created the illicit work ("the writings") to those who controlled the early printing press technology ("the publishing"). Article 17 does just this when it creates a prepublication licensing burden that is borne by the OCSSPs not the users.

Similarly, various proclamations made by the Catholic Queen Mary from 1553 to the end of her reign in 1558 banned printing, distributing, or possessing any books containing "wicked doctrine" and condemned "*the pryntynge of false fonde bookes, ballettes, rymes, and other lewd treatises.*"[88] As is well documented, Queen Mary's reign was one of constant struggle for control of the government. It is of little surprise, then, that these laws were passed to protect the royal government from criticism and, more importantly, to check the spread of the Protestant Reformation.

Again, it is interesting to note that these proclamations, banning particular types of works that shared particular messages (mostly anti-Catholic works), were representative of a shift in the target of the law's regulation: *the law was less concerned about the authors of such works, instead turning to those who controlled the technology.* In the historical example, the focus shifted from the authors of such "wicked" works to those who were capable of *publishing and distributing such works*. In the printing press regulations at the time, the author became less and less of a concern. This shifting liability, focusing on those who control the technology—subject to both privilege and regulation—was adapted later to the modern context. A similar shift is happening to hold social media platforms liable for the work of their users, as represented in the battle over Section 230 and the concerns arising from Article 17's spread through the EU (and eventually to any foreign jurisdiction that touches the EU over the internet).

However, evident from the record, laws that attempted to censor certain views—whether it be conservative bias or "wicked" works—were successfully utilized and merged with the growing use of licensing and the creation of copyright. Is our modern battle with Section 230 going to be subject to the same fate? Most likely not. With the rise of the Stationer's Company, those who controlled the printing technology, there was great concern with the economic control of the book trade. It was in the best interest of the Stationers to work hand in hand with the government. It is common sense to understand that the Stationers would not risk their lucrative government-sponsored monopoly by publishing heretical work by an antigovernment author. As a result, the Stationers could be biased and only publish works that were approved. The government was able to assert its control over the works disseminated, and the Stationers maintained their economic hold on the trade.

In the modern context, however, the "modern Stationers"—represented for purposes of this argument by the social media platforms and other internet companies that control technology platforms—do not have the same economic concerns that existed in the past. In fact, social media platforms and other internet companies, which all benefit from the safe harbor of Section 230, earn more than ever before. For example, in 2020 Facebook's annual revenue amounted to $85 billion,[89] while Twitter generated $3.7 billion.[90] These companies are less likely to work as closely as the Stationers did with the Crown to restrict publication of certain materials. More than likely, they will continue to advocate for continued protection under Section 230. Nor are they likely to secure the same type of government-sponsored economic monopoly like the Stationers. Arguably, these "modern Stationers" are already the monopoly themselves without the benefit of any government censorship agreement, like the royal printing prerogatives or the Licensing Act.[91]

However, the potential changes in Section 512's note-and-takedown provision and the EU's Article 17, are only just beginning. Changes in the law that favor the rightsholders, licensing agreements, and permission-based culture are already here. As technology is integrated into every aspect of the average copyright creator, we could all be subject to the same drastic reforms as seen from the Star Chamber decrees through the Stationers' internal regulations for publishing and access.

The emergence of copyright was a combination of many factors, though it was predominantly an attempt to control technology and implement the government's need for censorship. Licensing laws were developed as a means of accomplishing these goals. Additionally, the Stationers' Company, a government-approved organization, used licensing to maintain a monopoly over the printing technology. These licensing regimes limited the rights of authors and creators, and denied others access to the printing press as a form of technological censorship.

In the contemporary copyright system, we are beginning to see this same story play out again as power is being asserted over access to technology to control, censor, or suppress works on topics that are important in the modern era. Again, copyright is being weaponized to suppress criticism, commentary, research, and news reporting. The advent of licensing-only culture, prepublication licensing requirements, upload filters, DMCA takedown requests, and the latest challenge to CDA Section 230 immunity is remarkably similar to the historical narrative explored in this chapter. Distribution of copyrighted works is threatened via restrictive licensing and technological constraints. An overhaul of this system is due, much like the Statute of Anne initiated the move toward greater distribution and uses of works. It is critical that the cultural record—our books, articles, poems, music, and film—all copyrighted works—are no longer subject to the whim of third-party licensing terms, technological filters, or the government's concern over bias. Examination of historical copyright and technology narratives can help prevent our modern copyright system from succumbing to the threats posed in the past, including control and censorship, and thereby protect the ability of authors and creators to use modern technology for unrestricted sharing and distribution of creative expressions. If these problems are not addressed, the slide into licensing-only, permission-based technological control will continue, and this, as the US Supreme Court described, is "obnoxious to the public interest."[92]

NOTES

1. The term *copy right* first appears in the Stationers' Company records for the entry on May 31, 1701. Stationers' Company (London), George Edward Briscoe Eyre, and Roxburghe Club, *A Transcript of the Registers of the Worshipful Company of Stationers: From 1640–1708 A.D.*, vol. 3, *1675–1708* (London: privately printed, 1913), 494.

2. Venice, Italy, introduced the concept of granting printing licenses to access the new movable-type printers. In 1469 the Venetian Senate granted Johannes de Spira, a German printer, a five-year monopoly to print Cicero's letters. Many years later, the Venetian Senate granted the sole right to publish the book *Decade of Venetian Affairs* to its author who was the official historian of Venice. The Senate made a new modification to this grant, adding a fine for any infringement or unauthorized printing of the book.

3. Paul Goldstein, "Copyright and the First Amendment," *Columbia Law Review* 983 (1970): 70; B. MacDonald, *Copyright in Context: The Challenge of Change*" (Ottawa:

Economic Council of Canada, 1971); Zechariah Chafee, "Reflections on the Law of Copyright," *Columbia Law Review* 45, no. 5 (1945): 719–38; Wendy J. Gordon, "Toward a Jurisprudence of Benefits: The Norms of Copyright and the Problem of Private Censorship," *University of Chicago Law Review* 1009 (1990): 57.

4. The Constitution grants to Congress the power "To promote the Progress of Science and useful Arts, by securing for limited Times to Authors and Inventors the exclusive Right to their respective Writings and Discoveries." U.S. Const., art. I, § 8, cl. 8. See also *Feist Publications, Inc. v. Rural Tel. Serv. Co.*, 499 U.S. 340 ("The primary objective of copyright is not to reward the labor of authors, but to promote the Progress of Science and useful Arts").

5. Continental Congress, *Journals of the Continental Congress* 24 (1783): 326–538, 326, https://memory.loc.gov/cgi-bin/ampage?collId=lljc&fileName=024/lljc024.db&recNum=333&itemLink=?%230240334&linkText=1.

6. Continental Congress, *Journals of the Continental Congress* 24 (1783): 326.

7. Registration, although not required, does have certain benefits, and many legal systems, including the United States, encourage registration of copyrighted works. Registration creates a legal documentation of the ownership and is a prerequisite to filing a copyright lawsuit. Registration and Civil Infringement Actions, 17 U.S.C. § 411(a).

8. Exclusive Rights in Copyrighted Works, 17 U.S.C. § 106.

9. For more on contracting and author's rights, see Brianna L. Schofield and Robert Kirk Walker, *Understanding and Negotiating Book Publication Contracts* (Author's Alliance, 2018) at https://perma.cc/Q2ZG-YNL7.

10. "Right now, today, there are 650 million books that tax-paying citizens have paid to access that are sitting on shelves in closed libraries, inaccessible to them." Chris Freeland, "Internet Archive Responds: Why We Released the National Emergency Library," Internet Archive Blogs, http://blog.archive.org/2020/03/30/internet-archive-responds-why-we-released-the-national-emergency-library.

11. For example, many publishers created a new licensing regime at the height of library closures called "Read Aloud" licenses, requiring schools, libraries, or authors to seek permission before reading their books aloud online. Each publisher had different requirements for these permissions. See Kathy Ishizuka, "Abrams, HarperCollins, and Peachtree Extend Permission for Readalouds," *School Library Journal* (February 15, 2021), https://www.slj.com/story/remote-learning-still-the-norm-publishers-extend-permissions-for-read-alouds-COVID-19.

12. For more on this concept, see Aaron Perzanowski and Jason Schultz, *The End of Ownership*, Information Society Series (Cambridge, MA: MIT Press, 2016).

13. Augustine Birrell, *Seven Lectures on the Law and History of Copyright in Books* (Fred B Rothman & Co, 1971); Ronan Deazley, *On the Origin of the Right to Copy: Charting the Movement of Copyright Law in Eighteenth Century Britain (1695–1775)* (Hart Publishing 2004); John Feather, *Publishing, Piracy and Politics: An Historical Study of Copyright in Britain* (1994); Lyman Ray Patterson, *Copyright in Historical Perspective* (Nashville, TN: Vanderbilt University Press, 1968).

14. On modern licensing concerns, see the section on Licensing and Ownership.

15. See Eaton S. Drone, *A Treatise on the Law of Property in Intellectual Productions in Great Britain and the United States* (Boston: Little, Brown and Co., 1879) for the full list of declarations and decrees.

16. William Searle Holdsworth, Arthur L. Goodhart, Harold Greville Hanbury, and John McDonald Burke, *A History of English Law*, vol. 5 (London: Methuen, 1903); Frank Riebli, "The Spectre of Star Chamber: The Role of an Ancient English Tribunal in the Supreme Court's Self-Incrimination Jurisprudence," *Hastings Constitutional Law* 29, no. 4 (Summer 2002): 807–30.

17. Daniel L. Vande Zande, "Coercive Power and the Demise of the Star Chamber," *American Journal of Legal History* 50, no. 3 (2008–2010): 326–49.

18. Fred S. Siebert, *Freedom of the Press in England 1476–1776: The Rise and Decline of Government Controls* (Urbana: University of Illinois Press, 1952).

19. 28 Eliz., art 4 (1585).

20. Star Chamber Decree of 1586, reprinted in Edward Arber, ed., *A Transcript of the Registers of the Company of Stationers of London, 1557–1640* (London: Stationers Company, 1875), 2:807.

21. "A proclamation against the disorderly printing, vttering, and dispersing of bookes, pamphlets, &c," Early English Books Online Text Creation Partnership (2011), https://quod.lib.umich.edu/e/eebo2/B12865.0001.001/.

22. Star Chamber Decree of 1637, reprinted in Edward Arber, ed., *A Transcript of the Registers of the Company of Stationers of London, 1557–1640* (London: Stationers Company, 1877), 4:528.

23. Ronan Deazley, "Commentary on the Stationers' Royal Charter 1557," in *Primary Sources on Copyright (1450–1900)*, ed. Lionel Bently and Martin Kretschmer (2008), www.copyrighthistory.org.

24. Patterson, *Copyright in Historical Perspective*; Cyprian Blagden, *The Stationers' Company—A History 1403—1956* (George Allen, 1960).

25. For more on this period of hand copying as the means of distribution, see Leila Avrin, *Scribes, Script and Books: The Book Arts from Antiquity to the Renaissance* (American Library Association, 1991).

26. "For all I know, the monks had a fit when Gutenberg made his press." Justice Stephen Breyer, Transcript of Oral Argument at 11, *Metro-Goldwyn-Mayer Studios, Inc. v. Grokster, Ltd.*, 545 U.S. 913 (2005) (04–480).

27. Blagden, *Stationers' Company*.

28. Stationer's Charter 1557, reprinted in Edward Arber, ed., *A Transcript of the Registers of the Company of Stationers of London, 1557–1640* (London: Stationers Company, 1875), 1:xxviii.

29. Stationer's Charter 1557, 1:xxviii.

30. John Feather, *A History of British Publishing*, 2nd ed. (London, Routledge, 2006), 39.

31. On the early proclamation and decrees that predate the Stationer's Charter, see the next section on the Licensing Act.

32. Stationer's Charter 1557, 1:xxviii; Feather, *History of British Publishing*, 39.

33. See the section on the Statute of Anne below.

34. 3 and 4 Car.II, c.33 (1662).

35. However, much of this legislation was also based on the previous Star Chamber Decree of 1637, the notorious declaration that provided a more elaborate system for licensing both religious and secular works, and fully buttressed the Stationers' new copyright by requiring that no work could be printed without first being entered on the Stationers' Registry.

36. Arnold Plant, "The Economic Aspects of Copyright in Books," *Economica* 1, no. 2 (1934): 167–95.

37. Feather, *Publishing, Piracy and Politics*, 51–54.

38. 15 H.L. Jour. 280 (1693).

39. Augustine Birrell, *Seven Lectures on the Law and History of Copyright in Books* (London: Cassell, 1899).

40. Act for the Encouragement of Learning, 1710, 8 Ann., c. 19 (Eng.) ("Statute of Anne").

41. For example, Augustine Birrell, writing on copyright in 1889 stated that the Statute of Anne "butchered" copyright and was a "perfidious measure, rigged with curses dark" passed by "an ignorant Legislature." Birrell, *Seven Lectures*, 19, 22.

42. John Feather, "The Book Trade in Politics: The Making of the Copyright Act of 1710," *Publishing History* 8 (January 1, 1980): 19, 37.

43. Oren Bracha, "The Rise and Fall of Authorship-Based Copyright," in *Owning Ideas: The Intellectual Origins of American Intellectual Property, 1790–1909* (Cambridge, UK: Cambridge University Press, 2016), 54–123.

44. See *Millar v. Taylor* (1769) 98 Eng. Rep. 201 (K.B.).

45. U.S. Constitution, art. 1, § 8, cl. 8.

46. William Berkeley, *The Papers of Sir William Berkeley, 1605–1677*, ed. Warren M. Billings and Maria Kimberly (Richmond, VA: Library of Virginia, 2007).

47. John William Wallace, *An Address Delivered at the Celebration by the New York Historical Society, May 20, 1863, of the Two Hundredth Birth Day of Mr. William Bradford* (J. Munsell, 1863), 51.

48. Hugh Amory and David D. Hall, *A History of the Book in America*, vol. 1, *The Colonial Book in the Atlantic World* (Chapel Hill: University of North Carolina Press, 2016), 204.

49. Act of May 31, 1790, chap. 15, 1 Stat. 124 (1790).

50. Thomas Macaulay, *Speeches on Copyright* 11 (A. Thorndike ed., 1915)

51. *Golan v. Holder*, 565 U.S. 302, 345 (2012).

52. *Bobbs-Merrill v. Straus*, 210 U.S. 339 (1908).

53. *Bobbs-Merrill v. Straus*, 210 U.S. 339 (1908).

54. *Straus v. Victor Talking Mach. Co.*, 243 U.S. 490, 500–501 (1917). "Courts would be perversely blind if they failed to look through such an attempt as this 'License Notice.'" These attempted notices "have been hateful to the law from Lord Coke's day to ours, because obnoxious to the public interest."

55. *Bobbs-Merrill v. Straus*, 210 U.S. 339 (1908).

56. Limitations on Exclusive Rights, 17 U.S.C. § 109.

57. "A library may lend an authorized copy of a book that it lawfully owns without violating copyright laws." *Hotaling v. Church of Jesus Christ of Latter-Day Saints*, 118 F.3d 199, 203 (4th Cir. 1997).

58. A printing privilege was issued the Frenchman Richard Pynson as a two-year prohibition on others' reprinting a Latin sermon by the dean of St. Paul's Cathedral. Pynson was also the first in England to print and sell books of religious music.

59. Wendy Seltzer, "Free Speech Unmoored in Copyright's Safe Harbor: Chilling Effects of the DMCA on the First Amendment," *Harvard Journal of Law and Technology* 24, no. 2 (2010): 171–233.

60. The Communications Decency Act, 47 U.S.C. § 230, "CDA Section 230" (2021).

61. David R. Sheridan, "*Zeran v. AOL* and the Effect of Section 230 of the Communications Decency Act upon Liability for Defamation on the Internet," *Albany Law Review* 61 (1997): 147.

62. S. Rep. No. 105–190, at 2 (1998).

63. S. Rep. No. 105–190, at 2 (1998).

64. Communications Decency Act, 47 U.S.C. § 230.

65. See Seltzer, "Free Speech Unmoored"; Caroline Womack, "Revenge of the Retaliatory Takedown: Let's Plays, Fair Use, and an Unstable DMCA," *Quinnipiac Law Review* 37 (2019): 757; and Joel D. Matteson, "Unfair Misuse: How Section 512 of the DMCA Allows Abuse of the Copyright Fair Use Doctrine and How to Fix It" *Santa Clara High Technology Law Journal* 35 (2018): 1–22.

66. 1995 WL 323710 (N.Y. Sup. Ct. 1995).

67. 1995 WL 323710 (N.Y. Sup. Ct. 1995) at *1.

68. 1995 WL 323710 (N.Y. Sup. Ct. 1995) at *5.

69. H.R. Rep. No. 104–458 (1996).

70. Communications Decency Act, 47 U.S.C. § 230.

71. Communications Decency Act, 47 U.S.C. § 230.

72. "Section 230, which is a liability shielding gift from the U.S. to 'Big Tech' (the only companies in America that have it—corporate welfare!), is a serious threat to our National Security & Election Integrity. Our Country can never be safe & secure if we allow it to stand." @realDonaldTrump Twitter account on December 2, 2020 (account presently suspended; web archive at https://web.archive.org/web/20201202024527/https://twitter.com/realDonaldTrump/status/1333965375193624578).

73. It is not just the previous US administration that was attempting to reform Section 230. Several states have made similar attempts. For example, Florida recently passed SB 7072 which prohibits large social media platforms from willfully preventing a Florida political candidate from access to the platform. It also prohibited the social media platform from banning or limiting access to a user's posting without first giving notice and offering a rationale for the ban. A federal court ruled that this law was potentially in violation of the First Amendment. See Cat Zakrzewski, "Federal Judge Blocks Florida Law That Would Penalize Social Media Companies," *Washington Post*, June 30, 2021, https://www.washingtonpost.com/technology/2021/06/30/florida-social-media-law-trump/.

74. Donald J. Trump, Executive Order on Preventing Online Censorship (May 28, 2020), https://www.whitehouse.gov/presidential-actions/executive-order-preventing-online-censorship/.

75. U.S. Department of Justice, "Section 230—Nurturing Innovation or Fostering Unaccountability: Key Takeaways and Recommendations," June 2020, https://www.justice.gov/file/1286331/download.

76. Statement of Chairman Pai on Section 230, October 15, 2010, https://docs.fcc.gov/public/attachments/DOC-367567A1.pdf.

77. *Lenz v. Universal Music Corp.*, 815 F.3d 1145 (9th Cir. 2016).

78. Mitja Rutnik, "YouTube in Numbers," August 11, 2019, https://www.androidauthority.com/youtube-stats-1016070; Omnicore, "Instagram by the Numbers," February 28, 2023, https://www.omnicoreagency.com/instagram-statistics.

79. For more on these platforms' use of auto filters, see Maayan Perel and Niva Elkin-Koren, "Accountability in Algorithmic Copyright Enforcement," *Stanford Technology Law Review* 19, no. 3 (2016): 473–533.

80. Directive 2019/790 of the European Parliament and of the Council of 17 April 2019 on Copyright and Related Rights in the Digital Single Market and Amending Directives 96/9/EC and 2001/29/EC, 2001 O.J. (L 130) 92.

81. Directive 2019/790 at Article 2(6).

82. Note that the law also excludes many other types of platforms from Article 17's requirements, including nonprofit institutional repositories, nonprofit online encyclopedias, like Wikipedia, electronic communication service providers, cloud services, and more.

83. Art. 17(1) states "An [OCSSP] shall therefore obtain an authorisation from the rightholders . . . for instance by concluding a licensing agreement." Also, Art. 17(8) references "licensing agreements . . . concluded between service providers and rightholders."

84. U.S. Copyright Office, *Section 512 of Title 17: A Report of the Register of Copyrights*, 72, 198, May 2020, https://www.copyright.gov/policy/section512/section-512-full-report.pdf.

85. Relatedly, the use of printing patents was another form of government control. These were personal monopolistic grants by the Crown, slightly different from the early copyright and licensing laws. Again, much like proclamations, the royal printing patent grant might vest the ownership of a single book to an individual or it might reserve an entire discipline's publishing business (printing the law, for example) for the benefit of a particular person or group of persons. The most important of these were the royal printer's patent which was based on the similar early grants to William Facques.

86. An Act whereby Offences be made High treason and taking away all Sanctuaries for all manner of High Treasons 1534, 26 Hen.VIII, c.13, s.2 (emphasis mine).

87. There are numerous examples of these royal grants issued during the reign of Henry VIII. See Proclamation of July 8, 1546 ("Prohibiting Heretical Books; Requiring Printer to Identify Himself, Author of Book, and Date of Publication"), reprinted in *Tudor Royal Proclamations*, ed. Paul Hughes and James Larkin (New Haven, CT: Yale University Press, 1964), 1; Proclamation of June 22, 1530 ("Prohibiting Erroneous Books and Bible Translations"), *Tudor Royal Proclamations*, 193; Proclamation of March 6, 1529 ("Enforcing Statutes against Heresy; Prohibiting Unlicensed Preaching, Heretical Books"), *Tudor Royal Proclamations*, 181.

88. [Mary Proclamation]

89. "Facebook: Annual Revenue 2009–2020," 2021, Statista, https://www.statista.com/statistics/268604/annual-revenue-of-facebook.

90. "Twitter Announces Fourth Quarter and Fiscal Year 2020 Result," PRNewsWire, February 9, 2021, https://prn.to/3k7001V.

91. The most recent case addressing social media platforms as a monopoly is *Fed. Trade Comm'n v. Facebook, Inc.*, No. CV 20–3590 (JEB), 2021 WL 2643627, at *1 (D.D.C. June 28, 2021).

92. *Straus v. Victor Talking Mach. Co.*, 243 U.S. 490, 500–501 (1917).

Contributors

Joseph M. Adelman is an associate professor of history at Framingham State University where he is a scholar of media, communication, and politics in the Atlantic world, and an associate editor of the *New England Quarterly*. He is the author *of Revolutionary Networks: The Business and Politics of Printing the News, 1763–1789* (Johns Hopkins University Press, 2019) which was awarded an Honorable Mention for the 2019 St. Louis Mercantile Library Prize from the Bibliographical Society of America.

Benjamin Bankhurst is an associate professor and the Ray and Madeline Johnston Endowed Chair in American History at Shepherd University. Bankhurst is the author of *Ulster Presbyterians and the Scots Irish Diaspora, 1750–1764* (Palgrave Macmillan, 2013). He is the codirector of the Maryland Loyalism Project, a digital archive and database documenting the lives of Maryland Loyalists in the Revolutionary Atlantic.

Dorothy Berry is digital curator for the Smithsonian National Museum of African American History and Culture. Her work lies at the intersections of information discovery and African American history.

Gary Berton is an officer at the Thomas Paine National Historical Association and a program coordinator at the Institute of Thomas Paine Studies at Iona College. His work has appeared in *Thomas Paine Journal*, *Truth Seeker*, and the *Journal of Early American History*.

Mark Boonshoft is an associate professor and Conrad M. Hall '65 Chair in American Constitutional History at Virginia Military Institute. A historian of early America, he was the first executive director of the American Society for Eighteenth-Century Studies. Boonshoft is the author of *Aristocratic Education and the Making of the American Republic* (University of North Carolina Press, 2020) which was a finalist for the 2021 George Washington Book Prize.

Christian Boylston has an MA from Georgia Institute of Technology and is currently an associate at Loyal Health.

Lindsay M. Chervinsky is a senior fellow at the Center for Presidential History at Southern Methodist University. Previously, she was a historian at the White House Historical Association and a postdoctoral fellow at the Center for Presidential History at Southern Methodist University. She is the author of *The Cabinet: George Washington and the Creation of an American Institution* (2020), *Making the Presidency: John Adams and the Precedents That Forged the Republic* (2024), and coeditor of *Mourning the Presidents: Loss and Legacy in American Culture* (2023).

Kyle K. Courtney is a lawyer and librarian who serves as the director of Copyright and Information Policy for Harvard University, working out of Harvard Library. He also maintains a dual appointment at Northeastern University where he teaches in the interdisciplinary Cybersecurity Program at the Khoury College of Computer Science and teaches legal research and writing at the Northeastern University School of Law. In 2020 he cofounded the nonprofit organization Library Futures which empowers libraries to take control of their digital futures.

Sara Collini is a historian of early America studying slavery, women's history, the history of medicine, digital history, and public history. She is currently a research assistant professor at Clemson University where she is part of the research and community engagement team for the African American Burial Ground and Woodland Cemetery Historic Preservation project.

Michael Crowder is a public historian at Iona University. Dr. Crowder earned a PhD in the History Department at the Graduate Center, City University of New York in 2018. He is currently at work on a new history of Thomas Paine and the American Revolution, and in August 2020 he published "Remembering Revolution: Commemorating Thomas Paine and the Progressive Afterlives of the American Revolution," in *New York History*. He is an active public speaker on topics pertaining to the revolutionary era writ large, in particular revolutionary commemoration and the ongoing state and local organizing campaigns for America 250, the semiquincentennial commemorations of the American Revolution. His research interests also include the relationships between slavery, abolitionism, and capitalism in the revolutionary era.

Christy Hyman is a postdoctoral fellow for Freedom on the Move at Cornell University in the Department of History and an assistant professor of human geography in the Department of Geosciences at Mississippi State University with a joint appointment in the Program of African American Studies. Her scholarship focuses on African American cultural and political lives in the US South during the antebellum era.

Lubomir Ivanov is a professor of computer science and assistant chair of the Computer Science Department at Iona University. His scholarship focuses on authorship attribution, automatic music style/instrument recognition, chemical process modeling, and parallel image/video processing. His work has appeared in the *Journal of Computing Sciences in Colleges* and the *Journal of Technologies in Society*.

Maeve Kane is an associate professor of history at the University of Albany. Her research focuses on the social and economic history of gender, race, and material culture in early America and the early modern Atlantic World. She is the author of *Shirts Powdered Red: Haudenosaunee Gender, Trade, and Exchange across Three Centuries* (Cornell University Press, 2023).

Afshawn Lotfi has a degree in computer engineering from the Georgia Institute of Technology and is the founder and engineer of OpenOrion.org.

Molly Nebiolo is an assistant professor of history at Butler University, where she is a historian of the early Atlantic World, the history of health and medicine, and the digital humanities. Her scholarship has been supported by the McNeil Center for Early American Studies, the American Philosophical Society, the Huntington Library, and the John Carter Brown Library among others.

Marcus P. Nevius is an associate professor in the Department of History at the University of Missouri with a joint appointment at the Kinder Institute on Constitutional Democracy. His research focuses on slavery, the Revolution, confederation, and early republican periods in the early United States. Nevius is the author of *City of Refuge: Slavery and Petit Marronage in the Great Dismal Swamp, 1763–1856* (University of Georgia Press, 2020).

Jessica M. Parr is a professor of practice in digital humanities at Northeastern University and has held multiple fellowships, including from the Royal Historical Society. They have contributed to numerous history digital humanities projects, including the African Building Heritage Project, the Reckonings Project, the *Junto*, and *Black Perspectives*. Parr serves on the Board of Trustees of the *Programming Historian* and is the author of *Inventing George Whitefield: Race, Revivalism, and the Making of a Religious Icon* (University Press of Mississippi, 2016).

Smiljana Petrovic is an associate professor of computer science and chair of the Computer Science Department at Iona University. Her research focuses on the development of machine-learning techniques for authorship attribution and for solving constraint satisfaction problems. Her research has appeared in the *Journal of Technologies in Society*, the *International Journal on Artificial Intelligence Tools*, and *Constraint Programming Letters*.

Brad Rittenhouse is a research data facilitator in research computing at Stanford University. His scholarship traces the use of literary aesthetics as data management strategies in nineteenth-century American writing.

Kyle Roberts is the executive director of the Congregational Library and Archives. A prolific public historian and digital humanist, Roberts is the director of the Jesuit Libraries Provenance Project and codirector of the Maryland Loyalism Project. He is the author of *Evangelical Gotham: Religion and the Making of New York City, 1783–1860* (University of Chicago Press, 2016) as well as numerous anthologies and articles.

Cameron Shriver is a Myaamia research associate at the Myaamia Center and a visiting assistant professor of history at Miami University of Ohio. His research has appeared in *Early American Studies Journal*.

Nora Slonimsky is the director of the Institute for Thomas Paine Studies at Iona University. She is a scholar of political economy, legal history, and book history in the eighteenth-century anglophone world. Slonimsky serves as the social media editor for the *Journal of the Early Republic* and has published in *Early American Studies, The Junto,* and *Teaching US History*. Her first book, *The Engine of Free Expression: Copyrighting Nation in Early America*, is under contract with University of Pennsylvania Press.

Whitney Nell Stewart is an assistant professor of history at the University of Texas at Dallas. She is a historian of the US South, writing and teaching about slavery and plantations, material culture, and public history. She is the author of *This Is Our Home: Slavery and Struggle on Southern Plantations* (University of North Carolina Press, 2023) and coeditor with John Garrison Marks of *Race and Nation in the Age of Emancipations: An Atlantic World Anthology* (University of Georgia Press, 2018). Her work has also appeared in the *Journal of the Early Republic*, the *Journal of Social History*, and *Winterthur Portfolio*.

Jordan E. Taylor is digital projects editor at Colonial Williamsburg Innovation Studios and a historian of news, print, and politics in the American Revolution.

He is the author of *Misinformation Nation: Foreign News and the Politics of Truth in Revolutionary America* (Johns Hopkins University Press, 2022).

Ben Wright is an associate professor of history at the University of Texas at Dallas and a historian of the United States, history of religion, and slavery. Wright is the author of *Bonds of Salvation: How Christianity Inspired and Limited American Abolitionism* (Louisiana State University, 2020) and coeditor of *The American Yawp: A Massively Collaborative Open U.S. History Textbook* (Stanford University Press, 2019) with Joseph L. Locke and *Apocalypse and the Millennium in the American Civil War Era* (Louisiana State University, 2013) with Zachary W. Dresser.

Index

Aacimwahkionkonci, 7, 136–138, 143, 147, 149–150
Adams, Abigail Amelia Smith, 188–189
Adams, Abigail (Smith), 184, 188–189, 192
Adams, John, 21–22, 184, 188, 192–193, 214
Adams, John Quincy, 192–193
Adams, Louisa Catherine (Johnson), 188–189, 193–194
Age of Revolutions, 1–2, 83, 88, 115–116, 262
  rhetoric of, 160–161, 165–170, 175–178, 180–181
Aitchison, William, 220–222, 225, 228, 230
American Revolution, 1, 245–246
  African-Americans and, 222, 225–226
  commemoration of, 2, 11
  effects on women, 8, 185–187
  historiography of, 3–5, 18, 178, 239–240
  and Native people, 138–140, 152
  slavery and, 74–75
Antislavery, 51–56, 60–62, 74–75, 89–92, 172–173, 209–210
archives, 9, 50–51, 222
  administrative state and, 148–150
  Black presence in, 67–68, 92–98, 220, 223–228, 230–231
  collection practices, 189–191
  description of, 52–53
  digital access, 58–61, 62–63
"A Slave", 201–202, 207–208, 210–217
Assassin's Creed, 104
augmented reality. (See virtual reality).

Bache, Benjamin Franklin, 238, 242, 244
Berkeley, William, 237
Bleecker, Catharine, 53–55
Bleeding Kansas, 172–175
Blevins, Cameron, 6
Bobbs-Merrill v. Strauss, 282–284
Bon-Harper, Sarah, 27
Bonneville, Marguerite Brazier, 202, 215–216
"Book of Negroes", 35–36, 38–39, 42–43
Bourguinon, Jean Baptiste, 84
Brown, Vincent, 5, 87–88, 224
Brückner, Martin, 84, 114
Burr, Aaron, 194, 214

Burr, Theodosia, 194
Busa, Robert, 5

canals, 126–129
Carter, John, 242–243
censorship, 6, 267–268, 270–271, 273–281, 293–294
Chalmers, James, 43
Childs, Francis, 246–247
Chronological Bigram Analysis, 180
Colonial North America project (CNA), 51–52
common sense, 247
Communication technologies, 2, 240, 248, 254–256, 263, 286
Communications Decency Act (CDA), 268, 285, 287–289, 295
Conover, Peter, 56–58
conspiracy, 239–240, 245, 253, 257
Constitution, U.S., 24, 26, 74–75, 139–140, 238–239, 280
Continental Congress, 73, 268
copyright, 8, 267–270, 273–294
correspondence networks, 185–187, 191–193, 195–197
COVID-19, 10, 29, 50, 64, 196, 200, 253, 259, 262–264, 269–270
Cranch, Mary, 188, 192
Crummell, Alexander, 91–92

databases, 4, 6–7, 10, 36, 186–187, 259–260
  datafication of marginalized people, 41, 72
  methods of creating, 40–41, 68–69, 189–190
  relational, 7, 67, 69
  remediation, 39
Democratic-Republican Party, 208, 214, 243–244, 248–249
digital history, 2, 9
  Black DH, 10, 36, 39–40, 66–67, 69, 83–84, 224–225
digital humanities, 2, 5–6, 9, 36, 62, 104, 108
  historiography of, 5–6, 29–30, 66
Digital Millennium Copyright Act (DMCA), 268, 285–288, 290, 292, 295
digitization, 4, 8, 9, 39–40, 58–9, 141–142, 222

307

INDEX

dispossession, 136, 139–143, 149–150
Drinker, Elizabeth, 105, 107–109, 112–113
Duane, William, 208–209, 211, 214
Du Bois, W.E.B., 86–87, 89–91
Dunbar, Erica Armstrong, 17, 22–23
Dunlap, John, 246
Dunmore, Earl of (John Murray), 225–227, 258

Edelson, S. Mx, 5
Eden, Robert, 43
emancipation, 88–90
enslaved people, 7, 9, 21–26, 27–30
  Alfred, 43
  Ben, 43
  Bob, 43
  Christian, 43
  datafication of, 36, 43–46, 51–52, 72
  Eve, 132–133
  George, 131, 133
  Hannah, 28
  Hannah (Maryland), 43
  Ipheginia, 43
  James, 43
  Juda, 43
  Lewis, 55–58
  Manuel, 120–121, 129–131, 133
  Monimia, 43
  "a negroe man", 220Nelson, 28
  Plymouth, 43
  Queen, 43
  rebelling, 129, 258
  Rene, 43
  Sam, 43
  Sarah, 43
  Tom, 43
  Tom (New York), 54–55, 62
  Equiano, Olaudah, 92–93
EU Copyright Directive, 291–295

Facebook, 161, 248, 257, 262, 290, 294
fact-checking, 241, 247–249, 258
Federalists, 214, 243–245, 248–250
Fleet, Samuel, 57–59
Founder Online database, 4, 8, 184–188, 190–191, 195
founders, 8, 17, 184, 186
Franklin, Benjamin, 177, 184, 186–194
Franklin, Deborah, 192
French Revolution, 1, 179, 243–245, 247
Freneau, Philip, 244–245
Fuentes, Marissa, 9, 93–94

Gallon, Kim, 66, 83, 224–225
genealogy, 150–152
Geographical Information Systems (GIS), 6, 9, 137, 142–143
Gerry, Elbridge, 238–239
Glorious Revolution, 278
Goddard, Mary Katherine, 246
Goddard, William, 258
Google, 96, 102, 172, 262
Gordon-Reed, Annette, 17
Great Dismal Swamp, 8, 9, 126–128, 130–131, 133, 220–234
Gunston Hall, 65, 74

Haitian Revolution, 1, 90
Hall, Prince, 88
Hamilton, Alexander, 184, 186, 188–191, 193–194
Hamilton, Dr. Alexander, 103, 105, 107–109, 110–112, 114
Hartman, Saidiya, 9
Harvard University Libraries, 50–52, 59, 61–63
HathiTrust, 159, 163, 175
Haycorn, Cynthia, 9, 56–58, 59, 62
Highland, 6, 18, 26–28
Hsu, Wendy, 19
Hurley, Andrew, 21

Illuminati, 245
Indian Affairs (federal bureaucracy), 148–149, 151–152
information, 237–238, 240, 254, 259
Inspection Roll. See "Book of Negroes"
Institute for Thomas Paine Studies (ITPS), 8, 10, 209
International Image Interoperability Framework (IIIF), 60
Irving, Washington, 252–253

Jamaica, 87
January 6 insurrection, 170–172, 174–175, 253–254
Java Graphical Authorship Attribution Program (JGAAP), 208–209
Jefferson, Thomas, 27, 65, 67, 70–71, 74, 188–190, 201–202, 207–208, 214
  and *Notes on the State of Virginia*, 208, 214
Jefferys, Thomas, 84
Johnson, Jessica Marie, 36, 66, 69
Johnson v. M'Intosh, 139
JSTOR, 60
Judah, Moses, 9, 22–23, 55–56, 59
Judge, Ona/Oney, 23

Knox, Henry, 139–140
Kollock, Shepard, 258
Kramer, Michael, 39–40, 69

Latin American Revolutions, 1
Lee, Arthur, 188, 192–193
Library of Congress, 96, 260
licensing laws, 268–271, 276–282, 291–292
Longlois Reserve, 148–150
Loyalists, 35, 220, 246, 248
    Black Loyalists, 35, 42–46
    Claims Commission, 36–38, 41–42, 44–45
    "Inspection Roll. *See* "Book of Negroes"
    Loyalists, women, 42

Macauly, Catherine, 185
machine learning, 205–206, 209–210
Madison, Dolley, 24, 189, 193–197
Madison, James, 6, 18, 24–26, 190
mapping, 7–8, 83–84
    Black uses of, 86–87, 90–91
marronage, 8, 87, 90, 120, 124, 128, 227
Maryland Loyalist Project, 6, 35–36, 38–39
Mason, George, 65, 67, 74
McLuhan, Marshall, 2
media literacy, 239–241, 245–248, 263
metadata, 43, 46, 69–70, 75
Miami Tribe of Oklahoma, 7, 136, 141–143
    landownership of, 143–147, 150–152
midwives, enslaved, 6–7, 9, 66–67, 72–74
    Kate, 9, 65, 72
    Lucy, 71–72
    Nan, 9, 65
    Nell, 9, 65
    Pegg, 68, 71–72, 74
    Rachael, 9, 65, 70–71
Minkoff, Mary, 25
mobility, 122–123
Monroe, James, 6, 18, 26–29
Monticello, 27, 65, 70–71, 73
    Planation Database, 70
Montpelier, 6, 18, 24–26, 29
    Descendants Committee, 25
Moretti, Franco, 5
Morgan, Jennifer L., 9
Morris, Robert, 89–90
Mount Vernon, 23, 32, 65, 74
    Database of Mount Vernon's Enslaved Community, 69
Murrin, John M., 3
Myaamia Center, 136

National Endowment for the Humanities, 24–26, 260
Native Americans, 7
Natural Language Processing (NLP), 160–161
networks, 2, 4, 6–9, 92–93, 96–98, 113–115, 187–189, 191–195
news, 8, 241–244, 253
    anxiety, 8, 253–255, 257, 259, 263
    "bias" in, 240–241, 245–248, 259–260, 262–263
    business of, 253, 256–257, 263
    fake, 8, 238, 240–241, 243, 257
newspapers, 2, 4, 58, 67–68, 73, 95–96, 227, 233, 237–248, 255
New York
    abolition of slavery in, 54–55, 56
    Black people in, 51–58
New York Manumission Society, 51–52, 56
novels, 38, 160, 165–167, 255
North Carolina, 129
    1821 insurrectionary scare, 119–121, 124–126

Oakley, John, 55–57
Omeka, 40–43, 75
Omohundro Institute, 38
online content-sharing service providers (OCSSP), 291–292
Onslow County insurrection. See "North Carolina, 1821 insurrectionary scare"
open access, 10
Other[ed] Colonial Voices, 6, 53, 58–60, 62–63

Paine, Thomas, 8, 201–202, 209–210, 213–217, 261–262
Paris, Daniel, 53–54
Parker, James, 220–222, 225, 228, 230
Parkinson, Robert, 17
Parkland History Project, 19
Philadelphia, 105–106, 108–13
Pickering, Timothy, 190
Pigott, Charles, 214
post office, 238, 255–256
President's House, 6, 21–24
printing press, 165, 267–281
Privy Council, 271–272
public history, 6
    digital public history, 17–21, 29–30
public humanities, 137
Putnam, Lara, 223–224

QAnon, 240, 257
quantitative literary analysis, 159–160
"Quidnunc," 254–255

Readex, 4, 259–262
remediation, 4, 39–41
Rivington, James, 246
Robert David Lion Gardiner Foundation, 10
Robison, John, 245
Rosenzweig, Roy, 19

Saunders, Prince, 83
Sayer, Robert, 84
Scalar, 40–41, 48, 59, 62–63
Schuyler, Elizabeth, 189–190
Schuyler family, 190
Sherman, Roger, 238
slave narratives, 84, 95–96, 223
slavery
  pass laws, 123
  slave patrols, 123
slave trade, 121
social media, 268, 274, 289–290. *See also names of individual platforms*
"Star Chamber" (Court of Star Chamber), 271–273, 277, 288, 294
"Stationers" (Worshipful Company of Stationers), 271–280, 284–285, 287–288, 294–295
Statute of Anne, 279–282, 287
Steedman, Carolyn, 9
Steele, John, 238, 240
Stratton Oakmont, Inc. v. Prodigy Services, 288–289
students, 6, 22, 35–38, 45–46, 61–62
Supreme Court of the United States, 139, 283, 288, 295
swamps, 123–124

taverns, 37, 575, 70, 112–115
tea, 73, 110–113
term frequency-inverse document frequency (TF-IDF) analysis, 175–176
text attribution, 8, 202–205
Text Analysis Project (TAP), 202, 206–207, 217
Thelen, David, 19
Townshend Acts, 73
treaties, 139–141, 144–147
Trouillot, Michel-Rolph, 9
Twitter, 240, 257, 262–263, 290, 294

Ulrich, Laurel Thatcher, 9
University of North Carolina Libraries' Documenting the American South (DocSouth), 223
urban space, 7, 104–106, 111–113

virtual reality, 22–24, 27–28, 104–106, 114–116
Visualizing Colonial Philadelphia (VCP), 114–116

Warren, Mercy Otis, 185, 189, 193, 196–197
Washington, George, 18, 21–24, 65, 67, 74, 138, 140, 188–190, 214–215, 253, 261, 281
Washington, Martha, 185–186, 189–190, 193–197
Wheatley, Phillis, 92
Wikipedia, 240–241
Winkle, Rip Van, 252–253, 259, 262
women writers, 8–9, 98–99, 184–189
Wood, John, 131–132

YouTube, 262, 287, 291–292

www.ingramcontent.com/pod-product-compliance
Lightning Source LLC
Chambersburg PA
CBHW030734250426
43671CB00035B/328